Detection and Analysis of SARS Coronavirus

Detection and Analysis of SARS Coronavirus

Advanced Biosensors for Pandemic Viruses and Related Pathogens

Edited by
Chaudhery Mustansar Hussain
Sudheesh K. Shukla

WILEY-VCH

Editors

Professor Chaudhery Mustansar Hussain
Department of Chemistry and Environmental Sciences
151 D Tiernan, New Jersey Institute of Technology
161 Warren Street, University Heights
Newark, NJ 07102
USA

Dr. Sudheesh K. Shukla
Department of Biomedical Engineering
School of Biological Engineering and Life Science
Shobhit Institute of Engineering & Technology (Deemed-to-be University)
Modipuram
Meerut 250110
India

and

Department of Chemical Sciences
University of Johannesburg
Doornfontein Campus, P.O. Box 17011
Johannesburg, 2028
South Africa

Cover Design: Wiley
Cover Image: © Joao Paulo Burini/Getty Images

Library of Congress Card No.:
applied for

British Library Cataloguing-in-Publication Data
A catalogue record for this book is available from the British Library.

Bibliographic information published by the Deutsche Nationalbibliothek
The Deutsche Nationalbibliothek lists this publication in the Deutsche Nationalbibliografie; detailed bibliographic data are available on the Internet at <http://dnb.d-nb.de>.

Print ISBN: 978-3-527-34918-0
ePDF ISBN: 978-3-527-83250-7
ePub ISBN: 978-3-527-83251-4
oBook ISBN: 978-3-527-83252-1

Typesetting Straive, Chennai, India
Printing and Binding CPI Group (UK) Ltd, Croydon, CR0 4YY
Printed on acid-free paper

C113831_120721

We dedicate this book to all the CORONA WARRIORS who helped the entire GLOBE without caring themselves. A special dedication to my friend-cum-brother, ***Dr. Pawan Kumar****, Assistant Professor, Choudhary Charan Sing University, Meerut, India to whom we lost due to the CORONA virus infection.*

Contents

Preface

The world is witnessing the pandemic of severe acute respiratory syndrome coronavirus-2 (SARS-CoV-2), commonly known as 2019 novel coronavirus (COVID-19). The World Health Organization (WHO) has declared COVID-19 as pandemic. Scientists and researchers across the globe are working toward the development of prophylactics, therapeutics, and rapid diagnostic methods for the treatment and management of COVID-19. It is imperative to find that biological samples are appropriately analyzed using different tools and techniques. In addition, at this global emergency and urgent need, all the countries are facing a crisis in disposing of fast diagnosis of the disease, able to be deployed to an extensive number of persons. To address this issue, there is a need for fast, user-friendly, cheap, accurate, less time-consuming, and more reliable and sensitive tools and techniques to resolve the current global challenges to the analysis and treatment of COVID-19. Overall, this book highlights and gathers the current literature on advance analytical tools and techniques that would, in a rapid, selective, and efficient way, resolve this global and urgent issue.

This book provides a brief overview about the physical and chemical properties of analytical tools and techniques and puts forward the way to investigate the COVID-19 disease. This book will also demonstrate the different integration approaches for the development of sensor systems, along with design and commercialization guidelines. This book provides an up-to-date source of trusted information analytical tools and techniques and analysis for coronavirus. This book puts the strong recommendation based on the results in the current pandemic scenario that bio-based analytical tools and techniques could be adapted for standard protocols for permanent analytical solutions of the coronavirus.

Overall, this handbook is planned to be a reference handbook for researchers and scientists who are searching for new and advance analytical tools and techniques and analysis for coronavirus. The editor and contributors are famous researchers, scientists, and true professionals from academia and industry. On behalf of WILEY-VCH, we are very thankful to the authors of all the chapters for their wonderful and passionate efforts for compiling this handbook. A special thanks to Dr. Frank Weinreich, Dr. Martin Preuß, and entire team at WILEY-VCH for their dedicated support and help during compiling and publishing this handbook.

Chaudhery Mustansar Hussain, PhD
Sudheesh K. Shukla, PhD
(Editors)

About the Editors

Chaudhery Mustansar Hussain is Adjunct Professor, Academic Advisor, and Director of Chemistry & EVSc Labs in the Department of Chemistry & Environmental Sciences at the New Jersey Institute of Technology (NJIT), Newark, New Jersey, USA. His research is focused on the applications of nanotechnology & advanced materials, environmental management, analytical chemistry, and various industries. He is author of numerous papers in peer-reviewed journals as well as prolific author and editor of several scientific monographs and handbooks in his research areas.

Dr. Sudheesh K. Shukla works in the area of translational research and development of bioelectronics devices for disease alert and disorder with major focus on interfacing the chemistry (material science) and engineering for better healthcare (biology) and environmental applications. Currently, Dr. Shukla's research involves real-sample analysis of bio-chemical markers for personalized healthcare monitoring. In particular, Dr. Shukla is interested inintegrating biomaterials with micro- and nanosystems for sensing and actuation technologies. Dr. Shukla is author of numerous articles in peer-reviewed journals and as well as prolific author and editor of several scientific monographs in healthcare and environmental discipline published by Springer, John Wiley & Sons, and Elsevier. Dr. Shukla serves as an editor/editorial board member of several international peer-reviewed journals like *Chemosensor*, *Sensor International*, *Coronavirus*, *Reports in Electrochemistry*, etc.

Part I

Introduction

1

Current Diagnostic Approach for COVID-19

Nitika Thakur[1] and Rachit Sood[1]

[1] *Shoolini University of Biotechnology and Management Sciences, Faculty of Applied Sciences and Biotechnology, Department of Biotechnology, Solan, Himachal Pradesh, India*

1.1 Introduction

The ongoing spread of coronavirus has presented a threatening scenario globally because of the non-availability of accurate and rapid detection methods. However, on 30 January 2020, World Health Organization (WHO) has declared "COVID-19" (coronavirus disease 2019) as the largest threat under "public health emergency of global concern," as it is alone responsible for 250 000–260 000 deaths worldwide and across 3–3.5 million positive cases [1].

The detection and analysis procedure for this threatening virus started initially with a virus detection method, which somewhat has an advantage of non-detection of long culture cycles. Another way of detection is through the use of "nucleic acid profiling," which [2] can rapidly, sensitively, and accurately detect the pathogens in confirmed COVID patients, but large amounts of genetic variations, mismatches in primers, probes, and some target sequences may result in interpretation of false results. Detection via genomic sequence analysis and the point-of-care diagnosis have become popular in the detection of emerging viruses for finally detecting the specific antibodies IgM and IgG related to COVID [3].

Section 1.2 describes and highlights the current diagnostics and treatment strategies for COVID-19.

1.2 Recommended Laboratory Diagnosis for COVID-19

1.2.1 SARS-CoV-2 Testing: Detection Approach by Screening Suitable Specimen Cultures

The first and foremost step in diagnosis and identification is related to the appropriate collection of suitable specimens, which [4] are being collected from the upper and lower respiratory tracts, WBC's, and serum specimens. Furthermore, it has been mostly detected and screened from the swabs pertaining to nasopharyngeal area,

Detection and Analysis of SARS Coronavirus: Advanced Biosensors for Pandemic Viruses and Related Pathogens, First Edition. Edited by Chaudhery Mustansar Hussain and Sudheesh K. Shukla.

oropharyngeal, sputum, stool samples, urine, saliva, conjunctival area, and rectal swabs [5].

It is recommended that the samples and swabs should be strictly collected from the lower respiratory tract, for confirmatory diagnosis, even if the upper respiratory swab analysis is negative for COVID-19, as the receptor "AEC 2" is actively distributed in the alveolar lining of epithelial cells. Various studies compared [6] the viral loads from the lower respiratory tract specimen for the suspected and confirmed COVID patients. The study further stated that the average viral load differed in different collected samples [7], as the viral load detected in sputum was higher around 17 420 ± 6925 copies/test than the nasal swabs (655 ± 502 copies/test) and throat swabs (2555 ± 1965 copies/test). In addition, high viral load was also recorded in swabs collected from [8] the lower respiratory tract. Most of the cases were examined and confirmed positive through isolation and culturing techniques from oral swab on the first day, followed by a five [9–11] day diagnosis of anal swabs, indicating a shift from early period diagnosis to late period diagnosis. However, in asymptomatic conditions, it can be detected by analysis of urine sample, with no urinary irritation symptoms. Recently, it has also been detected in samples of saliva. In addition, it has been detected in nasopharyngeal swab, conjunctival tear swabs, and [12] oropharyngeal swabs. However, there still exist glitches in terms of monitoring and isolation process to screen conjunctival secretions for confirmatory diagnosis. Currently, the [13] virus has not been traced in many samples such as cerebrospinal fluid, semen, pericardial effusion, female reproductive tract, etc.

1.2.2 SARS-CoV-2 Detection: The Nucleic Acid Approach

For successful diagnostic strategies, identification of some specific primers and probes is important to screen out the target sequences. These target sequences for COVID-19 involve the "envelope – E," "the nucleocapsid – N," "spikes – S," "RNA-dependent RNA polymerase," and "open reading frame – ORF." WHO further recommends [14] reverse transcription polymerase chain reaction (RT-PCR) as a routine recommendation but lacks suitability in terms of time consumption, requirement of expensive equipment and biosafety conditions.

1.2.2.1 COVID-19 Detection Approach Through Real-Time PCR

The target gene sequences for detecting CoV-2 vary globally from China (ORF's), the United States (3 N gene), Germany (RdRp, N, and E genes) to France (two targets in RdRp). Center for Disease Control and Prevention (CDC) established a RT-PCR process for the detection and analysis of severe acute respiratory syndrome coronavirus 2 (SARS-CoV-2), with three specific primer sets to detect β forms of CoV-2 and the other two for SARS-CoV-2.

Different countries have a [15–18] large number of qRT-PCR (quantitative reverse transcription polymerase chain reaction) protocols provided by the official WHO website, which play the principal role in the detection of SARS-CoV-2. In recent time, different countries are following different protocols of gene targeting for the detection of SARS-CoV-2, for example, France (two targets in RdRp aka

RNA-dependent RNA polymerase), Japan (pancorona and numerous targets, spike protein), the United States (three targets in N gene), China (N genes and ORF1ab), Thailand (N gene), and [19–21] Germany (RdRp, N, and E genes). Different institutes use different RT-PCR primers or tests for the detection of SARS-CoV-2. A new RT-PCR panel has been rooted by the CDC for the universal detection of SARS-like β-CoVs and specific detection of SARS-CoV-2. For the N gene [22–25], three sets of distinct primers were devised – two sets of probes or primers were specific for identifying SARS-CoV-2 and the last set was universally used for detecting all β-CoVs. COVID-19 must be confirmed as positive for all the three individual targets. The Charite (Germany) developed two nucleic acid tests for the detection of E genes of the bat-like β-CoVs, SARS-CoV-2, and [26] SARS-CoV. If both of the tests are positive, only then it could enter the next level/step of detection, which is for the RdRp gene and is called the SARS-CoV-2-specific RT-PCR test [27].

Despite the various protocols developed by numerous institutions for SARS-CoV-2 testing, it is still not crystal clear whether the outcomes of the [28–31] nucleic acid tests based on the different targets can be compared or not. Various RNA transcripts that were extracted from a COVID-19 patient by Chantal et al. were used to study the detailed analytical sensitivities of the four qRT-PCR assays rooted in Hong Kong, Germany, China, and the United States. According to a study, in all the primer–probe sets enforced in the qRT-PCR tests, SARS-CoV-2 could be identified; however, there was a significant disparity in the ability to find the positives and negatives with a lesser viral load and in the detection limit. HKU-ORF1 (Hong Kong), 2019-nCV_N1 (United States), and E-Sarbeco (Germany) were found to have the highest sensitivity primer–probe sets, while RdRp-SARSr (Germany) had the lowest sensitivity, which can be due to the mismatching in the reverse primer. Also, the sputum samples or nasopharyngeal swab from the [32–34] COVID-19 patients (Germany) were used for comparing the qRT-PCR tests in a commercial reagent and different polymerase chain reaction (PCR) systems. A clear discrepancy in the analytical sensitivities among different PCR systems was detected when the same probes and primers were used. The results concluded that when a one-step qRT-PCR system was used, the RdRp [35] target was less sensitive than the E gene target. However, the test evaluation was not crystal clear as it was disturbed by the high background nature of the E gene target. The sensitivity may be improved by the additional optimization of the E gene assay [36].

1.2.2.2 Detection Approach Through Nested RT-PCR

To detect the low-copy-number SARS-CoVs present in the early stage of the disease, real-time nested RT-PCR assay [37] is the perfect choice as it bridges the real-time instruments (time-saving) with the high sensitivity of the nested PCR. The identification of the SARS-CoV-2 with the help of nested RT-PCR has already [38] been verified in countries like Japan during the initial days of the pandemic. This technique had already detected 20–25 COVID-19-positive patients in Japan, as of the first week of February 2020. A new OSN-qRT-PCR assay (one-step nested real-time RT-PCR) was recently devised by Ji et al. for targeting the N genes and SARS-CoV-2 ORF1ab genes. This assay had a difference in sensitivity (1 copy/test and 10-fold higher) with that of the commercial qRT-PCR assay (10 copies/test). The OSN-qRT-PCR confirmed the 14 samples with qRT-PCR negative, among the 181 clinical samples taken. Additionally, it also confirmed the seven samples as positive with qRT-PCR positive in the gray zone. In comparison with the qRT-PCR kit, it was clearly shown that nested RT-PCR analysis has both higher specificity and sensitivity, thus confirming that nested RT-PCR should be used for the clinical application for detecting [39–41] the SARS-CoV-2 whenever the viral load is low. However, there is a great chance of cross-contamination in nested RT-PCR, which may end in false-positive/negative results.

1.2.2.3 Detection and Analysis Approach via Droplet Digital PCR

For enhancing the accuracy of SARS-CoV-2 detection, sensitivity, and lower LOD (limit of detection), a technique called ddPCR (droplet digital polymerase chain reaction) has been implemented. By using the exact probe/primer sets issued by China CDC targeting ORF1ab or N gene, the utility of the ddPCR technique was studied by Suo et al. for the detection of SARS-CoV-2 RNA compared with the qRT-PCR. The importance [41–45] of ddPCR can be understood by the fact that 26 patients with negative RT-PCR results were re-confirmed as COVID-19 positive using this technique. There was a huge improvement in the accuracy and sensitivity from 47% and 40% of RT-PCR to 95% and 94% of ddPCR, respectively. Almost 43% of patients (42.9% to be exact as 6/14 patients) were tested positive by ddPCR within 5–12 days after the discharge. According to a study, a clear and large decrease can be observed in the number of false-negative results of qRT-PCR. The eight primers/probe sets [46] with the exact conditions and samples were used to further analyze the ddPCR and qRT-PCR performance. The results confirmed that qRT-PCR often gives us false results whenever the viral load is low as all the eight probes/primers that were used in qRT-PCR were not able to effectively [47–57] differentiate the positive and negative at a low viral load of 10–14 dilutions. qRT-PCR tests with false-positive results of US CDC-N1, N2, and China CDC-N probe/primer sets were identified. Although ddPCR was better than qRT-PCR in the overall performance, especially in the case of low viral load samples, however, it also had some limitations. Presently, ddPCR is more costly than qRT-PCR for each test performed by using consumables and suitable instruments [58–65]. Also, precise materials and gold standards still need to be effectively defined to ensure the commutuality between the molecular diagnostic laboratories.

1.2.2.4 Lab-on-chip Approaches Using Nucleic Acid as Chief Target Points

Loop-Mediated Isothermal Amplification Rapid amplification at a single temperature, which is highly effective and efficient in the rapid and safe diagnosis of coronaviruses, is the advantage of LAMP (loop-mediated isothermal amplification). The full LAMP primers that target the 5′ region of N genes and ORF1a genes of the SARS-CoV-2 and detected via colorimetric and visual RT-LAMP alongside a monetary RT-PCR assay were designed. In his experiment, total seven samples were taken among which six exhibited a visible [66] change in the color, thus depicting positive amplification, whereas one sample did not change its color and remained pink and thus was confirmed as negative. The RT-PCR results and the colorimetric RT-LAMP analysis were 100% consistent with each other across a range of Cq values (cycle quantification value) and matched with the RT-PCR in the point-of-care settings and field without any calibrant instrumentation. An isothermal LAMP-based detection method was designed by Yu et al. for the ORF1ab gene and is known as the isothermal LAMP-based method for COVID-19 (ILACO). The comparison of 11 respiratory viruses' sequences (2 normal CoVs, 2 influenza viruses, and 7 similar CoVs) was done using ILACO, which ultimately showed the species specificity. Moreover, the sensitivity of Taqman-based RT-PCR and ILACOs' was comparable to each other, which can detect as low as 10 copies of SARS-CoV-2. Another extremely sensitive, point-of-care test based on LAMP and Penn-RAMP (rapid isothermal amplification assay), nested-like amplification assay, was [64] designed. For the testing of purified targets, LAMP and RT-PCR sensitivity was 10 times lesser than that of RAMP, and for testing the samples that are minimally processed, it was 100 times lesser than that of RAMP. The method of RAMP is perfect for home use, point-of-care, and in the clinic with the least trained people and minimal instrumentation. It can also lessen the false-negative results from the normal nucleic acid tests.

1.2.2.5 Analysis Through Nanoparticle Amplification Process

In the nucleic acid amplification system, there is an important application of nanoparticles for enhancing the specificity and sensitivity of SARS-CoV-2 detection. A naked-eye colorimetric method that is based on AuNPs [67] (gold nanoparticles) along with thiol-modified ASOs (antisense oligonucleotides) targeting the SARS-CoV-2 N-gene was developed. In the test performed by Parikshit, the LOD calculated was 0.18 ng/μl of the SARS-CoV-2 viral load. Moreover, in a one-step nanoparticle-based biosensor (NBS) that was coupled with RT-LAMP, the LOD found was 12 copies/test. When the RNA templates from non-COVID-19 patients were studied, it was found that the specificity of the assay among the laboratory-confirmed COVID-19 patients was 100% (96/96) and the analytical sensitivity of SARS-CoV-2 was also 100% (33/33). The nanoparticles have some different properties that provide them an extra advantage over the classical conventional methods that are laborious and more expensive. For the diagnosis of the SARS-CoV-2 disease in first-line clinical laboratories, the nanoparticle-based amplification is an effective technique, particularly in the areas that have limited medical resources. The only limitations of this technique are that the pretreatment steps are quite complex, and also it is way more expensive as compared to qRT-PCR. Also, a high risk of photobleaching is there [68] that might end in false-negative/positive results and can also decrease the sensitivity, as it uses the conventional organic carriers [69].

1.2.2.6 Portable Methodology: The Concept of Benchtop-Sized Analyzer

One of the powerful, accurate, and highly sensitive methods for the rapid detection of SARS-CoV-2 has been provided by the [70] automated molecular diagnostic platform. This assay can easily achieve various technological innovations and instant decisions, even without any point-of-care testing or any PCR training in the laboratory. It has been observed that for the detection of SARS-CoV-2, the performance of different portable benchtop-sized analyzers was inconsistent. The QIAstat-Dx Respiratory SARS-CoV-2 panel was evaluated by Benoit et al. for SARS-CoV-2 detection. The sensitivity of this platform (LOD at 1000 copies/ml) was comparable to that of RT-PCR. The overall percentage [71–74] recommended by WHO of QIAstat-Dx SARS and RT-PCR was 97%, with a sensitivity of 100% (40/40), and a specificity of 93% (27/29). There was no observed cross-reaction of any other bacteria or respiratory virus in this assay. According to the results, the sensitivity of the QIAstat-Dx Respiratory SARS-CoV-2 panel was comparable to that of the RT-PCR assay (Table 1.1).

According to a recent research, the PPA (positive percent agreement) between the ID NOW tests and an improved test developed by CDC laboratory is 94%, whereas, according to other assessments, the PPA of ID NOW is [75–79] lower (75–87%), when compared with laboratory-developed reference methods. The detection time for each sample is the fastest for ID NOW (~17 minutes) when compared with the ePlex assay (~1.5 hours) and the Xpert Xpress assay (~46 minutes), but the limitation is the decrease in clinical and analysis performance, with the lowest PPA and highest LOD. Moreover, this assay had a specificity of 100% as shown by a research; however, among the 46 SARS-CoV-2-positive samples taken, 13 were found to be false negative; thus, the sensitivity got reduced to 71.7%. All the false negatives were actually the weak-positive samples [80–84]. Thus, it is clear that for the samples with [75–79] average or high viral RNA load, ID NOW has fair performance but shows low sensitivity in the case of weakly positive samples.

Xpert Xpress point-of-care assay (Cepheid GeneXpert systems) was evaluated by Femke et al. to target the SARS-CoV-2 E-gene and N2-gene in the medical laboratories of the Netherlands. It can detect SARS-CoV-2 with an LOD of 8.26 copies/ml in these laboratories. However, the Xpert Xpress test was reported for targeting the SARS-CoV-2 E-gene and the N2-gene with an LOD of 100 copies/ml [85–89]. The various methods were used to identify the input concentration that ultimately resulted in this difference, and it requires more verification. Compared to the LOD of ePlex (1000 copies/ml) and ID NOW (20 000 copies/ml), Xpert Xpress had the lowest LOD (100 copies/ml) and highest PPA (98.3%) when compared to ePlex (91.4%) and ID NOW (87.7%) – according to Wei's study.

1.3 Antigenic Approach for COVID-19 Diagnosis

There are various virus-encoded proteins such as E, S, M, and N proteins in SARS-CoV-2. The main antigenic targets of SARS-CoV-2 antibodies are N and S proteins. The S protein is spliced into two different polypeptides (S1 and S2) in most of the CoVs by the action of a host cell furin-like protease. Although the S protein exists on the viral surface and is also essential for viral entry, still the protein that is the most abundantly expressed immune dominant protein that interacts with the RNA is the N protein, and the N protein is also more conserved than the S protein.

Table 1.1 Techniques incorporated for diagnosis and detection for COVID-19.

Techniques incorporated for diagnosis and detection	Method employed for the specific technique	Advantages offered by the technique	Disadvantages offered by the technique
Sequencing of genome	Nanopore-assisted and targeted sequencing	Accuracy is very high, variation can be monitored, turnaround time is rapid, and detection range is wide	Costly, skill is required, and delicate instruments are used
	Sequencing based on hybrid capture	Intra-individual variations can be detected easily, highly sensitive	Recombinant viruses or highly diverse viruses cannot be sequenced using this method, use of sophisticated instruments
	Amplicon sequencing	Sensitivity is high, convenient, highly economical, low viral load samples can be detected easily	Recombinant viruses or highly diverse viruses cannot be sequenced using this method, use of sophisticated instruments
Polymerase chain reaction	Nested RT-PCR	Higher specificity than RT-PCR, viruses with low copy number can be detected easily, highly sensitive, requires less time	The pretreatment steps are quite complex, requires highly skillful people
	qRT-PCR	Equipments are not expensive, time-saving, and highly sensitive	Frequent problems of false-negative, pretreatment steps are quite complex, requires highly skillful people
	ddPCR	Low viral load samples can be detected easily, does not depend on standard curve, sensitive	Exogenous contamination can be seen, much expensive than qRT-PCR
Immunological diagnostic tests	ELISA	Risk of infection is low, simple, and convenient, detection is quantitative	Sensitivity is less, may encounter cross-reactivity, time-consuming, use of highly expensive monoclonal antibodies

The S1 subunit in the S protein [90] is more eminently specific to SARS-CoV-2 and is less conserved, thus proving that for the COVID-19 serologic identification, the S1 subunit is much more specific as an antigen when compared with the S2 subunit or full-length S protein. Moreover, the cross-reactivity of the RBD (receptor-binding domain) with other CoVs is very less. Also, in comparison with the full-length S or S1 subunit, the RBD domain of the S1 protein is much more conserved. The targets used are the various forms of S protein (RBD or S2 domain, full-length S, S1 domain) and N protein.

The most frequently used technique for SARS-CoV-2 antigen detection is the immunochromatographic assay. Bioassay, Liming bio, Savant, and RapiGEN (the four lateral flow antigen detection kits) were analyzed and compared by Thomas et al. for SARS-CoV-2 detection. There was an observable difference in the test performances. Out of all the four tests, the test with the highest accuracy (89.2%) and κ coefficient of 0.8 was the Bioassay test, and because of the poor performance, the Liming bio test was discontinued while testing. The sensitivity of the other kits varied from 16.7% for Sarvant assay to 85% for the Bioassay test.

For the detection of SARS-CoV-2, various highly sensitive biosensor-based tests have been established when compared to the lower sensitivity of immunochromatography. To detect the SARS-CoV-2 S1 protein, a rapid, portable cell-based biosensor with human chimeric spike S1 antibody was developed by Sophie et al., which permits tests completed within three minutes with a 1 fg/ml detection limit and a 10 fg to 1 μg/ml semilinear response range. Furthermore, for targeting the SARS-CoV-2 S1 protein, eCovSens [91] (a biosensor device) was designed, who correlated it with another commercial potentiostat biosensor. In the saliva samples, eCovSens had an LOD of 90 fM, while the LOD for the commercial potentiostat biosensor was 120 fM. Thus, these are helpful for monitoring SARS-CoV-2 antigen on large scale, thus providing hope of eventual control of the pandemic.

1.4 Antibody Diagnostic Strategies for Detection of COVID-19

The antibody test for the diagnosis of a specific antigen has become a preference in detecting the rising titers of individual antibodies such as IgM, IgG, and IgA. In addition, these antibody productions can be an indicative strategy that relies on the appearance of different antibodies indicating different infection situations, such as the rise of IgM that is produced within [92] 4–7 days is helpful in determining the frequency of recurrence of the infection, while the rise of IgG (10–15 days) provides a sure reason for easy detectability (Figure 1.1) of the viral infection, respectively. Further IgA is a useful indicator of mucosal immunity and can be easily detected in mucus secretions within five to eight days of onset of infection.

In situations where RT-PCR fails to demonstrate the results, serum analysis can be conducted during the important phases such as the acute and convalescent phase, which support validated serological procedures for rapid analysis of COVID-19.

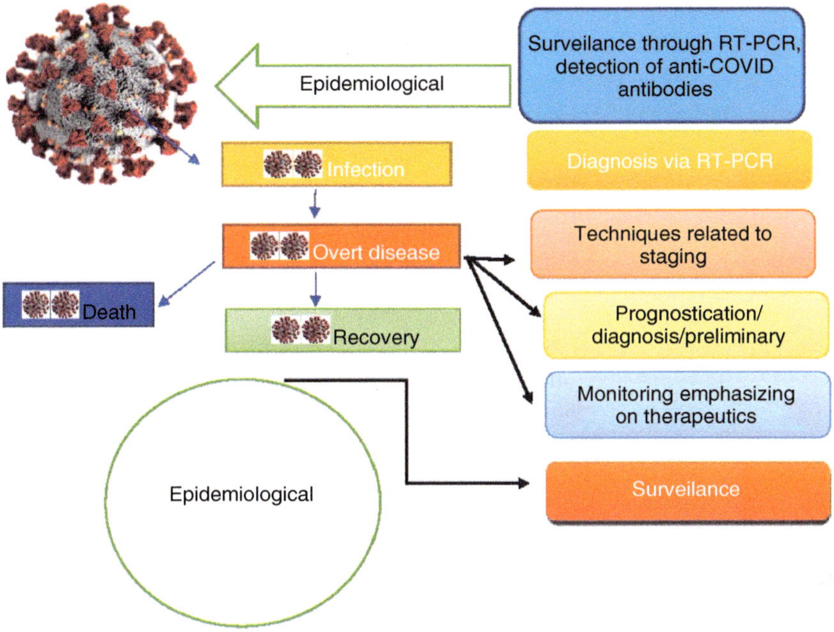

Figure 1.1 Epidemiological surveillance, monitoring, and prognostication of COVID-19. Courtesy of Alissa Eckert, MSMI, Dan Higgins, MAMS.

These antibody tests fall under two important diagnosis mechanisms, known as the "laboratory analysis and the point-of-care tests."

1.4.1 Enzyme-Linked Immunosorbent Strategies: The Vircell and Euroimmun ELISA

A study conducted highlights [93] the performance of different assays with a recombinant tagged N protein and S proteins corresponding to Vircell COVID-19 IgG and Euroimmun SARS-CoV-2 IgG.

The reports clearly indicated around 75–80% sensitivity in the case of Vircell enzyme-linked immunosorbent assay (ELISA) and around 50–60% in the case of Euroimmun ELISA on the 5–10th day of confirmation of COVID-19 by PCR. In addition, on the subsequent days, the sensitivity parameter was increased to 100% and 94% both in the case of Vircell and Euroimmun ELISA, respectively. Similarly, another research evaluated two specific diagnostic kits based on the N and S protein, where the sensitivity was reported around 90–95%. However, the overall sensitivity of both N- and S-based ELISA was around 65–70%, respectively.

1.4.2 Immunoassay-Based Detection Approach: Immunofluorescence and Chemiluminescence Assay

The first trial for detection through immunofluorescence technique was reported in Finland where the detection of SARS-CoV-2 IgM and IgG antibodies in serum was confirmed in a COVID-positive case [94].

The strategy involved the continuous dilution of patient's serum and a long incubation period in vero cell lines of 30 minutes for IgG and 2 hours for IgM. The sensitivity of IFA and the respective neutralization tests at different stages of COVID infection were recorded around 75–76%, which increased up to 100% by the 10th and 12th day of infection. Although this is a promising strategy for diagnostics, but in non-fluorescent cases, it may give false assumptions and results, research is needed to upgrade the specificity and sensitivity of the diagnostic strategies related to IFA. The advantage of quantitative detection and analysis belongs to the most popular diagnostic strategy known as the "chemiluminescence immunoassay (CLIA)." The recent research employs the benefit of detection against a specific target ranging from an ORF to N and S proteins. The individual sensitivity of both the antibodies (IgM and IgG) was recorded around 55–71%, whereas a combined sensitivity percentage increased to 82%. A similar research was carried out with four rapid tests known as "three LFA test and ELISA targeting IgM and IgG," for detecting COVID-19.

1.5 Point-of-care/Lab-on-chip Approaches: The LFA (Lateral Flow Assay)

The lateral flow assay (LFA) focuses on both in vitro semiquantitative and qualitative analysis and detection of SARS-CoV-2 IgM and IgG antibodies in plasma, serum, and venous blood samples [84–87].

Recent research studies highlight the working of three LFA tests known as the Quick Zen a Labo On Time and Avioq and in addition two [95] quantitative for the detection of SARS-CoV-2 IgG, IgM, and IgA antibodies in serum samples confirmed by RT-PCR. The test was recorded with 100% specificity in the analysis process, and further sensitivity of all these tests ranged from 90% to 95%, respectively. A related research also confirmed the three important factors (specificity, sensitivity, and seropositivity) in the diagnosis of COVID-19 patients. The recorded data show the sensitivity percentage of about 90% for IgG, followed by 91% in IgM and around 98% when seen in combination. A comparison between sensitivity percentage clearly depicts that a decrease has been witnessed in terms of sensitivity in IgG antibody as compared to IgM.

1.6 Miniaturization Detection Approach: Combining Microarray with Microfluidic Chip Technology

The concept of lab-on-chip technology has provided a strong microarray and microfluidic platform of technology where miniaturization can lead to automation and portability, high sensitivity, and high-throughput analysis in nano-based strategies.

The whole idea depends on the incorporation and integration of specific functions into small platforms known as "chips" for pathogen detection and diagnosis. Literature reports regarding a 65-microarray antigen concept for diagnosing respiratory viruses related to various species of SARS. It additionally includes the coated [96] antigens corresponding to the S, N, S2 MERS-CoV protein. This platform provides advantages related to low cost, high sensitivity, and high specificity and proves to be a potential valuable tool for sero-survillence of COVID-19 patients. The techniques allow easy and specific detection of COVID-19 but lacks in expressing in some important mammalian cells that generally needs to be optimized and standardized.

1.7 Neutralization Detection Approaches Toward COVID-19

The gold standard evaluation strategies include the discoveries related to viruses, their pathogenesis, and their ability to induce infection. In this context, neutralization tests have been recommended to evaluate the serum capacity derived from COVID-19-infected person to reduce down the CPE effects (cytopathic effects) [97] caused by SARS-CoV-2. A strong positive correlation is being reported between the neutralizing titers of antibodies and total CoV IgG (anti-S1 IgA, IgG, and IgM) antibodies. Neutralization assay stands as a specific choice of test to monitor patient immunity to the virus. However, when compared with serological tests, neutralization tests are marked as laborious and tedious and have a limitation of being SARS-CoV-2 restricted to only biosafety level 3 or BSL 2, respectively.

1.8 Genomic Sequencing Detection Approach: The Amplicon, Hybrid Capture, and Meta-transcriptomic Strategy

A powerful tool known as "genomic sequencing" for analyzing the evolution of virus, correlating genetic association to different diseases, tracing the outbreaks of diseases, and finally developing new strategies, therapies, and vaccines is the need of the hour. The first genomic sequence of SARS-CoV-2 was done by combining a meta-transcriptomic technique with Sanger's sequencing method [11, 33]. Research studies carried out by Lu et al. reported around 10 genome sequences of SARS-CoV-2 from targeted patients including BALF and culture samples with the help of meta-transcriptomic sequencing. The genome sequences in the study were similar; having 99% sequence similarity, the 10 genome sequences were nearly identical, displaying more than 99.98% sequence identity [13].

The genome analysis of the nasopharyngeal and oropharyngeal found somewhat identical to each other and similar to the SARS-CoV-2 sequences. The method involving the amplicon sequencing method and hybrid capture sequencing includes very high sensitivity with low accuracy. Further, these techniques cannot be applied to recombinant or hybrid viral strains because of the unavailability of probes and primers. Significant increase is being witnessed for the SARS reads, indicating enrichment efficiency that ranges from 5710- to 5595-fold in amplicon and hybrid sequencing patterns. However, lower frequency of sequencing for lower viral loads has been displayed by the alleles identified by hybrid capture sequencing when compared to amplicon and meta-transcriptomic sequencing [46, 98–109].

1.9 Conclusion

The global spread of COVID-19 has not only restrained the economic security but has also led to the health risks and threats to acerbate at a high rate. The detection and analysis thus become priority to diagnose the infection at an early stage and finally control its spread and transmission. RT-PCR routine confirmation has been widely accepted as the gold standard test for the screening and identifying SARS-CoV-2, which emphasizes the identification of conserved regions pertaining to the viral genome. However, some loopholes have been encountered regarding various mismatches in RNA primers, probes that can lead to poor performance ratios, and finally can result in negative or false results. In addition, serological tests can result in highlighting outbreaks and assessment of the percentage of viral attack in terms of antibody titer evaluation (IgG and IgM). However, these serological methods highlight diverse seroconversion processes that are not reliable for early detection. Further the use of immunological based strategies and molecular diagnosis are not suitable or not preferred for lab-on-chip or point-of-care detection strategies because of time consumption, expensive inputs, and strict biosafety regulations. Many upcoming strategies and methodologies that offer safety, quick diagnosis, efficacy, and sensitivity is into limelight (benchtop analyzers and lateral flow process). The demand of the present and future era is to provide and standardize some synergistic combinations of the techniques so that various advantages can be combined up for monitoring, detecting, screening, and diagnosing the current COVID-19.

References

1 Su, S., Wong, G., Shi, W. et al. (2016). Epidemiology, genetic recombination, and pathogenesis of coronaviruses. *Trends Microbiol.* 24 (6): 490–502.

2 Huang, P., Wang, H., Cao, Z. et al. (2018). A rapid and specific assay for the detection of MERS-CoV. *Front. Microbiol.* 9: 1101.

3 World Health Organization (2020). Coronavirus disease (COVID-19) situation reports. https://www.who.int/docs/default-source/coronaviruse/situation-reports/20200507covid-19-sitrep-108.pdf?sfvrsn=44cc8ed8_2 (accessed 7 May 2020).

4 World Health Organization (2019). Coronavirus disease (COVID-2019) situation reports. https://www.who.int/emergencies/diseases/novel-coronavirus-2019/situation-reports (accessed 02 March 2021).

5 Tahamtan, A. and Ardebili, A. (2020). Real-time RT-PCR in COVID-19 detection: issues affecting the results. *Expert Rev. Mol. Diagn.* 20 (5): 453–454.

6 Xie, X., Zhong, Z., Zhao, W. et al. (2020). Chest CT for typical 2019-nCoV pneumonia: relationship to negative RT-PCR testing. *Radiology* 296: E41–E45.

7 Mitchell, S.L. and George, K.S. (2020). Evaluation of the COVID19 ID NOW EUA assay. *J. Clin. Virol.* 128: 104429.

8 Zhou, P., Yang, X.L., Wang, X.G. et al. (2020). A pneumonia outbreak associated with a new coronavirus of probable bat origin. *Nature* 579 (7798): 270–273.

9 Maurier, F., Beury, D., Fléchon, L. et al. (2019). A complete protocol for whole-genome sequencing of virus from clinical samples: application to coronavirus OC43. *Virology* 531: 141–148.

10 Zhu, N., Zhang, D., Wang, W. et al. (2020). A novel coronavirus from patients with pneumonia in China, 2019. *N. Engl. J. Med.* 382 (8): 727–733.

11 Chen, L., Liu, W., Zhang, Q. et al. (2020). RNA based mNGS approach identifies a novel human coronavirus from two individual pneumonia cases in 2019 Wuhan outbreak. *Emerg. Microbes Infect.* 9 (1): 313–319.

12 Wu, C., Liu, Y., Yang, Y. et al. (2020). Analysis of therapeutic targets for SARS-CoV-2 and discovery of potential drugs by computational methods. *Acta Pharm. Sin. B* 10 (5): 766–788.

13 Lu, R., Zhao, X., Li, J. et al. (2020). Genomic characterisation and epidemiology of 2019 novel coronavirus: implications for virus origins and receptor binding. *Lancet* 395 (10224): 565–574.

14 Lam, T.T., Jia, N., Zhang, Y.W. et al. (2020). Identifying SARS-CoV-2-related coronaviruses in Malayan pangolins. *Nature* 583 (7815): 282–285.

15 Zhang, T., Wu, Q., and Zhang, Z. (2020). Probable pangolin origin of SARS-CoV-2 associated with the COVID-19 outbreak. *Curr. Biol.* 30 (8): 1578.

16 Liu, Y., Hu, G., Wang, Y. et al. (2020). Functional and genetic analysis of viral receptor ACE2 orthologs reveals a broad potential host range of SARS-CoV-2. *bioRxiv*. https://doi.org/10.1101/2020.04.22.046565.

17 Liu, Z., Xiao, X., Wei, X. et al. (2020). Composition and divergence of coronavirus spike proteins and host ACE2 receptors predict potential intermediate hosts of SARS-CoV-2. *J. Med. Virol.* 92 (6): 595–601.

18 (a) Riou, J. and Althaus, C.L. (2020). Pattern of early human-to-human transmission of Wuhan 2019 novel coronavirus (2019-nCoV), December 2019 to January 2020. *Eurosurveillance* 25 (4), 2000058; (b) Li, C., Zhao, C., Bao, J. et al. (2020). Laboratory diagnosis of coronavirus disease-2019 (COVID-19) *Clin. Chim. Acta; Int. J. Clin. Chem.* 510: 35–46. https://doi.org/10.1016/j.cca.2020.06.045.

19 Zhao, S., Lin, Q., Ran, J. et al. (2020). Preliminary estimation of the basic reproduction number of novel coronavirus (2019-nCoV) in China, from 2019 to 2020: a data-driven analysis in the early phase of the outbreak. *Int. J. Infect. Dis.* 92: 214–217.

20 Sanche, S., Lin, Y.T., Xu, C. et al. (2020). High contagiousness and rapid spread of severe acute respiratory syndrome coronavirus 2. *Emerg. Infect. Dis.* 26 (7): 1470–1477.

21 World Health Organization (2003). Consensus document on the epidemiology of severe acute respiratory syndrome (SARS). https://www.who.int/csr/sars/en/WHOconsensus.pdf (accessed 02 March 2021).
22 Jin, Y., Yang, H., Ji, W. et al. (2020). Virology, epidemiology, pathogenesis, and control of COVID-19. *Viruses* 12 (4): 372.
23 Guo, Z.D., Wang, Z.Y., Zhang, S.F. et al. (2020). Aerosol and surface distribution of severe acute respiratory syndrome coronavirus 2 in hospital wards, Wuhan, China, 2020. *Emerg. Infect. Dis.* 26 (7): 1583–1591.
24 van Doremalen, N., Bushmaker, T., Morris, D.H. et al. (2020). Aerosol and surface stability of SARS-CoV-2 as compared with SARS-CoV-1. *N. Engl. J. Med.* 382 (16): 1564–1567.
25 Zhang, W., Du, R.H., Li, B. et al. (2020). Molecular and serological investigation of 2019-nCoV infected patients: implication of multiple shedding routes. *Emerg. Microbes Infect.* 9 (1): 386–389.
26 Chen, W., Lan, Y., Yuan, X. et al. (2020). Detectable 2019-nCoV viral RNA in blood is a strong indicator for the further clinical severity. *Emerg. Microbes Infect.* 9 (1): 469–473.
27 Holshue, M.L., DeBolt, C., Lindquist, S. et al. (2020). First case of 2019 novel coronavirus in the United States. *N. Engl. J. Med.* 382 (10): 929–936.
28 World Health Organization (2019). Laboratory testing for 2019 novel coronavirus (2019-nCoV) in suspected human cases. https://www.who.int/publications-detail/laboratory-testing-for-2019-novel-coronavirus-in-suspected-human-cases-20200117 (accessed 19 March 2020).
29 Chan, J.F., Yuan, S., Kok, K.H. et al. (2020). A familial cluster of pneumonia associated with the 2019 novel coronavirus indicating person-to-person transmission: a study of a family cluster. *Lancet* 395 (10223): 514–523.
30 Kim, J.M., Chung, Y.S., Jo, H.J. et al. (2020). Identification of coronavirus isolated from a patient in Korea with COVID-19. *Osong Public Health Res. Perspect.* 11 (1): 3–7.
31 Yu, F., Yan, L., Wang, N. et al. (2020). Quantitative detection and viral load analysis of SARS-CoV-2 in infected patients. *Clin. Infect. Dis.* 71 (15): 793–798.
32 Rothe, C., Schunk, M., Sothmann, P. et al. (2020). Transmission of 2019-nCoV infection from an asymptomatic contact in Germany. *N. Engl. J. Med.* 382 (10): 970–971.
33 Wu, F., Zhao, S., Yu, B. et al. (2020). A new coronavirus associated with human respiratory disease in China. *Nature* 579 (7798): 265–269.
34 Guan, W.J., Ni, Z.Y., Hu, Y. et al. (2020). Clinical characteristics of coronavirus disease 2019 in China. *N. Engl. J. Med.* 382 (18): 1708–1720.
35 Peng, L., Liu, J., Xu, W. et al. (2020). SARS-CoV-2 can be detected in urine, blood, anal swabs, and oropharyngeal swabs specimens. *J. Med. Virol.*
36 To, K.K.-W., Tsang, O.T.-Y., Chik-Yan Yip, C. et al. (2020). Consistent detection of 2019 novel coronavirus in saliva. *Clin. Infect. Dis.* ciaa149.

37 To, K.K., Tsang, O.T., Leung, W.S. et al. (2020). Temporal profiles of viral load in posterior oropharyngeal saliva samples and serum antibody responses during infection by SARS-CoV-2: an observational cohort study. *Lancet Infect. Dis.* 20 (5): 565–574.

38 Chen, J.H., Yip, C.C., Poon, R.W. et al. (2020). Evaluating the use of posterior oropharyngeal saliva in a point-of-care assay for the detection of SARS-CoV-2. *Emerg. Microbes Infect.*: 1–14.

39 Tan, L.V., Ngoc, N.M., That, B.T.T. et al. (2020). Duration of viral detection in throat and rectum of a patient with COVID-19. *medRxiv*. https://doi.org/10.1101/2020.03.07.20032052.

40 Zhou, Y., Zeng, Y., Tong, Y., and Chen, C. (2020). Ophthalmologic evidence against the interpersonal transmission of 2019 novel coronavirus through conjunctiva. *medRxiv*. https://doi.org/10.1101/2020.02.11.20021956.

41 Xia, J., Tong, J., Liu, M. et al. (2020). Evaluation of coronavirus in tears and conjunctival secretions of patients with SARS-CoV-2 infection. *J. Med. Virol.* https://doi.org/10.1002/jmv.25725.

42 Hong, K.H., Lee, S.W., Kim, T.S. et al. (2020). Guidelines for laboratory diagnosis of coronavirus disease 2019 (COVID-19) in Korea. *Ann. Lab. Med.* 40 (5): 351–360.

43 Peng, L., Liu, J., Xu, W. et al. (2020). 2019 novel coronavirus can be detected in urine, blood, anal swabs and oropharyngeal swabs samples. *medRxiv*. https://doi.org/10.1101/2020.02.21.20026179.

44 World Health Organization (2020). Coronavirus disease (COVID-19) technical guidance: laboratory testing for 2019-nCoV in humans. https://www.who.int/emergencies/diseases/novel-coronavirus-2019/technical-guidance/laboratory-guidance (accessed 02 March 2021).

45 Centers for Disease Control and Prevention. Real-time RT-PCR panel for detection 2019-novel coronavirus. https://www.cdc.gov/coronavirus/2019-ncov/downloads/rt-pcr-panel-for-detection-instructions.pdf (accessed 4 February 2020).

46 Corman, V.M., Landt, O., Kaiser, M. et al. (2020). Detection of 2019 novel coronavirus (2019-nCoV) by real-time RT-PCR. *Eurosurveillance* 25 (3), 2000045.

47 Vogels, C.B.F., Brito, A.F., Wyllie, A.L. et al. (2020). Analytical sensitivity and efficiency comparisons of SARS-CoV-2 qRT-PCR assays. *medRxiv*. https://doi.org/10.1101/2020.03.30.20048108.

48 Konrad, R., Eberle, U., Dangel, A. et al. (2020). Rapid establishment of laboratory diagnostics for the novel coronavirus SARS-CoV-2 in Bavaria, Germany, February 2020. *Eurosurveillance* 25 (9), 2000173.

49 Lau, L.T., Fung, Y.-W.-W., Wong, F.-P.-F. et al. (2003). A real-time PCR for SARS-coronavirus incorporating target gene pre-amplification. *Biochem. Biophys. Res. Commun.* 312 (4): 1290–1296.

50 Jiang, S.S., Chen, T.-C., Yang, J.-Y. et al. (2004). Sensitive and quantitative detection of severe acute respiratory syndrome coronavirus infection by real-time nested polymerase chain reaction. *Clin. Infect. Dis.* 38 (2): 293–296.

51 Shirato, K., Nao, N., Katano, H. et al. (2020). Development of genetic diagnostic methods for novel coronavirus 2019 (nCoV-2019) in Japan. *Jpn. J. Infect. Dis.* https://doi.org/10.7883/yoken.JJID.2020.061.

52 Wang, J., Cai, K., Zhang, R. et al. (2020). Novel one-step single-tube nested quantitative real-time PCR assay for highly sensitive detection of SARS-CoV-2. *Anal. Chem.* 26 (8): 1076–1081.

53 Peiris, J.S.M., Chu, C.M., Cheng, V.C.C. et al., H.U.S.S. Group (2003). Clinical progression and viral load in a community outbreak of coronavirus-associated SARS pneumonia: a prospective study. *Lancet (London, England)* 361 (9371): 1767–1772.

54 Liu, X., Feng, J., Zhang, Q. et al. (2020). Analytical comparisons of SARS-CoV-2 detection by qRT-PCR and ddPCR with multiple primer/probe sets. *Emerg. Microbes Infect.* 9 (1): 1175–1179.

55 Suo, T., Liu, X., Feng, J. et al. (2020). ddPCR: a more accurate tool for SARS-CoV-2 detection in low viral load specimens. *Emerg. Microbes Infect.* 9 (1): 1259–1268.

56 Zhang, Y., Odiwuor, N., Xiong, J. et al. (2020). Rapid molecular detection of SARS-CoV-2 (COVID-19) virus RNA using colorimetric LAMP. *medRxiv*. https://doi.org/10.1101/2020.02.26.20028373.

57 Zhao, Z., Cui, H., Song, W. et al. (2020). A simple magnetic nanoparticles-based viral RNA extraction method for efficient detection of SARS-CoV-2. *bioRxiv*. https://doi.org/10.1101/2020.02.22.961268.

58 El-Tholoth, M., Bau, H.H., and Song, J. (2020). A single and two-stage, closed-tube, molecular test for the 2019 novel coronavirus (COVID-19) at home, clinic, and points of entry. *ChemRxiv*. https://doi.org/10.26434/chemrxiv.11860137.v1.

59 Zhu, X., Wang, X., Han, L. et al. (2020). Reverse transcription loop-mediated isothermal amplification combined with nanoparticles-based biosensor for diagnosis of COVID-19. *medRxiv*. https://doi.org/10.1101/2020.03.17.20037796.

60 Moitra, P., Alafeef, M., Dighe, K. et al. (2020). Selective naked-eye detection of SARS-CoV-2 mediated by N gene targeted antisense oligonucleotide capped plasmonic nanoparticles. *ACS Nano* 14 (6): 7617–7627.

61 Halfpenny, K.C. and Wright, D.W. (2010). Nanoparticle detection of respiratory infection. *Wiley Interdiscip. Rev. Nanomed. Nanobiotechnol.* 2 (3): 277–290.

62 (a) Visseaux, B., Le Hingrat, Q., Collin, G. et al. (2020). Evaluation of the QIAstat-Dx respiratory SARS-CoV-2 panel, the first rapid multiplex PCR commercial assay for SARS-CoV-2 detection. *J. Clin. Microbiol.* 58 (8): e00630-20; (b) Li, C., Zhao, C., Bao, J. et al. (2020). Laboratory diagnosis of coronavirus disease-2019 (COVID-19) *Clin. Chim. Acta; Int. J. Clin. Chem.* 510: 35–46. https://doi.org/10.1016/j.cca.2020.06.045.

63 Rhoads, D.D., Cherian, S.S., Roman, K. et al. (2020). Comparison of Abbott ID Now, Diasorin Simplexa, and CDC FDA EUA methods for the detection of SARS-CoV-2 from nasopharyngeal and nasal swabs from individuals diagnosed with COVID-19. *J. Clin. Microbiol.* 58 (8): e00760-20.

64 Zhen, W., Smith, E., Manji, R. et al. (2020). Clinical evaluation of three sample-to-answer platforms for the detection of SARS-CoV-2. *J. Clin. Microbiol.* 58 (8): e00783-20. https://doi.org/10.1128/JCM.00783-20.

65 Harrington, A., Cox, B., Snowdon, J. et al. (2020). Comparison of Abbott ID Now and Abbott m2000 methods for the detection of SARS-CoV-2 from nasopharyngeal and nasal swabs from symptomatic patients. *J. Clin. Microbiol.* 58 (8): e00798-20.

66 Hogan, C.A., Sahoo, M.K., Huang, C. et al. (2020). Five-minute point-of-care testing for SARS-CoV-2: not there yet. *J. Clin. Virol.* 128: 104410.

67 Wolters, F., van de Bovenkamp, J., van den Bosch, B. et al. (2020). Multi-center evaluation of cepheid xpert®xpress SARS-CoV-2 point-of-care test during the SARS-CoV-2 pandemic. *J. Clin. Virol.* 128: 104426.

68 Meyer, B., Drosten, C., and Müller, M.A. (2014). Serological assays for emerging coronaviruses: challenges and pitfalls. *Virus Res.* 194: 175–183.

69 Ward, S., Lindsley, A., Courter, J., and Assa'ad, A. (2020). Clinical testing for COVID-19. *J. Allergy Clin. Immunol.* 146 (1): 23–34.

70 Nisreen, M.A.O., Marcel, A.M., Wentao, L. et al. (2020). Severe acute respiratory syndrome coronavirus 2 – specific antibody responses in coronavirus disease 2019 patients. *Emerg. Infect. Dis.* 26 (7): 35–46.

71 Lv, H., Wu, N.C., Tsang, O.T. et al. (2020). Cross-reactive antibody response between SARS-CoV-2 and SARS-CoV infections. *bioRxiv*. https://doi.org/10.1101/2020.03.15.993097.

72 Mahari, S., Roberts, A., Shahdeo, D., and Gandhi, S. (2020). eCovSens-ultrasensitive novel in-house built printed circuit board based electrochemical device for rapid detection of nCovid-19 antigen, a spike protein domain 1 of SARS-CoV-2. *bioRxiv*. https://doi.org/10.1101/2020.04.24.059204.

73 Diao, B., Wen, K., Chen, J. et al. (2020). Diagnosis of acute respiratory syndrome coronavirus 2 infection by detection of nucleocapsid protein. *medRxiv*. https://doi.org/10.1101/2020.03.07.20032524.

74 Weitzel, T., Legarraga, P., Iruretagoyena, M. et al. (2020). Head-to-head comparison of four antigen-based rapid detection tests for the diagnosis of SARS-CoV-2 in respiratory samples. *bioRxiv*. https://doi.org/10.1101/2020.05.27.119255.

75 Mavrikou, S., Moschopoulou, G., Tsekouras, V., and Kintzios, S. (2020). Development of a portable, ultra-rapid and ultra-sensitive cell-based biosensor for the direct detection of the SARS-CoV-2 S1 spike protein antigen. *Sensors* 20 (11): 3121.

76 Padoan, A., Sciacovelli, L., Basso, D. et al. (2020). IgA-Ab response to spike glycoprotein of SARS-CoV-2 in patients with COVID-19: a longitudinal study. *Clin. Chim. Acta* 507: 164–166.

77 Kohmer, N., Westhaus, S., Rühl, C. et al. (2020). Clinical performance of SARS-CoV-2 IgG antibody tests and potential protective immunity. *bioRxiv*. https://doi.org/10.1101/2020.05.08.085506.

78 Liu, W., Liu, L., Kou, G. et al. (2020). Evaluation of nucleocapsid and spike protein-based enzyme-linked immunosorbent assays for detecting antibodies against SARS-CoV-2. *J. Clin. Microbiol.* 58 (6): e00461-20.

79 Haveri, A., Smura, T., Kuivanen, S. et al. (2020). Serological and molecular findings during SARS-CoV-2 infection: the first case study in Finland, January to February 2020. *Eurosurveillance* 25 (11), 2000266.

80 Long, Q.-X., Deng, H.-J., Chen, J. et al. (2020). Antibody responses to SARS-CoV-2 in COVID-19 patients: the perspective application of serological tests in clinical practice. *medRxiv*. https://doi.org/10.1101/2020.03.18.20038018.

81 Zhang, J., Liu, J., Li, N. et al. (2020). Serological detection of 2019-nCoV respond to the epidemic: a useful complement to nucleic acid testing. *medRxiv*. https://doi.org/10.1101/2020.03.04.20030916.

82 Cai, X.F., Chen, J., Hu, J.L. et al. (2020). A peptide-based magnetic chemiluminescence enzyme immune assay for serological diagnosis of coronavirus disease 2019 (COVID-19). *J. Infect. Dis.* 222 (2): 189–193.

83 Montesinos, I., Gruson, D., Kabamba, B. et al. (2020). Evaluation of two automated and three rapid lateral flow immunoassays for the detection of anti-SARS-CoV-2 antibodies. *J. Clin. Virol.* 128: 104413.

84 Jia, X., Zhang, P., Tian, Y. et al. (2020). Clinical significance of IgM and IgG test for diagnosis of highly suspected COVID-19 infection. *medRxiv*. https://doi.org/10.1101/2020.02.28.20029025.

85 Li, Z., Yi, Y., Luo, X. et al. (2020). Development and clinical application of a rapid IgM-IgG combined antibody test for SARS-CoV-2 infection diagnosis. *J. Med. Virol.* https://doi.org/10.1002/jmv.25727.

86 Pérez-García, F., Pérez-Tanoira, R., Romanyk, J. et al. (2020). All test rapid lateral flow immunoassays is reliable in diagnosing SARS-CoV-2 infection from 14 days after symptom onset: a prospective single-center study. *J. Clin. Virol.* 129: 104473.

87 Van Elslande, J., Houben, E., Depypere, M. et al. (2020). Diagnostic performance of seven rapid IgG/IgM antibody tests and the Euroimmun IgA/IgG ELISA in COVID-19 patients. *Clin. Microbiol. Infect.* 26 (8): 1082–1087.

88 de Assis, R.R., Jain, A., Nakajima, R. et al. (2020). Analysis of SARS-CoV-2 antibodies in COVID-19 convalescent blood using a coronavirus antigen microarray. *bioRxiv*. https://doi.org/10.1101/2020.04.15.043364.

89 Hedde, P.N., Abram, T.J., Jain, A. et al. (2020). A modular microarray imaging system for highly specific COVID-19 antibody testing. *bioRxiv*. https://doi.org/10.1101/2020.05.22.111518.

90 Tan, X., Lin, C., Zhang, J. et al. (2020). Rapid and quantitative detection of COVID-19 markers in microliter sized samples. *bioRxiv*. https://doi.org/10.1101/2020.04.20.052233.

91 Jiang, H.-W., Li, Y., Zhang, H.-N. et al. (2020). Global profiling of SARS-CoV-2 specific IgG/IgM responses of convalescents using a proteome microarray. *medRxiv*. https://doi.org/10.1101/2020.03.20.20039495.

92 Centres for Disease Control and Prevention (2019). Interim guidelines for COVID-19 antibody testing. https://www.cdc.gov/coronavirus/2019-ncov/lab/resources/antibody-tests-guidelines.html (accessed 02 March 2021).

93 Wang, P., Liu, L., Nair, M.S. et al. (2020). SARS-CoV-2 neutralizing antibody responses are more robust in patients with severe disease. *bioRxiv*. https://doi.org/10.1101/2020.06.13.150250.

94 Varnaitė, R., García, M., Glans, H. et al. (2020). Expansion of SARS-CoV-2-specific antibody-secreting cells and generation of neutralizing antibodies in hospitalized COVID-19 patients. *bioRxiv*. https://doi.org/10.1101/2020.05.28.118729.

95 Xiao, M., Liu, X., Ji, J. et al. (2020). Multiple approaches for massively parallel sequencing of HCoV-19 (SARS-CoV-2) genomes directly from clinical samples. *bioRxiv*. https://doi.org/10.1101/2020.03.16.993584.

96 Wang, M., Fu, A., Hu, B. et al. (2020). Nanopore target sequencing for accurate and comprehensive detection of SARS-CoV-2 and other respiratory viruses. *medRxiv*. https://doi.org/10.1101/2020.03.04.20029538.

97 World Health Organization (2019). Laboratory biosafety guidance related to coronavirus disease 2019 (COVID-19). https://www.who.int/publications-detail/laboratory-biosafety-guidance-related-to-coronavirus-disease-2019-(covid-19) (accessed 18 March 2020).

98 World Health Organization (2019). Coronavirus disease (COVID-19) advice for the public. https://www.who.int/emergencies/diseases/novel-coronavirus-2019/advice-for-public (accessed 02 March 2021).

99 World Health Organization (2020). Rational use of personal protective equipment for coronavirus disease 2019 (COVID-19). https://apps.who.int/iris/bitstream/handle/10665/331215/WHO-2019-nCov-IPCPPE_use-2020.1-eng.pdf (accessed 7 March 2020).

100 World Health Organization (2020). Advice on the use of masks in the context of COVID-19. https://www.who.int/emergencies/diseases/novel-coronavirus-2019/advice-for-public/when-and-how-to-use-masks (accessed 5 June 2020).

101 Long, Y., Hu, T., Liu, L. et al. (2020). Effectiveness of N95 respirators versus surgical masks against influenza: a systematic review and meta-analysis. *J. Evidence Based Med.* 13 (2): 93–101.

102 Jefferson, T., Jones, M., Al Ansari, L.A. et al. (2020). Physical interventions to interrupt or reduce the spread of respiratory viruses. Part 1 – face masks, eye protection and person distancing: systematic review and meta-analysis. *medRxiv*. https://doi.org/10.1101/2020.03.30.20047217.

103 Li, Q., Guan, X., Wu, P. et al. (2020). *Clin. Chim. Acta* 510: 35–4645.

104 Corman, V.M., Landt, O., Kaiser, M. et al. (2020). Detection of 2019 novel coronavirus (2019-nCoV) by real-time RT-PCR. *Eurosurveillance* 25 (3), 2000045. doi: https://doi.org/10.2807/1560-7917.ES.2020.25.3.2000045. Erratum in: Euro Surveill. 2020 25(14): Erratum in: Eurosurveillance. 2020 25(30): PMID: 31992387; PMCID: PMC6988269.

105 Institut Pasteu (2020). Real-time RT-PCR assays for the detection of SARS-CoV-2. France: Institut Pasteur. https://www.who.int/docs/default-source/coronaviruse/real-time-rt-pcr-assays-for-the-detection-of-sars-cov-2-institut-pasteur-paris.pdf (accessed 2 March 2020).

106 Naganori, N. and Shirato, K. (2020). *Detection of Second Case of 2019-nCoV Infection in Japan*. Japan: Department of Virology III, National Institute of Infectious Diseases.

107 Department of Medical Sciences, Ministry of Public Health, Thailand (2020). Diagnostic detection of novel coronavirus 2019 by real time RT-PCR. https://www.who.int/docs/default-source/coronaviruse/conventional-rt-pcr-followed-by-sequencing-for-detection-of-ncov-rirl-nat-inst-health-t.pdf (accessed 23 January 2020).

108 Poon, L., Chu, D., and Peiris, M. (2020). *Detection of 2019 Novel Coronavirus (2019-nCoV) in Suspected Human Cases by RT-PCR*. Hong Kong: School of Public Health, The University of Hong Kong.

109 (a) Liu, T., Hsiung, J., Zhao, S. et al. (2020). High-accuracy multiplexed SARS-CoV-2 antibody assay with avidity and saliva capability on a nano-plasmonic platform. *bioRxiv*. https://doi.org/10.1101/2020.06.16.155580; (b) Li, C., Zhao, C., Bao, J. et al. (2020). Laboratory diagnosis of coronavirus disease-2019 (COVID-19) *Clin. Chim. Acta; Int. J. Clin. Chem.* 510: 35–46. https://doi.org/10.1016/j.cca.2020.06.045.

2

COVID-19 Diagnostics: Current Approach, Challenges, and Technology Adaptation

Prama Bhattacherjee[1], Santanu Patra[*2,3], Abhishek Mishra[4,5], Trupti R. Das[6], Hemlata Dewangan[7], Rajgourab Ghosh[8], Sudheesh K. Shukla[9], and Anshuman Mishra*[*2,3]*

[1] *Indian Institute of Technology (ISM) Dhanbad, Department of Chemistry, Jharkhand 826004, India*
[2] *Institute of Advanced Materials, IAAM, Gammalkilsvägen 18 Ulrika 590 53, Sweden*
[3] *VBRI, 7/16 Kalkaji Extension, New Delhi 110019, India*
[4] *Invertis University, Bareilly, Uttar Pradesh 243123, India*
[5] *Oriental Institute of Science & Technology, Bhopal 462021, India*
[6] *Veer Surendra Sai University of Technology (VSSUT), Department of Physics, Burla, Odisha 768018, India*
[7] *Shreyansh Hospital and Research Centre, Raipur, 493441, India*
[8] *Amity University, Amity Institute of Biotechnology (AIBNK), Kolkata, West Bengal 700135, India*
[9] *Department of Biomedical Engineering, School of Biological Engineering and Life Science, Shobhit Institute of Engineering & Technology (Deemed-to-be University), Modipuram, Meerut 250110, India*

2.1 Introduction

Humanity is challenged by many pandemics throughout its long history, which are caused by many deadly pathogens such as bacteria, protozoa, and viruses. Among the most recent deadly outbreaks are of Ebola, Zika, and the human coronaviruses (SARS-CoV and MERS-CoV). The present pandemic coronavirus disease 2019 (COVID-19) caused by novel coronavirus severe acute respiratory syndrome coronavirus-2 (SARS-CoV-2) has caused more mortality because of its high infectivity, absence of specific drugs, treatment, and standard healthcare practices.

The causative infectious agent SARS-CoV-2 is also called as COVID-19 disease, globally resulted in 38 002 699 confirmed cases, including 1 083 234 deaths, reported to World Health Organization (WHO; as of 4:14 p.m. CEST, 14 October 2020) (Figure 2.1a,b) [1]. This disease was declared pandemic on 11 March 2020 by WHO. SARS-CoV-2 is a highly contagious disease. The virus spreads mainly by a droplet from saliva and respiratory secretions of an infected person. Once the virus enters the respiratory tract, it causes upper respiratory infection and subsequently it infects the bronchial system and lungs causing pneumonia.

While immune responses are different between severely and moderately infected persons, uncontrolled levels of cytokines and chemokines cause overactive inflammatory responses or cytokine storm. Hyperactive immune response along with

*Corresponding authors: patra.santanu82@gmail.com (Santanu Patra), dr.anshuindia@gmail.com (Anshuman Mishra)

Detection and Analysis of SARS Coronavirus: Advanced Biosensors for Pandemic Viruses and Related Pathogens, First Edition. Edited by Chaudhery Mustansar Hussain and Sudheesh K. Shukla.

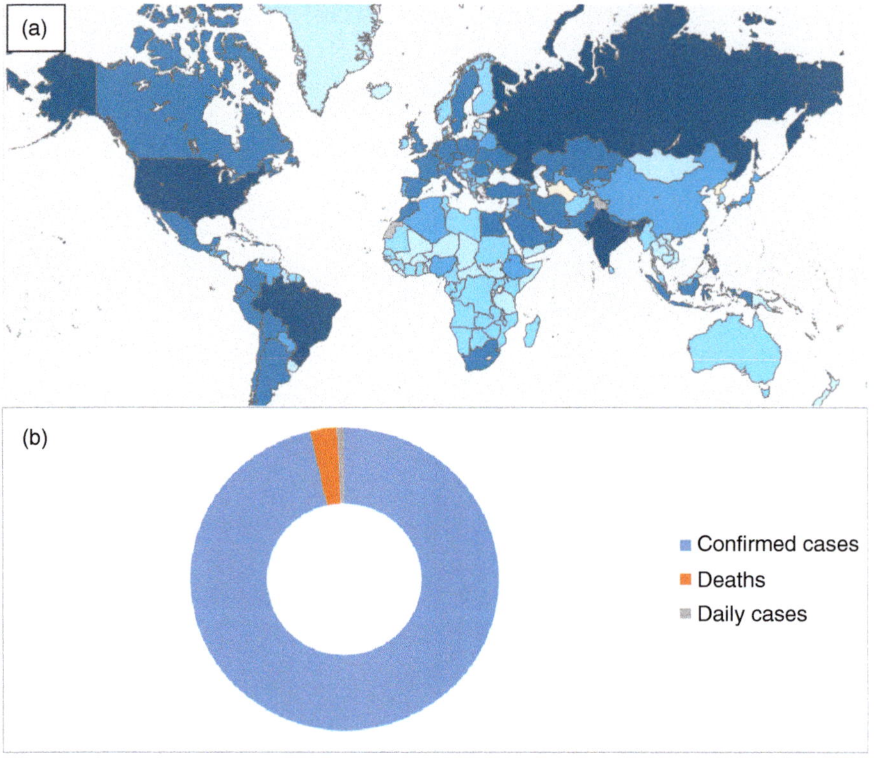

Figure 2.1 (a) COVID-19-confirmed cases in different countries. Source: World Health Organization, WHO Coronavirus Disease (COVID-19). (b) Globally, as of 4:14 p.m. CEST, 14 October 2020, there have been 38 002 699 confirmed cases of COVID-19, including 1 083 234 deaths and 287 031 daily cases, reported to WHO.

impaired adaptive immune response may trigger pulmonary injury, ARDS, viral sepsis, and organ failure-like complications and eventually death [2]. Study on angiotensin-converting enzyme 2 (ACE2) approved that COVID-19 is the gateway to multiple organ failure syndromes [3].

The etiologic diagnosis confirmation of COVID-19 is done primarily by genetic and immunological methods. Molecular diagnosis based on genetic material (detection of ribonucleic acid [RNA] of the SARS-CoV-2 virus) and immune assay (measuring antigens and antibodies in human serum) is a well-standardized method for clinical decision. Diagnostic tests covering molecular and antigen tests detect a virus for infected persons, while antibody tests or serology tests detect the status of past infections.

Studies discuss the critical research needs for better improvements in RT-PCR, development of alternative nucleic acid amplification techniques, incorporating CRISPR technology for point-of-care (POC) applications, and standardized genetic analysis of viral genome for more sensitive and specific results because of several challenges [4].

The challenges of immunological testing have many issues such as differentiation in viral loads during the course of the development of infection. This is due to the wrong protocol of sample collection, handling, or less specificity. The genetic methods have challenges because of changing virus genome, high rate of mutations, and different adaptability of virus in various geographical regions. Until now, human susceptibility and resistance to virus infections are not understood fully. Overall, worldwide crisis of COVID-19 resulted in shortage of resources, public health drawbacks, socioeconomic loss, and risk of increasing infections. Presently, the world needs increasing testing capacity along with rapid, reliable, and widely accepted diagnosis. A well-defined public health model having early diagnosis and control method approaches is considered the best at this stage.

In this perspective, this review article summarizes the update for the molecular diagnosis of COVID-19 diagnostic, highlighting the opportunities and challenges. The choice of an appropriate public health control model of the pandemic in terms of fast, cost-effective diagnosis, and treatment at the mass level is the correct lead.

2.2 Diagnosis of COVID-19

COVID-19 is an irresistible infection brought about by SARS-CoV-2. It was first detected in December 2019 in Hubei, Wuhan, and has brought about a progressing pandemic [5, 6]. The common symptoms of this disease are fever, hack, weakness, windedness, and loss of breathing [7]. Prescribed measures to prevent the infection include frequent hand washing and maintaining physical distance from others. Well-being authorities likely expressed that clinical evaluation, face covers, for example, N95 covers, face shields, and for those who care the infected people directly, PPE ought to be utilized distinctly by social insurance laborers [8]. There are no demonstrated antibodies or explicit antiviral medicines for COVID-19. Management involves the treatment of symptoms, supportive care, isolation, and experimental measures. When there is no vaccine and specific medicine in this place, the antiviral plasma becomes the treatment for COVID-19. Before the diagnosis, this COVID-19 can be detected by some treatment described below.

2.2.1 Clinical Diagnosis

Clinical diagnosis is based on symptoms and clinical findings. For an asymptomatic patient, generally lab investigation is not required. Table 2.1 refers to clinical feature-based information and test recovered from persons. The symptoms of COVID-19 infection varied and belong to the below mentioned clinical features. The symptoms and severity vary from person to person, and clinical findings may be of similar to pneumonia, septic shock, etc.

- *Fever may or may not be associated with chills and rigor*
- *Cough*
- *Shortness of breathing*
- *Respiratory distress*
- *Tiredness and weakness*
- *Headache and bodyache*

Table 2.1 Laboratory-based clinical diagnosis. Depending on clinical features categories of subject such as mild, moderate and severe, subject are refer to several clinical based biochemical, molecular and immunological tests on clinical practitioner recommendation.

Timing of test	Mild	Moderate	Severe/critical
At admission	CBC RBS ECG HbA1C for DM RFT LFT D-dimer	CBC with N/L ratio RBS ECG HbA1c for DM RFT LFT D-dimer CRP S. LDH S. Ferritin IL-6 ABG PT/INR Procalcitonin Chest X ray-PA HRCT-Thorax If cardiac patient Trop-T and I Echocardiography	CBC N/L ratio Ratio RBS ECG HbA1c for DM RFT LFT D-dimer CRP S. LDH S. Ferritin IL-6 S. Cotisol S. Magnesium S. Calcium ABG PT/INR Procalcitonin Chest X ray-PA HRCT-Thorax If cardiac patient Trop-T and I NTProBNP Echocardiography
Repeat daily		CBC, RFT, LFT, ABG	CBC, RFT, LFT, ABG
Repeat every 72 h	If initial D-dimer is high	CRP, D-dimer, S. Ferritin, LDH, CxR-PA	CRP, D-dimer, S. Ferritin, LDH, CxR-PA
At the time of discharge		CRP, D-dimer, S. Ferritin, LDH, CxR-PA	CRP, D-dimer, S. Ferritin, LDH, CxR-PA RT-PCR-Nasal & Throat

This table refer to clinical feature-based information's recovered from persons.
N/L, neutrophil/lymphocyte ratio.

- *Flu-like symptoms*
- *Loss of smell or taste*
- *Diarrhea and nausea*

2.2.2 Sample Collection and Testing

The samples tested vary from the region and testing methods. RT-PCR is the most overwhelmingly utilized technique for diagnosing COVID-19 utilizing respiratory samples [9]. Upper respiratory samples are comprehensively suggested in spite of the fact that lower respiratory samples are suggested for patients displaying gainful cough [10]. Upper respiratory lot tests incorporate nasopharyngeal swabs, oropharyngeal swabs, nasopharyngeal washes, and nasal suctions. Lower respiratory lot tests incorporate sputum, BAL liquid, and tracheal suctions. Generally, the respiratory specimens preferred for the diagnosis purposes are as follows:

- Nasal swabs
- Nasopharyngeal swabs
- Oropharyngeal swabs
- Bronchoalveolar lavage specimens

2.3 Understanding Genetic Consequences

Genetic information of the viruses is important for the medical and biological purposes and useful for the prevention, diagnosis, and therapy of infectious diseases [11].

2.3.1 SARS-CoV-2 Genome and Database

SARS-CoV-2 is an enveloped positive-sense single-stranded RNA (ssRNA) virus. The viral capsid is enclosed within a lipid bilayer, and the viral genome present inside encodes viral proteins. After infection in the host cells, the virus genome (RNA) of SARS-CoV-2 is replicated in 4 structural (E, M, S, and N) and 25 nonstructural proteins. SARS-CoV-2 genome sequences (~100 nm in diameter) range from 26 to 32 kb in length (~29 903 nucleotide [nt] long). More than 46 000 SARS-CoV-2 genome sequences have been publicly shared under the Global Initiative on Sharing All Influenza Data (GISAID) up to mid-June 2020 [12].

2.3.2 Infection and Genetic Diagnosis

Infection in host (human) is initiated by binding of the spike protein of virus to ACE2 receptors on the host cell surface. Later on, packaging of the genomic RNA with the structural proteins in the host cells results in the formation of new SARS-CoV-2 viruses. SARS-CoV-2 genetic tests are designed to target one or more common regions, including proteins (E, N, and S) and the open reading frame (ORF) region [13].

2.3.3 Real-Time PCR

Centers for Disease Control and Prevention (CDC) approved the qRT-PCR test as the standard for SARS-CoV-2 detection (https://www.cdc.gov/coronavirus/2019-ncov/lab/rt-pcr-panel-primer-probes.html).

Design a nucleic acid test for SARS-CoV-2. Nucleic corrosive testing is an essential technique for diagnosing COVID-19 [14]. Various converse record polymerase chain response (RT-PCR) packs have been planned to recognize SARS-CoV-2 hereditarily. RT-PCR includes the converse record of SARS-CoV-2 RNA into reciprocal DNA (cDNA) strands, trailed by enhancement of explicit areas of the cDNA. The design cycle for the most part includes two primary advances: (i) succession arrangement and groundwork design and (ii) measure advancement and testing. Corman et al. adjusted and examined various SARS-related viral genome arrangements to plan a lot of preliminaries and tests.

RT-PCR can be acted in either a one-venture or a two-venture measure [15–17]. In a one-venture examination, switch record and PCR intensification are merged into one response. This test organization can give quick and reproducible outcomes to high-throughput examination. The test is the trouble in enhancing the converse record and intensification steps as they happen at the same time, which prompts lower target amplicon age. In the two-venture measure, the response is done consecutively in independent cylinders. This test design is more delicate than the one-venture measure, yet it is also more time consuming and requires advancing extra parameters [9, 10]. Finally, controls should be painstakingly chosen to guarantee the dependability of the test and to recognize exploratory mistakes.

2.4 Understanding Immunological Consequences

The presence of acute antibody responses to COVID-19 in patients or asymptomatic persons generated as antiviral immunoglobulin-M (IgM) and antiviral immunoglobulin-G (IgG). The seroconversion for IgG and IgM occurred simultaneously or sequentially. The study suggests that seroconversion is approximately 13 days (median) for IgM and IgG [18]. However, this seroconversion started on day 5 after disease onset and IgG level rose earlier than IgM [19]. Thus, infection status and severity can be detected by antiviral immunoglobulin as described below:

- IgM may be a useful marker of more recent infection
- IgG is a reliable marker of past infection

Comparison between patients with different disease severities suggested that early seroconversion and high antibody titer were linked with less severe clinical symptoms [19]. The importance of serologic tests is more because of cost-effectiveness, rapidness, simplicity, and usefulness for field-based setup for confirming the status of infection. The antibody- or antigen-based serological testing can also be useful for suspected subjects having negative RT-PCR results.

2.4.1 Role of Immunological Test

Developed kits vary in collection methods, sample handling, amount of detection, sensitivity, and specificity. Immunological tests detect antibodies against most immunogenic proteins such as spike protein (S) and/or nucleoprotein (N) of COVID-19 virus [20]. The S protein has a receptor binding domain (RBD), through which virus makes connections with the ACE2 receptor present on the human cells.

Thus, serological tests detecting anti-spike protein antibodies (against spike protein) are considered as the marker of the human immune response against virus spike protein [21]. Antigen tests identify the most infectious people. Although, antigen assays are much faster and cheaper than PCR, however, they are less sensitive and require specific amounts of virus antigen for the detection.

2.4.2 Rapid Antigen Testing

The tests detect specific SARS-CoV-2 proteins called as antigens and indicate the presence of virus amount in the body and can identify those who are at greatest risk of spreading the disease. Antigen tests give results in less than 30 minutes, required thousands of virus particles per microliter to produce a positive result so, and a critical amount is required to detect it [22].

2.4.3 Rapid Antibody Tests

Antibody tests those sense molecules that people produce when they have been infected by the virus. SARS-CoV-2 infection antibodies can be developed against four structural proteins: spike (S), envelope (E), membrane (M), and nucleocapsid (N). Out of these, N and S protein are considered more immunogenic and suitable targets for antibody assays.

Antibodies, also called immunoglobulins are produced by a human in response to virus antigens. Immunological test kits detection IgG and IgM antibodies in human serum as rapid methods.

Development of antibodies can take time (several days) after an infection and present even after recovery. In SARS-CoV-2, antibody tests have limited use in diagnosis, and it can be used for sero-prevalence studies to understand the population exposure and immunity protection studies.

2.5 Protein Testing

Viral protein antigens and antibodies that are made because of a SARS-CoV-2 contamination can be utilized for diagnosing COVID-19. Changes in viral burden over the course of the disease may make viral proteins hard to identify. For instance, Lung et al. indicated high salivary viral burdens in the main week after beginning of indications, which progressively declined with time [23]. Interestingly, antibodies created in reaction to viral proteins may give a bigger window of time for in a roundabout way identifying SARS-CoV-2. Immunizer tests can be especially valuable for reconnaissance of COVID-19.

2.5.1 Computed Tomography

Because of the deficiency of packs and false-negative pace of RT-PCR, the Hubei Province, China, briefly utilized computed tomography (CT) filters as a clinical conclusion for COVID-19 [24]. Chest CT filters are non-intrusive and include taking numerous X-beam estimations at various edges over a patient's chest to create cross-sectional images [25, 26]. The pictures are dissected by radiologists to search for irregular highlights that can prompt a diagnosis. The imaging highlights of COVID-19 are various and rely on the phase of contamination after the beginning of indications.

2.6 Challenges

Challenges started from the collection, diagnosis methods, and detection of viral sample with sensitivity and specificity during the pandemic. Most importantly, false-negative or false-positive results from either detection of human antibodies or viral RNA are another challenge. Variation in viral loads, changing virus morphology, and genetic structure due to mutation also limiting factors in diagnosis and treatment. Varied clinical features and asymptomatic infections need critical research discussion and improvements in the existing diagnostic technology, procedures, and development of new technology for better solutions. Unfortunately, test sensitivity in asymptomatic patients is not established. Improved assays in molecular research via adding more biomarkers, incorporating CRISPR technology, and addition of POC tests are missing. Understanding epidemiology via pathogen diversity, human susceptibility, and risk factors study through environmental surveillance the is need of hours for better control strategies.

2.6.1 Challenges of Developing COVID-19 Tests

The pandemic has reminded us to make strategies for the mass community, i.e. development of technology for rapid and cost-effective diagnosis tests. The detection of viral RNA does not mean that it is transmissible in the community. The genetic material of dead viral particles can remain within the epithelial cells and can be detected [27]. Coronavirus test based on RT-PCR has its own disadvantages. qRT-PCR test requires complex equipment, extensive training for potential users, and multiple hours to complete the procedure [28]. There are three issues that have emerged with RT-PCR. To start with, the accessibility of PCR reagent packs has not kept up with request. Second, people group emergency clinics outside of metropolitan urban communities do not have the PCR foundation to oblige high test throughput. In conclusion, RT-PCR depends on the presence of noticeable SARS-CoV-2 in the sample collected.

Most importantly, positive results do not rule out the possibility of other infection or co-infection with other viruses because of sequence similarity or complementary primer binding or non-specific binding of biological components or interference of reactions during genetics assay. In an event where an asymptomatic patient was contaminated with SARS-CoV-2 yet has since recouped, PCR would not recognize this earlier disease; furthermore, control measures would not be authorized. In immunological assays, one potential challenge with creating exact serological tests incorporates expected cross-reactivity of SARS-CoV-2 antibodies with antibodies produced against different COVID-19. Lv et al. tried plasma tests from 15 COVID-19 patients against the S protein of SARS-CoV-2 and SARS-CoV and saw a high recurrence of cross-reactivity [29].

2.6.2 Sample Collection and Tests

SARS-CoV-2 has been detected in saliva from a nasopharyngeal aspirate via a non-invasive method of the diagnosis. Different diagnostic strategies of COVID-19 will be useful in asymptomatic, symptomatic, and healthy subjects. Varied clinical features and multiorgan failure are major concerns in COVID-19. Even in symptomatic patients, various severities are reported. Diagnosis of viral RNA in patients from another region of human body may lead to different clinical decisions and better understanding of severity of the diseases. It is a well-known fact that the receptor of SARS-CoV-2, ACE2, has been found highly expressed in several organelles. Diagnosis in gastrointestinal (GI) epithelial cells, and the stool specimens of infected patients studied and concerned for potential oral-fecal transmission of SARS-CoV-2 [30].

2.7 Advanced Diagnosis Technologies and Adaptation

COVID-19 pandemic has created worldwide panic and forced everyone to adopt hygienic and healthy life with more awareness about pathogen infections. Now onwards, many gadgets amplifying day by day in human life having new technologies and innovation changing living standards of the human. Currently, major technologies identified included telemedicine and mobile care based on robotics and artificial intelligence (AI) for complete healthcare packages.

2.8 Adaptation of a New Approach

In the present scenario, RT-PCR considered as gold standard tests and the nasopharyngeal swab considered as the gold standard specimen for COVID-19 diagnosis. There are many areas where future science and technology should be focused for better results in COVID-19 diagnosis. Adaptation of new approaches is a good option as it carries below benefits:

- Development of a more innovative method of sample source after validation.
- Adaptation of more advanced technology for genetic tests having high sensitivity.
- Coronavirus is at present determined to have RT-PCR and has been screened for with CT checks; however, every procedure has its own disadvantages.
- Different combinations need to be used in the SARS-CoV-2 diagnosis considering the inadequacies in the test sensitivity and specificity.
- In the interim, genetic RT-PCR is costly, require confidence of testing based on other earlier tests results such as phenotype-based clinical features or other immunological or biochemical based cheaper and specialized tests, and cannot explicitly analyze COVID-19.

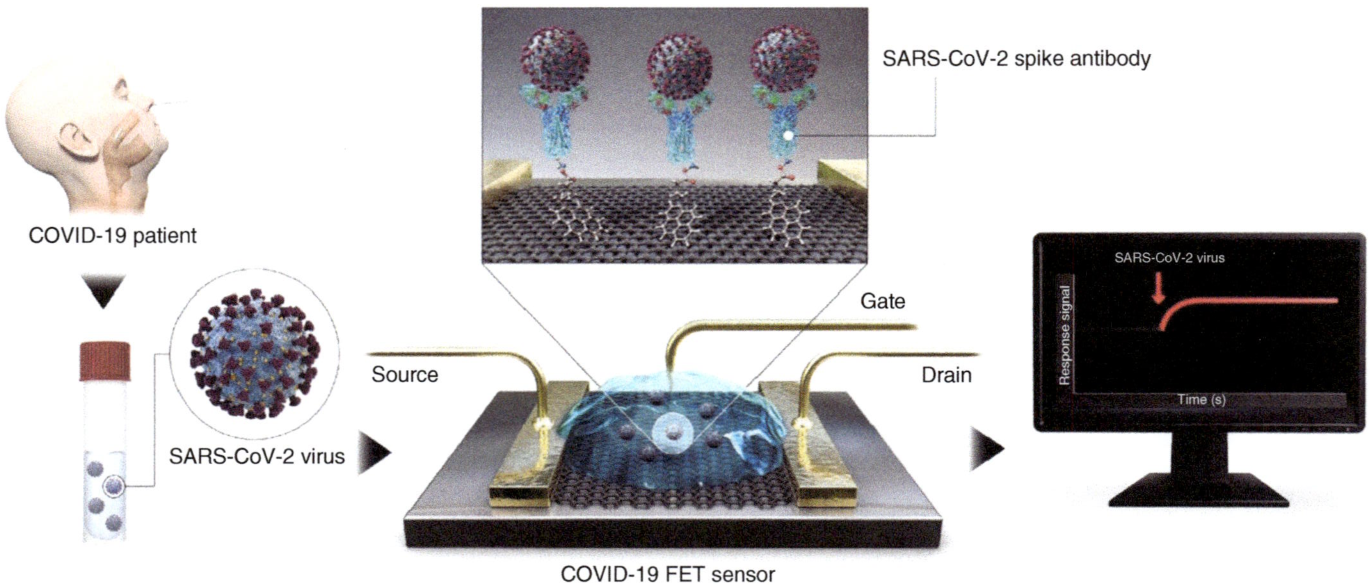

Figure 2.2 Schematic diagram of COVID-19 FET sensor operation procedure. Graphene as a sensing material is selected, and SARS-CoV-2 spike antibody is conjugated onto the graphene sheet via 1-pyrenebutyric acid *N*-hydroxysuccinimide ester, which is an interfacing molecule as a probe linker. Source: Seo et al. [32].

- At mass level, immunological test with "pooled testing" methods can be adopted.
- Clinical features should be monitored through technology and emergency use of genetic methods adopted to save resources.
- During a surge in infections, multiple tests should be performed for random samples to identify any other mechanism, pathogens, or infections in the region.

2.8.1 Emerging Diagnostic Tests for COVID-19

As indicated by the WHO, the quick need for COVID-19 diagnostics research is the improvement of nucleic corrosive and protein tests what's more, identification at the purpose of-care.2 The more drawn-out term need is to coordinate these tests into multiplex boards. So, as to improve reconnaissance endeavors, serological tests utilizing proteins are required notwithstanding nucleic basic analyses. These tests have the advantages of discovery after recuperation, not at all like nucleic basic analyses. This empowers clinicians to follow both wiped out and recuperated patients, giving a superior gauge of all out SARS-CoV-2 diseases. Purpose of-care tests are financially savvy, hand-held gadgets used to analyze patients outside of brought together offices. These can be worked in zones like network focuses to diminish the weight on clinical laboratories [31]. In Figure 2.2, the operation procedure has been demonstrated [32].

2.8.2 Role of siRNA, Nanoparticle Toward COVID-19

siRNA called short meddling RNA is a class of twofold abandoned non-coding RNA particles. siRNA controls quality articulation through a wonder called RNA interference [33]. As of late numerous investigations continuous for siRNA in the field of prostate and bosom malignant growth yet it ought to be engaged toward the COVID-19 on the grounds that As we probably am aware Coronaviruses are having a place with the family coronaviridae. Dissimilar to MERS-CoV and SARS-CoV, COVID-19 is the seventh individual from the COVID family that taints people [34]. In addition, it is a positive-abandoned RNA infection whose glycoprotein tops distend from the external film of the infection, introducing the presence of COVIDs or halogen infections [35]. For this situation, siRNA-based treatment could give a viable answer for COVID-19. In an event where there is a particular procedure that utilizes NPs to convey the siRNA-based COVID-19 to a particular site of disease, it might thump down the COVID-19-positive RNA. More direct, it can control the infection's generation. A large number of the medications that have been tried, for example, CQ, and HCQ hence, this medication has been tried in clinical preliminaries and end up being viable in COVID-19 [36]. This medication can likely be moved to explicit areas with NPs to expand its restorative impact with the assistance of NPs. In this way, a novel COVID treatment dependent on siRNA can be created, which can hit the exceptionally monitored district of COVID-19. Notwithstanding, nano-transporter conveyance frameworks will be more effective. Accordingly, these restrictions should be overwhelmed by the improvement of viable conveyance frameworks that can give exceptional advances in the region of breathed-in siRNA dose structures. This methodology can help accomplish better treatment objectives and hence decrease the pandemic danger of COVID-19.

2.8.3 RT-LAMP Nucleic Acid Testing

A reverse transcriptase loop-mediated isothermal amplification (RT-LAMP) was standardized using universal and serotype specific genes in a single tube. This test does not require sophisticated equipment and can detect virus at serotype level in an hour. In addition, the sensitivity and specificity of this test are comparable to conventional reverse transcriptase PCR and real-time PCR (rRT-PCR) [37]. LAMP possesses few advantages such as amplification at a constant temperature, exclusion of a thermal cycler, and rapid result, with good sensitivity and specificity, thus making it more suitable than RT-PCR [28]. RT-LAMP utilizes DNA polymerase and four to six preliminaries to tie to six particular locales on the objective genome. Nucleic analyses (RT-LAMP) utilizing isothermal enhancement for SARS-CoV-2 detection do not require specific lab hardware [38–41].

2.8.4 Point-of-care Testing

POC tests are utilized to analyze patients without sending tests to laboratory or incorporated offices, consequently empowering healthcare management to identify patients. Here, a paper-like film strip is covered with two lines: gold nanoparticle neutralizer forms are present in one line and catch antibodies in the other. The patient's sample (e.g. blood, swab, and pee) is kept on the layer, and the proteins are drawn over the strip by slim activity. As it passes the principal line, the antigens tie to the gold nanoparticle neutralizer forms and the complex streams together through the film. As they come to the second line, the complex is immobilized by the catch antibodies and a colored line is formed.

2.8.5 FNCAS9 Editor-Limited Uniform Detection Assay

The test, called FNCAS9 Editor-Limited Uniform Detection Assay (FELUDA), addresses the urgent need for accurate mass testing. FELUDA uses CRISPR-Cas technology for the detection of genes specific to SARS-CoV-2 virus (https://health.economictimes.indiatimes.com/news/medical-devices/indias-feluda-covid-19-test-cheaper-faster-alternative-to-rt-pcr/78381106). Here, a protein called FnCas9 and a guide RNA (gRNA) which helps in recognizing the viral genes is used. The test is rapid, low cost, and easy to use compared to the rRT-PCR test. If the patient sample has the viral gene, this gRNA–FnCas9 complex binds to the gene, and using a paper strip, this binding can be visualized. It had a 96% sensitivity, and a 98% specificity comes during the trials of 2000 patients. This test kit has been developed by CSIR-Institute of Genomics and Integrative Biology (CSIR-IGIB) scientists Souvik Maiti and Debojyoti Chakaraborty.

2.8.6 Development of a Novel Technology for COVID-19 Rapid Test

In a remarkable exertion, researchers at IIT Kharagpur have advanced a novel portable diagnostic device to recognize COVID-19 disease. This first-of-its-sort device will bring the testing for COVID-19 out from walls of costly labs and RT-PCR machines and empower testing at moderate expenses for the underserved network over the world. This whole test with the RNA separated from the patient spit tests can be directed in an ultra-low-cost portable enclosure in area as an option in contrast to particular research facility laboratories. A similar compact unit can be utilized for an enormous number of tests, on unimportant substitution of the paper cartridge after each test. The device has been proven to create no outcome with amazing precision and affectability good to standard RT-PCR tests. This test has an unprecedented minimal effort of not exactly Rs. 400/– per test, considering all parts of costs and plan of action. The institute is available to tie-ups, including a mode where the administration mediates with respect to meeting our minimal effort medicinal services objective for the underserved network as an arrangement measure to ensure the enthusiasm of general well-being in the midst of the pandemic circumstance, rather than simply building up a solid benefit-situated model [42–44].

2.8.7 Specific High-Sensitivity Enzymatic Reporter Unlocking

The diagnostic is based on the technology of specific high-sensitivity enzymatic reporter unlocking (SHERLOCK), a qualitative test for the detection of nucleic acid from SARS-CoV-2 in samples [44]. The test works via the programming of a CRISPR molecule designed to detect the genetic signature of the SARS-CoV-2 virus; if the signature is found, the CRISPR enzyme is activated and releases a detectable signal. Here, SHERLOCK enables rapid identification of a single alteration in a DNA or RNA sequence in a single molecule. The precision, coupled with its capability to be deployed to multiplex over 100 targets or as a simple POC system, will make it a rapid diagnostic for COVID-19.

2.9 Digital Healthcare Technologies

COVID-19 has demonstrated the impact of a technology-oriented healthcare consultancy using digital platform, which makes sure hygienic environment of healthcare professionals and other non-infected family members. Under health emergency, round the clock consultancy available from distant locations also and getting best suggestions for relieve. Hospital-at-home care, remote monitoring services, and persons under investigation in home quarantine all these concepts are reality, where physicians and health systems may need to track large populations of patients on a daily basis easily with digital technology from remote locations [45, 46].

Digital healthcare technologies and robot-induced instruments would assist for healthcare monitoring and generate real-time patient data and services under critical care conditions. Various advanced wearable devices can be used to monitor patient's physicality changes in a real-time manner. WHO recognized such field and consider it successful solutions for population screening, tracking the infection, prioritizing the use and allocation of resources, and designing targeted responses [47].

2.9.1 Artificial Intelligence and Mass Healthcare

AI significantly influences the practice of medicine and the delivery of public healthcare at mass level. With AI algorithms, healthcare professionals analyzed a much wider scope of data, new drug combinations for personalized healthcare to public healthcare. AI methods, potential benefits, challenges, and futuristic vision in an everyday medical practice are enormous [48].

The era of AI used inherited intelligence of a computer machine, which the computer learns by repeated use of end users (humans). This is a non-human intelligence that can do wonders with 90–96% accuracy and reliability [49].

AI and the underlying algorithms can predict the health issues and potentially save lives. The applications involve pattern recognition, robotics, speech recognition, etc., in various areas of medicine and healthcare.

The AI-driven technology (computational, ML, IoT, sensors, software, etc.) can help people with disease (infectious, Alzheimer's, lifestyle, etc.), evaluating signs and symptoms of the patients and mass community health data. Digital healthcare technologies and robot-induced instruments help extensively and transform the delivery of healthcare. However, AI adaptation required substantial training and expertise.

A risk-based approach to autonomy in medical AI systems should inform regulation, clinical validation, and clinician engagement, and it is the duty of providers to understand the ways in which AI systems might function autonomously [50].

2.9.2 Standard Healthcare Management During Pandemic Crisis

Successful disease management requires a rapid and sensitive diagnosis method that can recognize early infection even before the manifestation of its clinical signs. Standard healthcare management has a significant role to manage pandemic burden. During pandemic crisis, pathogens present a significant risk to the human health and generate health emergencies and shortages of resources. Rapidly spreading pathogens make people ill and weaken their immunity, which in turn increase the chances of other infections with related flu strains. In such case, phenotype-based diagnosis or early diagnosis is best lead. Antigen tests are easier to interpret; however, molecular tests are fast and highly sensitive compared to other tests for confirmatory results to start immediate treatment. Adaptation of regional and standard public healthcare model for control covering sanitization, diagnosis, and drug practices with a follow-up strategy is best health management. Consequently, control and diagnosis is the best precautionary strategy adopted for mass community. Under limited resources, genetic test should be directed toward those who need them most.

2.10 Implications of Technology-Based Diagnosis and Testing

As a general concept, antibody tests identify prior infection, while genetic diagnostic identifies current active infection. Considering the widespread impact of this pandemic, we have to develop multiple test options and treatment method systems as a diagnostic strategy in mass community, as given below:

- Asymptomatic testing – immunological and serological tests
- Symptomatic testing – genetic testing as RT-PCR

There are several factors that can lead to false results; in particular, false positives or false negatives are based on pathogen and host diversity. A test's sensitivity indicates the likelihood of false-negative results; its specificity indicates the likelihood of false-positive results. Asymptomatic persons were tested preferentially as they can be the carriers and spread the disease in the community. So, understanding test result and clinical feature correlations through follow-up is must. Sometimes, sampling time influences the result because of the delayed response of the human immune system or wrong methodology. Because of the aforementioned complication in testing, combined assay of immunological and genetic mechanism can be deployed as a POC test. Public healthcare management should adopt defined and structured diagnostic strategies to support a disease-free and healthy system.

2.10.1 Benefit of Diagnosis

Any testing (genetic and immunological) for diagnosis purpose is important for the implementation of effective public health management models to minimize the pandemic impact. Table 2.2 indicates the characteristics and usefulness of the diagnosis of the below methods.

- By using an antigen test, we can slow down the outbreaks, and the focus should be on identifying those who are at risk of spreading SARS-CoV-2 to other people [51].
- RT-PCR test is the gold standard test for the diagnosis of COVID-19. By detecting viral RNA, it is useful in detecting the active infection cases.
- The human immune system produces antibodies against virus proteins or antigens that are useful to fight and clear the virus.
- The immunity generation and lasting through antibody are not yet fully understood in the research studies, so antibody testing is not recommended until at least 14 days after the onset of the symptoms.
- Another benefit of accurate antibody testing is that people who have recovered from COVID-19 may be eligible to donate plasma, a part of their blood. This plasma could be used to treat others with severe disease and boost the ability to fight the virus. Doctors call this convalescent plasma.

2.11 Conclusion

Coronavirus has been tried on numerous medications, and a few outcomes are empowering. Notwithstanding, it is absolutely impossible to dispose of them, and if there are no successful antiviral medications to battle COVID-19, transformations, it might cause an emergency soon. Antiviral mixes have been tried

Table 2.2 Characteristics and usefulness of the diagnosis. Based on top three test such as RT PCR, Antigen and Antibody tests clinical professional start correlating with clinical features. However, RT PCR is most confirmatory test to start treatment.

S. no.	Category	RT-PCR	Antigen test	Antibody test
1	Detection	Nucleic acids of SARS-CoV-2	SARS-CoV-2 protein	Human immunoglobulin
2	Indication	Active cases	Active cases	Exposure status
3	Method	Swab	Swab	Serum or plasma
4	Positive results	Matching with genetic sequence	Binding with tagged antibody	Qualitative test for IgG/IgM
5	Cost effectiveness	High	Low	Low
6	Time	Longer, more than hour	Rapid, less than hour	Rapid, less than hour
7	Type	Genetic test	Immunological assay	Serology test
8	Usefulness	Information of active infection	Risk of spreading	Donate convalescent plasma
9	Sensitivity	High	Middle	Low

in numerous clinical preliminaries previously, however have not yet been utilized clinically and require pressing COVID-19 testing. Nanotoxicology is as yet a moderately new subject and should be additionally evolved in displaying and investigation. Without an inside and out comprehension of the well-being impacts of NPs, large-scale manufacturing, and applications in different fields will not be completely considered. The primary objective of COVID-19 diagnosis research is to find and focus on innovations of genomic approaches and technology adaptations. The digital medicine practices have less history of diagnosis and treatment of pathogen illness with less sensitivity. The integrated diagnostic workflow of molecular detection (detects currently active infection) and antibody testing (prior infection/immunity) should provide a better lead.

2.12 Future Prospects

The best solution of diagnosis lies in combined approaches where both biological diagnosis and digital diagnosis should be integrated with caution. If we avoid medical emergency, then this kind of practices should be well spread and serve common people in a cost-effective manner. Using more clinical research studies and advanced technologies in such kind of traditional practices will help more in the public healthcare management of emerging pathogens.

Acknowledgment

Dr. Anshuman Mishra acknowledge the Prof. Herbert Gleiter Fellowship from IAAM, Sweden. Abhishek Mishra acknowledge the DNA Consultancy Services, Bhopal 462041, Madhya Pradesh, India.

References

1 WHO Data. https://covid19.who.int (accessed 14 October 2020).
2 Prompetchara, E., Ketloy, C., Palaga, T. et al. (2020). Immune responses in COVID-19 and potential vaccines: lessons learned from SARS and MERS epidemic. *Asian Pac. J. Allergy Immunol.* 38: 1–9.
3 Loganathan, S., Kuppusamy, M., Wankhar, W. et al. (2020). Angiotensin-converting enzyme 2 (ACE2): COVID 19 gate way to multiple organ failure syndromes. *Respir. Physiol. Neurobiol.* 283: 103548. https://doi.org/10.1016/j.resp.2020.103548.
4 Feng, W., Newbigging, A.M., Le, C. et al. (2020). Molecular diagnosis of COVID-19: challenges and research needs. *Anal. Chem.* 92 (15): 10196–10209.
5 Lai, C.-C., Shih, T.-P., Ko, W.-C. et al. (2020). Severe acute respiratory syndrome coronavirus 2 (SARS-CoV-2) and coronavirus disease-2019 (COVID-19): the epidemic and the challenges. *Int. J. Antimicrob. Agents* 55: 105924.
6 Singhal, T. (2020). A review of coronavirus disease-2019 (COVID-19). *Indian J. Pediatr.* 87: 281–286.
7 Rothe, C., Schunk, M., Sothmann, P. et al. (2020). Transmission of 2019-nCoV infection from an asymptomatic contact in Germany. *N. Engl. J. Med.* https://doi.org/10.1056/NEJMc2001468.
8 Hafeez, A., Ahmad, S., Siddqui, S.A. et al. (2020). A review of COVID-19 (coronavirus disease-2019) diagnosis, treatments and prevention. *Eurasian J. Med. Oncol.* 4: 116–125.
9 Wong, M.L. and Medrano, J.F. (2005). Real-time PCR for mRNA quantitation. *BioTechniques* 39 (1): 75–85.
10 Bustin, S.A. (ed.) (2004). *A–Z of Quantitative PCR*. La Jolla, CA: International University Line.
11 Khailany, R.A., Safdar, M., Ozaslan, M. et al. (2020). Genomic characterization of a novel SARS-CoV-2. *Gene Rep.* 19: 100682. https://doi.org/10.1016/j.genrep.2020.100682.
12 Global Initiative on Sharing All Influenza Data (GISAID). www.gisaid.org (accessed 08 May 2020 and 15 June 2020).
13 Wu, F., Zhao, S., Yu, B. et al. (2020). A new coronavirus associated with human respiratory disease in China. *Nature* 579: 265–269.
14 US Centers for Disease Control and Prevention: Atlanta, GA (2020). CDC 2019-Novel Coronavirus (2019-nCoV) Real-Time RT-PCR Diagnostic Panel. Division of Viral Diseases, CDC-006-00019, Revision: 05.

15 Freeman, W.M., Walker, S.J., and Vrana, K.E. (1999). Quantitative RTPCR: pitfalls and potential. *BioTechniques* 26 (1): 124–125.

16 Kageyama, T., Kojima, S., Shinohara, M. et al. (2003). Broadly reactive and highly sensitive assay for Norwalk-like viruses based on real-time quantitative reverse transcription-PCR. *J. Clin. Microbiol.* 41 (4): 1548–1557.

17 Corman, V., Bleicker, T., Brünink, S., and Zambon, M. (2020). *Diagnostic Detection of Wuhan Coronavirus 2019 by Real-Time RT-PCR*. Geneva: World Health Organization.

18 Long, Q., Liu, B., Deng, H. et al. (2020). Antibody responses to SARS-CoV-2 in patients with COVID-19. *Nat. Med.* 26: 845–848.

19 Kong, W.H., Zhao, R., Zhou, J.B. et al. (2020). Serologic response to SARS-CoV-2 in COVID-19 patients with different severity [published online ahead of print, 2020 Jul 23]. *Virol. Sin.*: 1–6. https://doi.org/10.1007/s12250-020-00270-x.

20 Zhang, W., Du, R.H., Li, B. et al. (2020). Molecular and serological investigation of 2019-nCoV infected patients: implication of multiple shedding routes. *Emerging Microbes Infect.* 9: 386–389.

21 Van Elslande, J., Decru, B., Jonckheere, S. et al. (2020). Antibody response against SARS-CoV-2 spike protein and nucleoprotein evaluated by four automated immunoassays and three ELISAs. *Clin. Microbiol. Infect.* https://doi.org/10.1016/j.cmi.2020.07.038.

22 Larremore, D.B., Wilder, B., Lester, E. et al. (2020). Test sensitivity is secondary to frequency and turnaround time for COVID-19 surveillance. *medRxiv*. https://doi.org/10.1101/2020.06.22.20136309.

23 To, K.K.-W., Tsang, O.T.-Y., Leung, W.-S. et al. (2020). Temporal profiles of viral load in posterior oropharyngeal saliva samples and serum antibody responses during infection by SARS-CoV-2: an observational cohort study. *Lancet Infect. Dis.* https://doi.org/10.1016/S1473-3099(20)30196-1.

24 Yang, W. and Yan, F. (2020). Patients with RT-PCR confirmed COVID-19 and normal chest CT. *Radiology* 295: 200702.

25 Whiting, P., Singatullina, N., Rosser, J.H. et al. (2015). Computed tomography of the chest: I. Basic principles. *Contin. Educ. Anaesth. Crit. Care Pain* 15 (6): 299–304.

26 Lee, E.Y.P., Ng, M.-Y., Khong, P.-L. et al. (2020). COVID-19 pneumonia: what has CT taught us? *Lancet Infect. Dis.* 20: 384.

27 Wölfel, R., Corman, V.M., Guggemos, W. et al. (2020). Virological assessment of hospitalized patients with COVID-2019. *Nature* 581 (7809): 465–469.

28 Thompson, D. and Lei, Y. (2020). Mini review: recent progress in RT-LAMP enabled COVID-19 detection. *Sens. Actuators Rep.* 2 (1): 100017.

29 Lv, H., Wu, N.C., Tsang, O.T.-Y. et al. (2020). Cross-reactive antibody response between SARS-CoV-2 and SARS-CoV infections. *bioRxiv*. https://doi.org/10.1101/2020.03.15.993097.

30 Guan, W.-j., Ni, Z.-y., Hu, Y. et al. (2020). Clinical characteristics of coronavirus disease 2019 in China. *N. Engl. J. Med.* 382: 1708–1720.

31 Mahmoudi, T., de la Guardia, M., and Baradaran, B. (2020). Lateral flow assays towards point-of-care cancer detection: a review of current progress and future trends. *TrAC, Trends Anal. Chem.* 125: 115842.

32 Seo, G., Lee, J., Kim, M.J. et al. (2020). Rapid detection of COVID-19 causative virus (SARS-CoV-2) in human nasopharyngeal swab specimens using field-effect transistor-based biosensor. *ACS Nano* 14 (4): 5135–5142.

33 Ueda, M., Uchimura, K., Narita, Y. et al. (2015). The serine protease inhibitor camostatmesilate attenuates the progression of chronic kidney disease through its antioxidant effects. *Nephron* 129 (3): 223–232.

34 Zhou, Y., Vedantham, P., Lu, K. et al. (2015). Protease inhibitors targeting coronavirus and filovirus entry. *Antiviral Res.* 116: 76–84.

35 Uno, Y. (2020). Camostatmesilate therapy for COVID-19. *Intern. Emerg. Med.* 29: 1–2.

36 Cantini, F., Niccoli, L., Matarrese, D. et al. (2020). Baricitinib therapy in COVID-19: a pilot study on safety and clinical impact. *J. Infect.* 81 (2): 318–356.

37 Farooq, U., Latif, A., Irshad, H. et al. (2015). Loop-mediated isothermal amplification (RT-LAMP): a new approach for the detection of foot-and-mouth disease virus and its sero-types in Pakistan. *Iran. J. Vet. Res.* 16 (4): 331–334.

38 Lamb, L.E., Bartolone, S.N., Ward, E. et al. (2020). Rapid detection of novel coronavirus (COVID-19) by reverse transcription-loop-mediated isothermal amplification. *medRxiv*. https://doi.org/10.1101/2020.02.19.20025155.

39 Yu, L., Wu, S., Hao, X. et al. (2020). Rapid colorimetric detection of COVID-19 coronavirus using a reverse transcriptional loop-mediated isothermal amplification (RT-LAMP) diagnostic plat-form. *Clin. Chem.* https://doi.org/10.1101/2020.02.20.20025874.

40 Zhang, Y., Odiwuor, N., Xiong, J. et al. (2020). Rapid molecular detection of SARS-CoV-2 (COVID-19) virus RNA using colorimetric LAMP. *medRxiv*. https://doi.org/10.1101/2020.02.26.20028373.

41 Yang, W., Dang, X., Wang, Q. et al. (2020). Rapid detection of SARS-CoV-2 using reverse transcription RT-LAMP method. *medRxiv*. https://doi.org/10.1101/2020.03.02.20030130.

42 Sorabh, N. (2020). IIT Kharagpur researchers develop novel technology for COVID-19 rapid test. *DD News* (25 July). https://www.medrxiv.org/content/10.1101/2020.03.02.20030130v2.

43 ANI (2020). IIT Kharagpur researchers develop novel technology for COVID-19 rapid test. *ANI* (25 July).

44 Asutosh, A. (2020). IIT(ISM) techies develop anti-viral nano-coating for clothes. *Times of India* (17 June).

45 Keesara, S., Jonas, A., and Schulman, K. (2020). COVID-19 and health care's digital revolution. *N. Engl. J. Med.* 382: e82.

46 Vozella, J. (2020). https://www.fda.gov/media/137747/download (accessed 03 September 2020).

47 Digital technology for COVID-19 response. https://www.who.int/news/item/03-04-2020-digital-technology-for-covid-19-response (accessed 02 November 2020).

48 Meskó, B. and Görög, M. (2020). A short guide for medical professionals in the era of artificial intelligence. *npj Digital Med.* 3 (126) https://doi.org/10.1038/s41746-020-00333-z.

49 Bhattacharya, S., Pradhan, K.B., Bashar, M.A. et al. (2019). Artificial intelligence enabled healthcare: a hype, hope or harm. *J. Family Med. Prim. Care* 8 (11): 3461–3464. https://doi.org/10.4103/jfmpc.jfmpc_155_19.

50 Bitterman, D.S., Aerts, H.J.W.L., and Mak, R.H. (2020). Approaching autonomy in medical artificial intelligence. *Lancet Digital Health* 2 (9): 447–449. https://doi.org/10.1016/S2589-7500(20)30187-4.

51 Guglielmi, G. (2020). Fast coronavirus tests: what they can and can't do. https://www.nature.com/articles/d41586-020-02661-2 (accessed 03 November 2020).

3

Current Scenario of Pandemic COVID-19: Overview, Diagnosis, and Future Prospective

Bindu Mangla[1], Shinu Chauhan[1], Shreya Kathuria[1], Prashant[1], Mohit[1], Meenakshi[2], Santanu Patra[3], Sudheesh K. Shukla[4], and Chaudhery Mustansar Hussain[5]

[1] *Department of Chemistry, J.C. Bose University of Science and Technology YMCA, Faridabad 121006, India*
[2] *Ben-Gurion University of the Negev, The Jacob Blaustein Institutes for Desert Research, The Swiss Institute for Dryland Environmental and Energy Research, Midreshet Ben-Gurion 8499000, Israel*
[3] *Institute of Advanced Materials, IAAM, Ulrika 59053, Sweden*
[4] *Department of Biomedical Engineering, School of Biological Engineering and Life Science, Shobhit Institute of Engineering & Technology (Deemed-to-be University), Modipuram, Meerut 250110, India*
[5] *New Jersey Institute of Technology, Department of Chemistry and Environmental Science, NJ 07102, USA*

3.1 Introduction

Coronaviruses belong to the novel β-coronavirus section of the Coronaviridae family of Sarbecovirus sub-genus and the Nidovirales order distributed in animals and in humans at large scale [1]. The two β-coronaviruses reported earlier were epizootic disease extreme coronavirus causing severe acute respiratory syndrome coronavirus named SARS-CoV [2] and MERS-CoV leading to illness and Middle East respiratory syndrome coronavirus [3]. Over the past two decades, more than 10 000 cases have risen, 10% of death rates were recorded for SARS-CoV and 37% for MERS-CoV [4]. The previously reported coronaviruses are severe but not worst, with even more newly and extreme zoonotic incidents to be discovered. Several issues with undisclosed cause of pneumonia emanate in Wuhan, Hubei, China, in December 2019 result in highly similar to viral pneumonia through diagnostics [5]. Analyses of clinical data resulted in a novel coronavirus that was reported from lower respiratory tract samples, which later termed officially as COVID-19 (coronavirus 2019) [6]. WHO advisors concentrated on the type of virus that triggers the disease when choosing the name. Co and Vi are originated from coronavirus, and Tedros clarified the year the author announced the first cases of D, representing the disorder and 19 representing the disease in 2019 [7].

However, the initial investigation suggested a correlation between a single national fish market and the wildlife market, with most infections, suggesting possible communication between animals and humans [8, 9]. Studies have shown that SARS-CoV-2 is increasingly transferred from human-to-human through physical communication of virus [10, 11]. COVID-19 infection is caused by virus entry and

Detection and Analysis of SARS Coronavirus: Advanced Biosensors for Pandemic Viruses and Related Pathogens, First Edition. Edited by Chaudhery Mustansar Hussain and Sudheesh K. Shukla.

is prone to both the immunosuppressed and normal population, where patients are mostly found between age 35 and 55. Children and infants have less reported cases [12, 13]. An initial transmission dynamics analysis of the virus showed that the median age of patients ranged from 15 to 89 years, with the majority (59%) being male [14]. With these studies, it was confirmed that individuals with weak immune function, such as elderly people and people with kidney and hepatic diseases, could be the possible category, which is at higher risk [15].

On 29 December 2019, those who linked to a seafood market, the first four cases of acute respiratory syndrome of unknown origin were registered in Wuhan City, Hubei Province (China) [16]. In acute respiratory distress syndrome (ARDS)in all first 41 cases reporting 2019-nCoV infections, a viral pneumonia was detected in one-third of patients needing intensive care and six patients (14.6%) died initially [17]. According to the data reported by WHO on 13 March 2020 as per the national authorities, globally, 132 758 cases were reportedly confirmed including 4955 deaths, where in China, 80 991 cases were confirmed including 3180 deaths, and outside of China 51 767 were confirmed in 122 countries including 1775 deaths [18]; the risk assessment rate is very high not only in China but globally, and emergency has been declared in many countries around the world because of this pandemic. According to the report of BBC news in India on 13 March 2020, 82 cases of COVID-19 were reported, including 15 Italian tourists, with 2 deaths reported; the cases reported were transmitted from either imported cases or transmitted locally [19]. Four coronaviruses infecting humans including 229E, NL63, OC43, and HKU1 are widespread in immunocompetent individuals and typically triggers signs of common cold. SARS-CoV that affects severe acute respiratory syndrome (SARS) is exceptional in pathogenesis because it induces inflammation of the upper and lower respiratory tract. The 2019-nCoV genome sequence is about 89% close to the SARS-like CoV-ZXC21 bat and 82% identical to human SARS-CoV comparably [20]. 2019-nCoV is reported to use the similar SARS-CoV cell entry-receptor, ACE2, to infect humans [21], so clinical correlation is expected in between the two viruses, mainly in normal cases. It is a spherically wrapped fragment comprising single-stranded (positive-sense) RNA linked with a nucleoprotein within a mirid consisting of matrix protein (Figure 3.1). The wrapper holds club-shaped glycoprotein projections. Some coronaviruses also have a hemagglutinin esterase protein [8].

Many public health measures have been established that may deter or delay the transmission of 2019-nCoV, including quarantine, contact identification and surveillance, ambient disinfection, and the use of personal protective equipment [22]. R_0 is a relevant vial transmissibility-related threshold and is the basic reproduction number. The epidemiological definition of R_0 is the estimated figure of individual's one infectious individual may receive a disease. This refers in particular to a group of people who were previously infectious and had not been vaccinated. There are three reasons for the potential propagation or regression of a disease based on its R_0 meaning: Firstly, if R_0 is less than 1, increasing current infection has triggered new infection. In this case, the illness can gradually disappear with initial decline. Secondly, disease cannot be an outbreak but can stay alive if R_0 is equal to 1. Finally, if R_0 is greater than 1, cases could increase exponentially, triggering an outbreak or even a

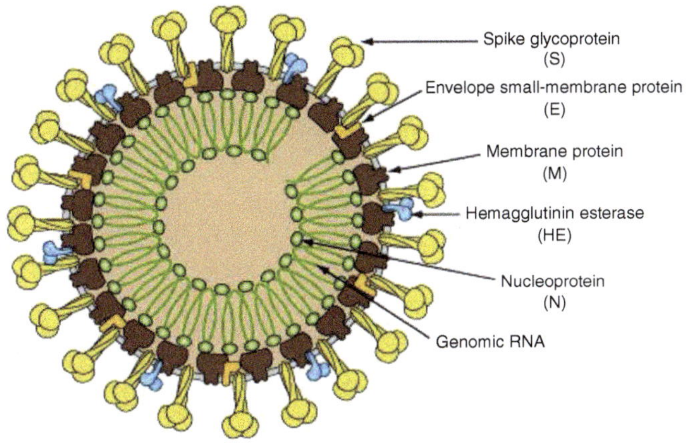

Figure 3.1 Structure of COVID-19 virus. Source: Mousavizadeh and Ghasemi [8].

pandemic [23]. The 2019-nCoV transmissibility and pandemic risk were considered to be higher than the 2019-nCoV SARS-CoV, showing a valid reproductive number (R) (2.9), which is predicted to be higher at the early stage than the recorded effective SARS R (1.77) [24]. The estimated incubation time for 2019-nCoV was 4.8 ± 2.6 and went from 2–11 to 5.2 days 95%. The normal duration of the incubation may be 7 days, varying from 2 to 14 days according to Chinese health authorities' current guidance [25].

The signs most frequent at the onset of the illness are fever, cough, and myalgia/weakness; the occurrence of sputum, hemoptysis, and diarrhea are less common. More than half of the patients reported with developed dyspnea. Blood counts revealed leucopenia (white blood cells [WBCs] below 4/109/l) and lymphopenia (lymphocyte counts $< 1 \cdot 0 \times 10^9$/L) in infected individuals. For severe cases, the length of prothrombin and the levels of D-dimer were higher [26] and have significantly increased levels of protein transferase from aspartate. There was a significant increase in the diagnosis of heart injury-related virus in patients with hypersensitive troponin I (hs-cTnI). In infected patients, the concentrations were higher for IL1B, IL1RA, IL7, IL8, IL9, IL10, GMCSF, IFNγ, IP10, MCP1, MIP1A, MIP1B, PDGF, TNFα, and VEGF. For severe cases, the plasma concentration had higher levels of IL2, IL7, IL10, GCSF, IP10, MCP1, MIP1A, and TNFα. IL1B, IFNγ, IP10, and MCP1, which likely resulted in T-helper-1 (Th1) cell responses being triggered, also suggested that the cytokine storm was related to seriousness of disease [26]. All confirmed patients had pneumonia. ARDS contained severe complications, followed by RNAaemia, acute heart injury, and secondary infection [27].

Between COVID-19 and earlier reported infections of β-coronavirus, similarities in clinical characteristics have been shown. Some patients experienced diarrhea, dry cough, dyspnea, and bilateral defects of the ground glass in a group of infected patients on chest computed tomography (CT) scans. These COVID-19 signs have been comparable to SARS-CoV infections and MERS-CoV infections [28]. In comparison, certain individuals suffering with COVID-19 infection displayed clear symptoms and signs of the upper respiratory tract (e.g. sore throat, rhinorrhea, or sneezing), demonstrating that the cell localizing in the lower airway could be targeted cells. In comparison, patients with 2019-nCoV hardly reported intestinal symptoms such as diarrhea, while around 20–25% of MERS-CoV or SARS-CoV patients suffer from diarrhea [29]. The pathophysiology for SARS-CoV/MERS-CoV with unusually high pathogenicity has not been clearly understood. However, 2019-nCoV infection has similarly led to increased development of T-helper-2 (Th2) cytokines (e.g. IL4 and IL10) that inhibit infection differing from SARS-CoV [30]. Advance research is needed in 2019-nCoV infections to identify the reactions Th1 and Th2 and to explain pathogenesis. Autopsy/biopsy tests help to understanding the disease.

Viruses show seasonality, as viruses tend to peak in winters and slower down in warm summers; there were few reports in which COVID-19 was compared to follow the similar pattern that warm season kills coronavirus [31]. As stated by the Chinese Center for Disease Control and Prevention, COVID-19 remains stable at 4 °C and can live at 60 °C below zero for several years. Its tolerance decreases at higher temperatures, but the temperature just influences the survival time of the virus, not its ability to kill, says the center. Similar findings have been identified with coronavirus SARS. As per the study at China's Academy of Military Medical Sciences, SARS remains stable at 4 °C but will lose its operation in three days at 37 °C and can live for only 15 minutes at 70 °C below freezing [32]. Similar to COVID-19, the first case affecting SARS was documented late in the year. The number of definite cases peaked in April and July 2003 saw no further patients. A treatment for SARS has never been found by scientists, and it is widely believed that the infection dies off during warm weather. However, variations between SARS and COVID-19 make the resistance of the new virus hard to predict. Like SARS, which can only affect others by fever patients, people with COVID-19 can spread the disease deprived of showing any symptoms, rendering the infection considerably more difficult to control. A latest research has shown that COVID-19 is significantly very infectious than SARS. Published by the medical research journal bioRxiv on 15 February, the study leads through a team managed by University of Texas researcher Jason McLellan, who has been studying various types of coronaviruses, such as SARS and MERS, for years [33, 34].

Not all coronavirus meets the same regulations. For example, the one causing MERS has not shown the capacity to quickly spread from person-to-person, says Amesh Adalja. An infectious disease specialist and high-ranking researcher at the Johns Hopkins Centre for Health Security: "It doesn't have any seasonality because it's just a human virus infection and not what you see as a seasonal phenomenon causing disease." Yet he says COVID-19 looks more like the cold of the season [35]. Yet coronaviruses affect as much as one third of common colds. "We've seen exponential dissemination of person-to-person infection inside China, essentially, and – in that way – it's actually behaving like a common cold coronavirus," Adalja says.

3.2 Diagnosis and Treatment

Early diagnosis can be done by specific molecular test on respiratory samples; virus can be detected in stool and in severe cases in the blood sample. Suspect cases can be detected a positive molecular test after checking travel history and the symptoms, but later pandemic spread via travel history becomes irrelevant. Samples were sent to various labs for investigation and found lower count of white cells than normal. The current diagnostic workflow for COVID-19 is described in Figure 3.2 [36]. Various techniques used for the detection include CT, polymerase chain reaction (PCR), etc. [37]. Researchers have also developed isothermal amplification tests, serological tests, and lateral flow assays for the diagnosis.

Reverse transcription polymerase chain reaction (RT-PCR) test is considered as one of the most efficient with reliable results but not 100%. Sometimes, a person with negative RT-PCR report may show symptoms in CT imaging. In a case study at the *Shaanxi province of China*, a 56 year old patient was reported who traveled to be admitted due to hyperthermia [38, 39]. Laboratory reports for WBC count, lymphocyte cell count, and serum procalcitonia came out to be normal while some tests including C-active protein, erythrocyte sedimentation rate, and alanine aminotransferase ware abnormal (Figure 3.3). The CT scan showed multiple ground-glass opacities in both lungs. Three RT-PCR tests of the oropharyngeal swab sample came out to negative. After further symptomatic treatment, CT scans showed increase of multi-focal ground-glass opacification and mixed consolidation mainly along the periphery of the lungs. Keeping this in view, fourth SARS-CoV-2 nucleic acid test was done that came out to be positive.

The positive result shown in the fourth swab test was due to increase in the amount of the virus that leads to the deteriorating condition of the infected [38]. The viral load is very important in determining the severity of the infection particularly for the throat swab test.

Computed tomography or simply CT (or temporal changes, also screening tool) has been an important imaging modalities (modality is the term used in radiology to refer to one form of imaging) in the diagnosis and management of patient with COVID-19 and reports on the radiological appearances of COVID-19 pneumonia are emerging.

The predominant CT findings include ground-glass opacification, consolidation, bilateral involvement, and peripheral and diffuse distribution. Asymptomatic (subclinical) group of patients showed early CT changes. However, the evolution of the disease on CT is not well understood. It is also not very clear whether the threshold for performing CT evaluation of potential lung changes should be lower when chest radiographs are normal.

In a case study, an asymptomatic novel coronavirus pneumonia laboratory-confirmed COVID-19 patient demonstrated that abnormal CT findings can precede clinical symptoms, and these abnormal CT findings may include bilateral pleural effusions (fluid can accumulate around the lungs due to poor pumping by the heart or by inflammation), previously not reported in association with COVID-19. The imaging manifestations, high-resolution CT, may be helpful for the diagnosis

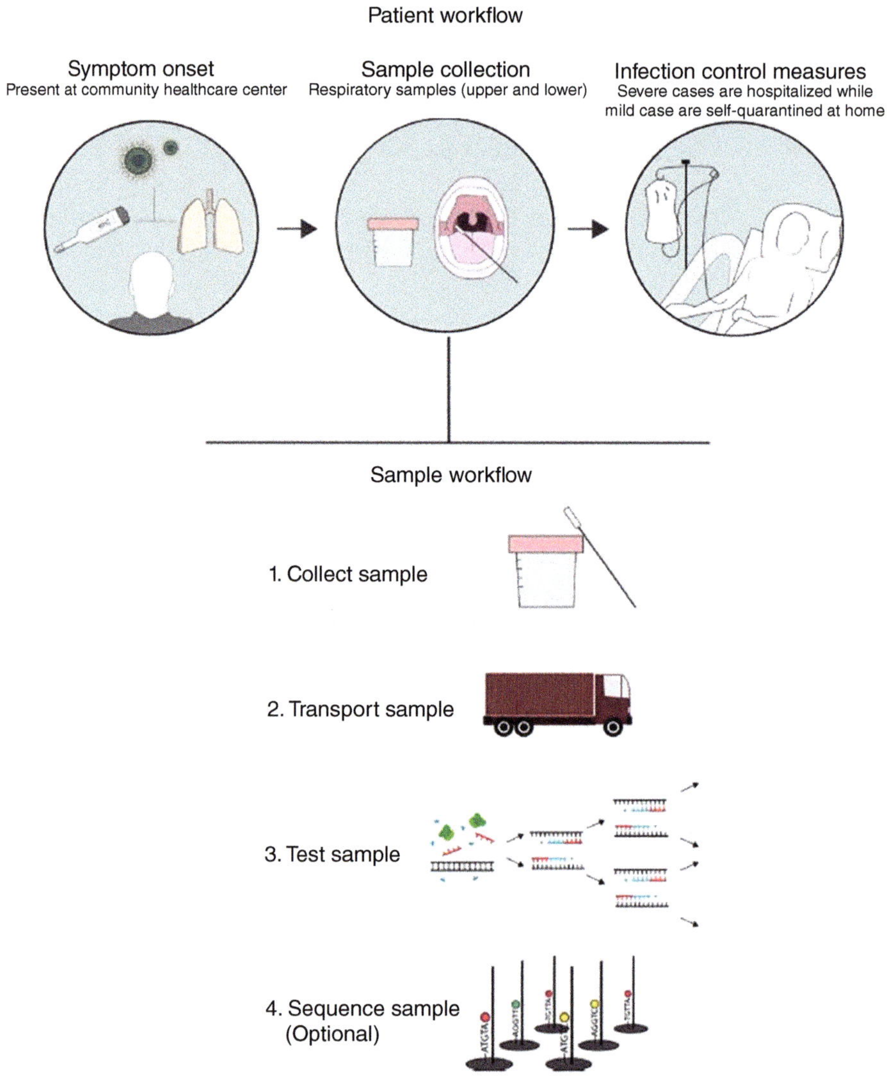

Figure 3.2 Example of patient and sample workflow during the COVID-19 outbreak. Patients present at a healthcare facility for triage. The collected samples are tested on-site if possible or transported for molecular testing and sequencing. Patients are then managed appropriately. Source: Udugama et al. [36].

and observation in the course of this unfortunately exponentially growing disease (Figure 3.4) [40].

Several artificial intelligence (AI) tools build upon machine learning algorithms are employed for analyzing the data and decision-making processes. Such tools can also be used to identify and predict the nature of spread of virus across the different parts of the world. However, such tools require adequate training data, which consumes a lot of time. However, in the present scenario, when pandemic has claimed many lives, we do not have much time to train the machines. Instead of having

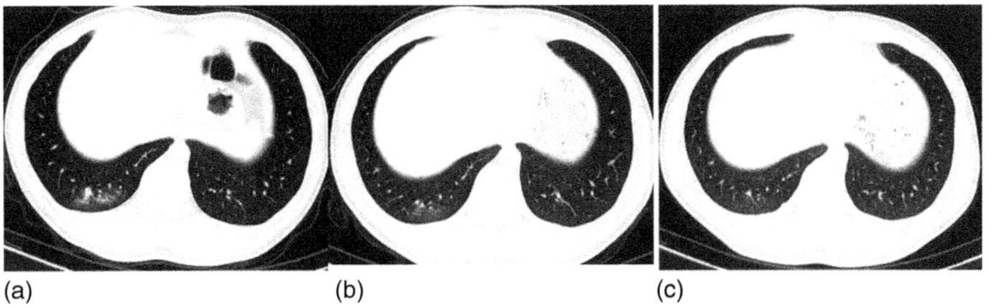

Figure 3.3 A 28 year old male who had a fever for 10 days. CT scan examination was performed on 3 February 2020. Ground-glass opacity was observed in the lower lobe of right lung. (a) Thickening of the blood vessel shadow was observed; (b) after seven days, the CT scan examination showed that the area and density of lesions had decreased and was in the recovery period; (c) on day 17, the follow-up examination showed that the lesions were fully absorbed, and chest CT scan examination showed no abnormality. Source: Deng et al. [39].

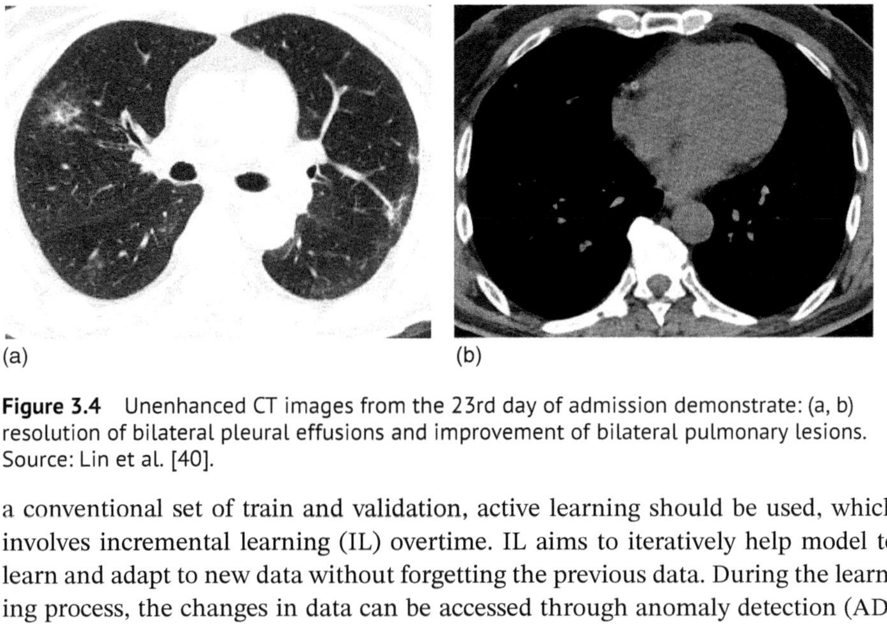

Figure 3.4 Unenhanced CT images from the 23rd day of admission demonstrate: (a, b) resolution of bilateral pleural effusions and improvement of bilateral pulmonary lesions. Source: Lin et al. [40].

a conventional set of train and validation, active learning should be used, which involves incremental learning (IL) overtime. IL aims to iteratively help model to learn and adapt to new data without forgetting the previous data. During the learning process, the changes in data can be accessed through anomaly detection (AD) techniques that can help to find rare items, events [41]. Nanotechnology holds great promise in offering innovative solutions to a wide range of problems regarding the prevention, diagnosis, and treatment of COVID-19. The potential of nanoparticles to be used as novel adjuvants and the use of nanoemulsions as colloidal vaccine carriers are being explored in the field of immunization.

3.3 Infection and Control

Because there is no approved treatment, prevention is crucial; isolation for confirmed and suspected cases is highly recommended. Patients should wear surgical face mask and practice cough hygiene, and health workers should be very careful too (provided with fit nested N95), negative molecular test should not prerequisite discharge. Avoid crowded places and further traveling should be avoided. If an individual is facing the symptoms associated with the COVID-19 infection, then the person should stay in isolation and undergo the treatment even if the RT-PCR test reports comes out to be negative. AI-driven tools can be used for the detection and treatment of this virus, thus minimizing the contact with the infected and reducing the risk of further spread among the healthcare workers. Nanotechnology and nanomaterial may also play a major role in fight against the COVID-19. Because of small size and large surface ratio, nanoparticles are widely used for the diagnostic and therapeutic purposes.

Nanomaterials are being used for the development of

i) *Rapid point-of-care diagnostics* – The availability of cost-effective, rapid, point-of-care diagnostics can help in the timely detection and treatment of the disease.
ii) *Surveillance and monitoring* – Mass surveillance can help the concerned departments in monitoring the spread of virus, identify the hotspots, and take appropriate measures for controlling the further spread.
iii) *Therapeutics* – Studies related to nano–bio interactions could be adapted to understand how the virus infects the cells that can lead to new therapeutic agents and design.

Nano-fabricated structures, coated with elements such as gold that have affinity for biomolecules, are incorporated into the biosensors that are employed for detection of various infections in the past. The rapid COVID-19 test based on IgG–IgM combined antibody and gold nanoparticles has great potential benefit for the fast screening of COVID-19 infections within 15 minutes. Nanomaterials have also shown the capability to inactivate viruses, including SARS and bird flu. For example, specially developed nanoparticle clusters can reduce virus levels by 80–100% through direct contact. Nanomedicine research, about the role of chloroquine in nanoparticle uptake, in cells showed promise in developing an effective treatment for COVID-19. Different carbon quantum dots (CQDs) are reported for the treatment of human coronavirus infections. Quantum dots intervene with the cellular binding sites and thus prevent the virus to use the cellular machinery of the host to grow and reproduce. The glycoprotein S of MERS-CoV is important for initial interactions with host cell; similarly spike protein S1 of n-CoV is found to have initial biochemical binding interactions with the host cell. Quantum dots inhibit the entry owing to the interactions of functional groups such as boronic acid and thiazide [42].

3.4 Current Status of COVID-19

A number of patients associated with seafood and live animal stock market in Wuhan, China, were reported to have acute severe lower respiratory tract illness.

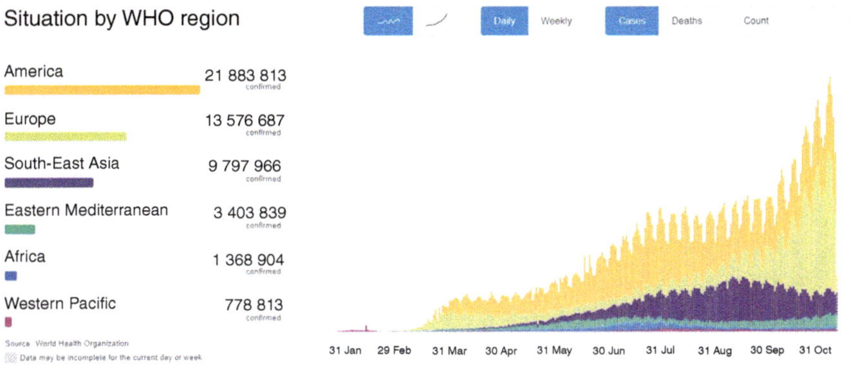

Figure 3.5 Geographical distribution of COVID-19 cases across the world as of 11 November 2020 at 16.00 CET. Source: Adapted from World Health Organization [45].

On investigating these patients, an unknown virus was isolated from the respiratory epithelial cells. This virus was found to belong to the coronavirus family. Seven coronaviruses are known to infect humans of whom SARS-CoV and MERS-CoV having zoonotic origins are linked with severe respiratory illness outbreaks. COVID-19 is also assumed to have zoonotic origin.

The COVID-19 virus has claimed many lives since it was identified in November last year. Initially confined to China, this virus has spread all over the world. The novel coronavirus belongs to the lineage B of the genus β-coronavirus of the coronavirus family of which SARS-CoV and MERS-CoV are also a part. In comparison with the 10% fatality rate of SARS-CoV and 37% fatality of MERS-CoV, 2019-nCoV is the third most deadly virus [4]. The novel coronavirus (COVID-19) is serving as a continuous threat to the whole world because it was identified in late 2019. This virus is mainly spreading among people through respiratory mode [43]. The symptoms shown by an infected individual include fever, cough, and shortness of breath. However, these symptoms take 2–14 days to occur from the day of exposure. According to WHO, by 9 March 2020, China has reported 80 904 cases and 3123 deaths [44, 45]. Outside China, the virus among America affects almost every continent and Europe is most badly hit by this pandemic. Considering the vast spread of the virus, WHO has already declared a public health emergency? Keeping in view of this, there is an urgent need to work on the treatment of this pandemic (Figure 3.5).

3.5 Recommendation

The whole world is suffering from the havoc of the novel coronavirus. Originating from China, this virus has spread across the globe. The number of the infections is increasing day by day. It deals with the acquired immune response, variant viruses, and to work against their chances of recovery, prevention measures, and use of modern technology to control COVID-19 for good human health. For the fast observation of disease outbreaks and understanding its process, there is a need of advanced technology with high sensitivity, specificity, and cost effectiveness to detect virus.

Modern and advanced technologies such as AI, IOT, Cloud computing, etc., can be used for fast progress and to get better results. AI promises to be a paragon for the healthcare. Along with the use of active learning in machine learning, cross-population trains must be used so that the AI-driven model can perform automatic detection. Further, the collected data can be used to train the model over time, which is based on the decisions. Also the use of multi-dimensional and multi-model data helps in decision-making process with greater confidence.

Nanotechnology has the potential to offer advances to current attitudes for immunization, drug design and drug delivery, diagnostics, and cross-infection control and is also unexpectedly delivering many new tools and competencies. Nanofiber-enriched facemasks and respirators supported with accelerated copper oxide not only mechanically intercept viruses and bacteria but they actively kill them as well.

3.6 Conclusion

Genome sequence of new virus strain SARS-CoV-2 is not found identical but similar to other coronavirus outbreaks namely SARS and MERS. However, species classification based on amino acid sequence suggests that both the viruses belong to the same species SARS-CoV. RNA-dependent RNA polymerase from bat coronavirus (RaTG13) shares high sequence similarity throughout the 2019-nCov. S-gene that codes for receptor binding spike protein also shows divergence from other coronavirus. Globally, SARS-CoV-2 has caused huge negative impacts on population health and economy. Effective preventive measures must be implemented to control it from global spreading. Continuous efforts are made on the development of vaccine and antiviral drugs to prevent and control the outbreak of this and the other novel viruses in future.

The swab test has however proven to be a milestone in the detection of COVID-19 infection, but because of its time-consuming procedure and lack of the viral substance in the samples, sometimes, it is unable to fulfill the present need. CT is useful in studying temporal changes in both symptomatic and asymptomatic COVID-19 patients, but more exploration is desired in selecting the patients for CT over other diagnostic techniques. CT is basically more easy to use and requires less intravenous contrast agent. CT besides chest radiography serves the purpose better, but application of AI can make the process easy and efficient. Active learning must be employed in the tools. AI-driven tools are also expected to work as cross-population train/test models. This novel coronavirus-2 has become a big impediment for future development of the world. Worldwide knowledge sharing for designing better and efficient models has become the need of the day to fight the ongoing and future pandemics. More focus on new sciences and technologies (like AI-enabled drug designing) is required than short-term technological advancement.

Because the viruses are mutating at unmatchable pace, they can easily bypass drug therapies, so more efficient ways such as CQDs, which do not interact with virus but interact with the cellular protein or other factors, thus making the cell incompatible for the virus to bind. CQDs are a possible option for inhibitor of virus infection cycle. More therapeutic options can help us to fight and eradicate the future fierce viral pandemics.

References

1 Llanes, A., Restrepo, C.M., Caballero, Z. et al. (2020). Betacoronavirus genomes: how genomic information has been used to deal with past outbreaks and the COVID-19 pandemic. *Int. J. Mol. Sci.* 21: 4546.

2 Lai, C.-C., Shih, T.-P., Ko, W.-C. et al. (2020). Severe acute respiratory syndrome coronavirus 2 (SARS-CoV-2) and coronavirus disease-2019 (COVID-19): the epidemic and the challenges. *Int. J. Antimicrob. Agents* 55: 105924.

3 Hajjar, S.A., Memish, Z.A., and McIntosh, K. (2013). Middle East respiratory syndrome coronavirus (MERS-CoV): a perpetual challenge. *Ann. Saudi Med.* 33 (5): 427–436.

4 Raoult, D., Zumla, A., Locatelli, F. et al. (2020). Coronavirus infections: epidemiological, clinical and immunological features and hypotheses. *Cell Stress* 4 (4): 66–75.

5 She, J., Jiang, J., Ye, L. et al. (2020). 2019 novel coronavirus of pneumonia in Wuhan, China: emerging attack and management strategies. *Clin. Transl. Med.* 9: 19.

6 Wang, L., Wang, Y., Ye, D., and Liu, Q. (2020). Review of the 2019 novel coronavirus (SARS-CoV-2) based on current evidence. *Int. J. Antimicrob. Agents* 55 (6): 105948.

7 WHO names deadly viral disease from China as 'COVID-19'. https://www.thehindu.com/news/international/who-names-deadly-virus-disease-from-china-as-covid-19/article30797838.ece (accessed 01 October 2020).

8 Mousavizadeh, L. and Ghasemi, S. (2020). Genotype and phenotype of COVID-19: their roles in pathogenesis. *J. Microbiol. Immunol. Infect.*. https://doi.org/10.1016/j.jmii.2020.03.022.

9 Mackenzie, J.S. and Smith, D.W. (2020). COVID-19: a novel zoonotic disease caused by a coronavirus from China: what we know and what we don't. *Microbiol. Aust.*: MA20013. https://doi.org/10.1071/MA20013.

10 Chang, D., Lin, M., Wei, L. et al. (2020). Epidemiologic and clinical characteristics of novel coronavirus infections involving 13 patients outside Wuhan, China. *JAMA* 323 (11): 1092–1093.

11 Huang, C., Wang, Y., Li, X. et al. (2020). Clinical features of patients infected with 2019 novel coronavirus in Wuhan, China. *Lancet* 6736 (20): 30183–30185.

12 Medical Expert Group of Tongji Hospital (2020). Quick guide to the diagnosis and treatment of pneumonia for novel coronavirus infections (third edition). *Her. Med.* http://kns.cnki.net/kcms/detail/42.1293.r.20200130.1803.002.html.

13 Wang, C. and Wang, X. (2020). Prevalence, nosocomial infection and psychological prevention of novel coronavirus infection. *Chin. Gen. Pract. Nurs.* 18: 2–3.

14 WHO (2020). Novel coronavirus – China. https://www.who.int/csr/don/12-january-2020-novel-coronavirus-china/en (accessed 1 February 2020).
15 COVID research updates: many surfaces carry coronavirus RNA – but not much of it (2020). *Nature*. https://doi.org/10.1038/d41586-020-00502-w.
16 Hamid, S., Mir, M.Y., and Rohela, G.K. (2020). Novel coronavirus disease (COVID-19): a pandemic (epidemiology, pathogenesis and potential therapeutics). *New Microbes New Infect.* 35: 100679.
17 D'Abramo, A., Lepore, L., Palazzolo, C. et al. (2020). Acute respiratory distress syndrome due to SARS-CoV-2 and influenza a co-infection in an Italian patient: mini-review of the literature. *Int. J. Infect. Dis.* 97: 236–239.
18 World Health Organization (2020). Novel coronavirus (2019-nCoV), situation report-4, 24 January 2020. https://www.who.int/docs/default-source/coronaviruse/situation-reports/20200124-sitrep-4-2019-ncov.pdf?sfvrsn=9272d086_8 (accessed 04 November 2020).
19 Coronavirus: second death confirmed in India. https://www.bbc.com/news/world-asia-india-51866903 (accessed 04 November 2020).
20 Chan, J.F., Kok, K.H., Zhu, Z. et al. (2020). Genomic characterization of the 2019 novel human-pathogenic coronavirus isolated from a patient with atypical pneumonia after visiting Wuhan. *Emerging Microbes Infect.* 9: 221–236.
21 Zhou, P., Yang, X.L., Wang, X.G. et al. (2020). A pneumonia outbreak associated with a new coronavirus of probable bat origin. *Nature* 579: 270–273.
22 Bauchner, H., Fontanarosa, P.B., and Livingston, E.H. (2020). Conserving supply of personal protective equipment – a call for ideas. *JAMA* 323 (19): 1911.
23 Chen, J. (2020). Pathogenicity and transmissibility of 2019-nCoV – a quick overview and comparison with other emerging viruses. *Microbes Infect.* 22: 69–71.
24 Liu, T., Hu, J., Kang, M. et al. (2020). Transmission dynamics of 2019 novel coronavirus (2019-nCoV). *bioRxiv*. https://doi.org/10.1101/2020.01.25.919787.
25 National Health Commission C (2020). Prevent guideline of 2019-nCoV. www.nhc.gov.cn/xcs/yqfkdt/202001/bc661e49b5bc487dba182f5c49ac445b.shtml (accessed 04 November 2020).
26 Huang, C., Wang, Y., and Li, X. (2020). Clinical features of patients infected with 2019 novel coronavirus in Wuhan, China. *Lancet* 395: 497–506.
27 Zhang, X., Li, S., and Niu, S. (2020). ACE2 and COVID-19 and the resulting ARDS. *Postgrad. Med. J.* https://doi.org/10.1136/postgradmedj-2020-137935.
28 WHO (2020). Coronavirus. https://www.who.int/health-topics/coronavirus (accessed 03 November 2020).
29 Hajifathalian, K., Mahadev, S., and Schwartz, R.E. (2020). SARS-CoV-2 infection (coronavirus disease 2019) for the gastrointestinal consultant. *World J. Gastroenterol.* 26 (14): 1546–1553.
30 Su, S., Wong, G., Shi, W. et al. (2016). Epidemiology, genetic recombination, and pathogenesis of coronaviruses. *Trends Microbiol.* 24 (6): 490–502.
31 Coronavirus: how will winters impact COVID-19 pandemic infection?. https://www.financialexpress.com/lifestyle/health/coronavirus-how-will-winters-impact-covid-19-pandemic-infection/2100689 (accessed 03 November 2020).

32 Darnell, M.E.R., Subbarao, K., Feinstone, S.M., and Taylor, D.R. (2004). Inactivation of the coronavirus that induces severe acute respiratory syndrome, SARS-CoV. *J. Virol. Methods* 121 (1): 85–91.

33 Will warming weather kill off COVID-19? Scientists aren't sure. https://www.caixinglobal.com/2020-02-19/will-warming-weather-kill-off-covid-19-scientists-arent-sure-101517340.html (accessed 03 November 2020).

34 Scudellari, M. (2020). The sprint to solve coronavirus protein structures – and disarm them with drugs. *Nature* 581: 252–255.

35 Amesh Adalja: taking pandemic preparation seriously. https://www.who.int/bulletin/volumes/98/5/20-030520/en (accessed 03 November 2020).

36 Udugama, B., Kadhiresan, P., Kozlowski, H.N. et al. (2020). Diagnosing COVID-19: the disease and tools for detection. *ACS Nano* 14 (4): 3822–3835.

37 Kim, H., Hong, H., and Yoon, S.H. (2020). Diagnostic performance of CT and reverse transcriptase-polymerase chain reaction for coronavirus disease 2019: a meta-analysis. *Radiology* 17: 201343.

38 Hao, W. and Li, M. (2020). Clinical diagnostic value of CT imaging in COVID-19 with multiple negative RT-PCR testing. *Travel Med. Infect. Dis.* 34: 101627.

39 Deng, L., Khan, A., and Zhou, A. (2020). Follow-up study of clinical and chest CT scans in confirmed COVID-19 patients. *Radiol. Infect. Dis.* https://doi.org/10.1016/j.jrid.2020.07.002.

40 Lin, C., Ding, Y., Xie, B. et al. (2020). Asymptomatic novel coronavirus pneumonia patient outside Wuhan: the value of CT images in the course of the disease. *Clin. Imaging* 63: 7–9.

41 Santosh, K.C. (2020). AI-driven tools for coronavirus outbreak: need of active learning and cross-population train/test models on multitudinal/multimod data. *J. Med. Syst.* 44: 93. https://doi.org/10.1007/s10916-020-01562-1.

42 Loczechin, A., Séron, K., Barras, A. et al. (2019). Functional carbon quantum dots as medical countermeasures to human coronavirus. *ACS Appl. Mater. Interfaces* 11: 42964–42974.

43 Modes of transmission of virus causing COVID-19: implications for IPC precaution recommendations. https://www.who.int/news-room/commentaries/detail/modes-of-transmission-of-virus-causing-covid-19-implications-for-ipc-precaution-recommendations (accessed 03 November 2020).

44 Report of the WHO-China joint mission on coronavirus disease 2019 (COVID-19). https://www.who.int/docs/default-source/coronaviruse/who-china-joint-mission-on-covid-19-final-report.pdf (accessed 03 November 2020).

45 World Health Organization (2020). WHO coronavirus disease (COVID-19) dashboard. https://covid19.who.int/?gclid=CjwKCAiAtK79BRAIEiwA4OskBntjlCp_TaMofSIsrqq73sPyYQcj0e0t-NUrAKLP76VYenM67Cf_ORoClDMQAvD_BwE (accessed 11 November 2020).

Part II

Bio-analytical Strategies for SARS-CoV-2/COVID-19

4

COVID Detection via Nanotechnology: A Promising Field in the Diagnosis and Analysis of Novel Coronavirus Infection

Nitika Thakur[1], Sudheesh K. Shukla[2], and Chaudhery M. Hussain[3]

[1] *Shoolini University, Faculty of Applied Sciences and Biotechnology, Solan, Himachal Pradesh 173229, India*
[2] *Department of Biomedical Engineering, School of Biological Engineering and Life Science, Shobhit Institute of Engineering & Technology (Deemed-to-be University), Modipuram, Meerut 250110, India*
[3] *New Jersey Institute of Technology, Department of Chemistry and Environmental Science, NJ 07102, USA*

4.1 Introduction

4.1.1 Pandemic Outbreak of COVID-19: A Tour Around the Globe from Wuhan

The pandemic outbreak of coronavirus disease 2019 (COVID-19) was initiated in Wuhan, (Hubei) Province, China. The outbreak was associated with a similar strain of severe acute respiratory syndrome (SARS) virus and was possibly related to SARS-CoV-2 (severe acute respiratory syndrome coronavirus 2) consisting of a single-stranded RNA virus (25–38 kb genome) with a diameter ranging from 60 to 130 nm. The virus has been traced to follow a path starting and including soft tissues such as eyes, nose, and mouth [1].

The disease has spread globally within a short period of time from China as the center of origin of the infection and finally it was tagged as COVID-19, a novel coronavirus currently spreading through 212–250 countries, areas, and regions [2, 3].

The diagnosis and treatment process generally highlight two important criteria that should be taken into consideration while controlling the COVID-19 pandemic: the first important criteria is to reduce and overcome the rate of acerbating infections and the second point is to decrease the rate of date. Keeping in view the above points, many vaccine trials have been initiated including screening of potential therapies that includes antivirals and antibiotics to treat primary and secondary infections related to bacteria, sepsis including use of corticosteroids finally reducing the inflammation. However, the main loopholes behind these therapies are their failure because of the initiation of "cytokine storm induction" in lungs by COVID-19 [4].

However, the ongoing therapies have been successful in inhibiting various processes related to virus entry, replication inhibitors, heterocyclic antivirals, etc., thus displaying a moderate hand in handling. The viral infection causes a deadly threat because of the novel respiratory disease (SARS). Examination and early detection are only needed to restrict the spread of COVID pandemic further [5].

Detection and Analysis of SARS Coronavirus: Advanced Biosensors for Pandemic Viruses and Related Pathogens, First Edition. Edited by Chaudhery Mustansar Hussain and Sudheesh K. Shukla.

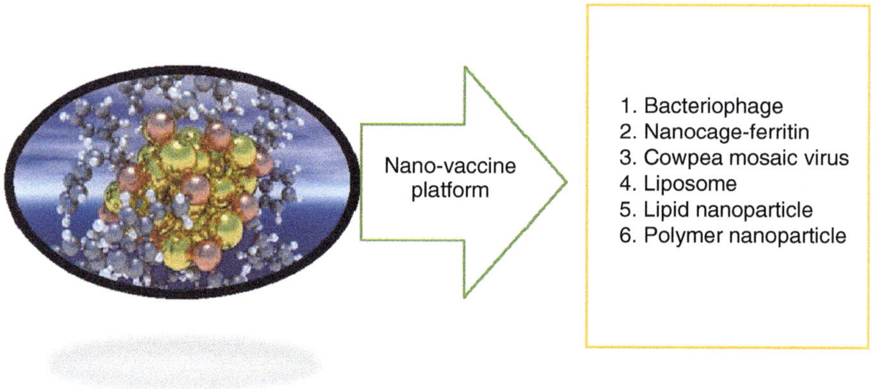

Figure 4.1 The nanoplatform integrating different potential nanocandidates for vaccine development.

4.1.2 Nanotech Solutions for Faster Detection Analysis of COVID-19

Nanotechnology highlights the effectiveness of nanoparticles because of their small size and large surface-to-volume ratio, which make them promising tools for medicine, sensors, DNA/RNA labeling and detection, and drug delivery [6]. They have been continuously exploited extensively to improve the antiviral drug delivery, efficacy, particularly of analogs related to nucleoside, having potential application in terms of drug-resistant against viral infections [5, 7]. Further, the nanoformulations have an efficient mechanism for drug delivery, which makes them potential for faster detection analysis in COVID-19 (Figure 4.1).

Thus, vaccines are the most promising solutions to the current COVID-19 scenarios to prevent it from spreading further.

4.2 Methodologies from Lab to People: Advantages of Nanovaccines in Providing Point-of-care Diagnosis

There are numerous nanotechnology research labs who are directing their efforts toward the integration of nanoscience with therapeutics for the development of nanovaccine or nanoformulation for the treatment of COVID-19, which can be initially screened beforehand at preclinical and clinical stages. Nanotechnology thus provides a potential advantage of nanoformulation as the most promising detection tool that may assist in nanomaterial-based viral agents/disinfectants [8].

Further, they may play an important role in integral aspects of vaccine designing, its accurate and specific delivery process, and finally its administration. In addition, nanoparticles incorporate an excellent strategy of "multiple antigen presentation" and finally stabilization of these antigens when they are administered. Nanoparticles do act as excellent adjuvants and carriers to boost up the immune process and to serve as vectors/carriers for the target delivery of the specified antigen [9, 10].

Furthermore, a liposomal nanoparticle is being tagged for the delivery of an mRNA vaccine is serving as a potential candidate, which is being currently preferred in ongoing clinical trials against SARS-CoV-2. However, still it is questionable as until now, no mRNA or DNA vaccine has been currently preferred and approved because of the glitches in terms of the mechanism of delivery of nucleic acids with some modification or a tagging of nanodevice to generally prevent their degradation [11]. Furthermore, the science of nanotechnology has been upcoming as the science of new innovation through the most important breakthrough of cancer nanovaccines to boost up the host immune responses. This breakthrough has provided a solid imprint of diverting these immune-assisted approaches against the current pandemic (SARS-CoV-2).

The most important aspect of technologies related to nanotechnology in COVID diagnosis is that these are point-of-care (POC) technology or rather lab-on-chip (LOC), generally targeting the administration and distribution of vaccines to resource poor population or on the other hand the densely populated areas. It provides ease in walking toward a mode of "self-administration," in terms of "SDSRI" (single-dose slow release-assisted implants), microneedles, film-based vaccines, plant viral-based nanovaccines for effective delivery, and administration process.

4.3 An Overview: The Potential Strategies Related to Nanotechnology for Combating COVID-19

There are many loopholes regarding the conventional detection strategies, infrastructure related to healthcare, which puts a questionable halt to the check on viral surveillance and screening methods for improved diagnostics. Therefore, there is a quick need to access diagnostics and therapeutics that can be molded toward quick screening, specified viral surveillance strategies, and tagging out the positive cases at initial level in COVID-19 pandemic.

These quick diagnostics for COVID-19 may prove a boost in terms of rapid implementation of preventive measures and precautions to finally restrict the spreading nature of this virus and further march in the direction of "contact tracing" (tracing the chain of infection and the people involved). The COVID-infected respiratory tract can be easily differentiated from healthy one, as the infected lungs are more profuse, denser, and concomitant [8].

In the current scenario, COVID-19 is being diagnosed through reverse transcription polymerase chain reaction (RT-PCR) and computed tomography (CT) scans has been used for screening purposes. Further molecular methods can be used in diagnosing diseases more rapidly as they can easily classify and differentiate a wide range of pathogens, when compared with CT scans. The approaches have specific drawbacks associated with them [9].

The upcoming urge of detection and diagnosis brings a new insight to nanotechnology, which showcase newer opportunities in terms of specific, cost-effective, and specific detection methods with safer equipment and successful medications.

In this context, nanosensors come up with a very promising strategy of detecting bacterial and viral loads at very low concentration, so that this can be used as a tool for pre-diagnosis in asymptomatic conditions. Recently, several viral strains have been sequenced and tagged up with illumina, which is a "sequence-by-sequence synthesis" using solid-phase amplification process to finally translocate a nucleoside through protein tunneling and screening out the sequence analysis by witnessing the changes in voltage. In addition to this, oxford nanopore technology is coming up with promising results in terms of accurate and cost-effective methodology [10].

The scenario of technology coupled with research serve as the strongest weapon in order to fight against this current pandemic type. Nanotechnology thus provides effective solutions to treat, diagnose, and prevent diseases, thus adding an important flavor to the area of therapeutics [11].

Nano-based strategies and methods are thus, coming into limelight due to their ease of use, cost effectiveness, specificity, enhancing power and finally out ruling the conventional methods. The upcoming potential advanced methods are being discussed in the following section.

4.3.1 Loop-Mediated Isothermal Reverse Transcriptase Coupling with Nanobiosensors

COVID-19 is posing a great threat in terms of its rapid spread, which has created panic among people globally. The current scenario utilizes the use of RT-PCR [12] and some gene amplification strategies but has been discarded because of some drawbacks in relation to its complex nature, requirement of experts and skilled staff, and above all these are time-consuming in nature.

A modified form of RT-PCR by tagging it with LAMP and a nanosensor has been used in 2003 for easy detection of the targeted genes in diagnostic pathway of SARS-CoV-2 analysis [13]. This modified technique merges the combination of the conventional RT-PCR, LAMP amplification strategy for target sequences, and a nanoparticle-based biosensor for a single-way detection (25–30 minutes time slot) of COVID-19 [14]. This technique ensures precision, as it aims at targeting sequences through six to eight different locations using RNA as a preferred template and can be easily witnessed through color changes by the microscope [8]. Furthermore, these methodologies include amplification of involved polymerases, amplification assisted by helicase, and loop-assisted amplification [8, 13].

A one-tube RT-LAMP-NBS assay has been successfully utilized for detection analysis of COVID-19 because of its simplicity in operational methodologies, basic structure requirement, and finally the need of inexpensive tools and equipment to maintain a standard temperature around 60–63 °C [15]. It has been found way better than the conventional RT-PCR in visualization of accurate results and also in terms of non-requirement of expensive and complicated processes and tools. Research is still going on for increasing the efficacy, optimizing the amplification conditions and finally the feasibilities.

4.3.2 Nanopoint-of-care/Lab-on-chip Diagnosis: A Strategy to Reach out the Resource-Poor Areas

The nanopoint-of-care diagnosis has an advantage in treating patients at resource-poor areas and decentralized hospital systems [16]. Therefore, the POC diagnosis in nanotesting methodology for COVID-19 is through "the lateral flow-assisted antigen detection" [17]. The available commercial kit for lateral flow assays basically contains a simple paper like strip or membrane coated usually with two different lines having connected purpose. The one line is incorporated with nanoparticles synthesized from gold and conjugated to an antibody, while the other line holds up the antibody firmly.

The samples such as blood and urine from the patient are collected on the incorporated membrane that is taken into consideration for analysis. The movement of the first line results in attaching the antigen to the complex known as the "antibody–nanoparticle conjugate" and pulls this complex across the membrane. The antibodies caught by the second line is generally for the immobilization of the complex; as a result, the coupled plasmon bands makes the nanocoated complex to turn from red to blue confirming positive detection [13]. Furthermore, many POC technologies such as paper-based nanosystem and nano-electrosystems are being developed for providing an effective mode of POC or LOC technologies for resource poor areas.

4.3.3 Tagging up the Biosensor with Optics for Reducing the Long Detection Time

To reduce the constrains regarding the time consumption criteria, the science of nano-optics provides a quick solution, resolving detection analysis around a time span of 25–30 minutes approximately without requiring the laboratory systems that are working in a centralized mode. This new upcoming technology can efficiently differentiate between a normal flu "Influenza flu" and a more severe "Corona" infection. In addition, this technology will sense and analyze the potential reservoir animals such as bats and birds generally to track the potential evolutionary lineage of virus, so that the potential future outbreaks can be finally avoided [18]. This would further open doors regarding the pathogenicity of the particular virus, which can be easily traced through the evolutionary pattern, and finally, it would be easier to screen these viruses, their potential vectors, reservoirs, and degree of disease transmission. This opens future venue for new diagnosis and therapeutics needed to develop effective vaccination patterns along with easy detection and analysis procedure to restrict the dangerous viruses from spreading globally.

4.3.4 Sequencing Strategy Involving the Nanopore-Assisted Target Sequencing (NTS)

The more advanced method that provides additional analysis of other respiratory viruses too, along with the specific detection of SARS-CoV-2, is considered as a time-effective strategy where results can be obtained within a time period of 5–10 hours. Basically, this detection method focuses on the genes that are related to virulence (11 SARS-CoV-2), followed by quick amplification of targeted genes on a nanopore-assisted platform [19]. The major advantage lies in the screening and detection of long nucleoside chains or fragments with a result output recorded in real time, thus allowing a clear pattern of sequences by mapping and matching it with sequence reads of SARS-CoV-2 genome. Thus, this test includes one mini-chip sequencer for all test samples with different protein tags (S, orf3A, E, M, ORF6, ORF7a, ORF8, and ORF10) to detect various virulence genes. The technology still contains glitches (qPCR) regarding deciding the positive sample by taking into consideration probably two different sites [20, 21].

4.4 Screening of Potential Agents for Restricting the Rapid Spread of COVID-19

The burning spread of COVID-19 really needs an effective strategy incorporated into the therapeutics to combat with this particular virus and finally put a full stop to its rapid spread globally. In this context, many potential agents have been tested to overcome this deadly virus, but no drug has been found effective in treating it completely. A series of antiviral agents and immunomodulators have been tested, such as remdesivir, favipiravir, cobicistat, mycophenolate, oseltamivir, interferon, convalescent plasma therapy, and ribavirin [22]. However, the use of hydroxychloroquine was labeled a necessary and emergency step by the Food and Drug authorities on 28 March 2020, but according to the latest recommendation, "remdesivir" has been found as the drug of choice in Switzerland and other parts of the country, whereas in countries like China and Japan, the use of another antiviral agent called as "favipiravir" has been approved for influenza and is under examination for utilizing the same for COVID-19.

The literature shows and reports the use of many immunomodulators and antiviral agents, and some are still under clinical examination [23–27]. The Chinese traditional ways of treating infections have come up with new liquids "Shuanghuanglian oral liquid" [28, 29] containing acids such as chlorogenic acid, which are antiviral and antibacterial in nature.

The current scenario is still searching ways for manufacturing correct vaccine for COVID-19, which can be used globally. The studies are ongoing for the vaccines, which are still under clinical trials [30]. Currently, no vaccines have been approved for COVID-19; however, some newly developed vaccines are in clinical trials.

The application of nanoscience can be a useful tool in treating COVID-19 complications as they serve as potential antibacterial and antiviral agents and can very well reduce severity and risks associated with COVID-19. However, still we lack the nano-based vaccines today, and the work is in progress for these effective nanoformulations and vaccines where tagging of some potential candidates such as zinc can be utilized. Zinc nanoconjugations or the tagged Zn nanoparticles can increase the host resistance. The studies have been already conducted against some common cold viruses and influenza viruses. In addition to the conventional methods the upgraded advanced ones have been found promising in terms of rhinoviruses. Therefore, it can be a potential nanoparticle that can work against COVID-19 and can help to lower the disease severity and burden among populations [11, 30].

It has already been proven that Zn-coated nanoparticles reduce viral replication effectively in in vitro analysis and studies [31–33] and have reduced the prevalence of pneumonia infections in children globally. It may help in inhibiting the initial binding of viruses to the potential target sites and also induces the production of antiviral agents and interferons (interferon α and γ) [30]. Further, Zn has been known to activate potential enzymes for different cell-mediated and humoral functions. It is clear from the recent study that Zn deficiency can lead to non-activation of humoral response or very low activation, which results in low production of B antibodies and lesser phagocytic action against parasites [30, 34, 35]. Zn supplementation is needed for the boosting of immune response (50–55 mg/d), thus preparing host to fight against potential pathogenic invaders. These potential characters make it a potential candidate to be tagged up as a nanoparticle, providing additional defense against COVID-19 [30, 36, 37]. It has been reported that face masks having Zn oxide coating possibly prevent and kill pathogens on contact because of its antibacterial and antiviral efficacy [38, 39].

4.5 Potential New Generation Vaccines: A Journey from Nucleoside, Subunit, Peptide Analogs to Nanoformulation

4.5.1 Nucleoside Analog Vaccines: Searching Potential Candidates Among DNA, RNA, and mRNA

The use of DNA or RNA vaccine to elicit immune responses by triggering the cell-mediated immunity through important helper T cells: CD4+ [40] and toxic CD8 cells have resulted in boosting up the immune reaction and finally eradication of virus. The utility of this typical vaccine was tested, and various clinical process was initiated by ovio pharmaceutical and Entos pharmaceuticals on 6 April 2020 for the phase 1 clinical trials [41, 42]. The advance mRNA-based methodology was tested on 16 March 2020 for phase 1 clinical trials in US43. In addition, BioNTech currently approved four lead mRNA in the phase 1 and 2 clinical trials [43] (Figure 4.2).

These DNA vaccines do offer higher ratio of stability when compared to the higher risk posing mRNA vaccines, which can result in higher mutational rates. However, the stability, the associated half-life, and factors such as immunogenicity can be modified in mRNA vaccines [44].

The nanoscience has been integrated for the effective delivery of these nucleoside analogs by directing the vaccine to appropriate cellular and subcellular locations.

The use of synthetic nanoparticles/nanocarriers have been found to incorporate the polymeric and cationic liposomal particles for the effective delivery process across the targeted cell membrane which can further help in acerbating the translocation process of the selected plasmid DNA [44, 45].

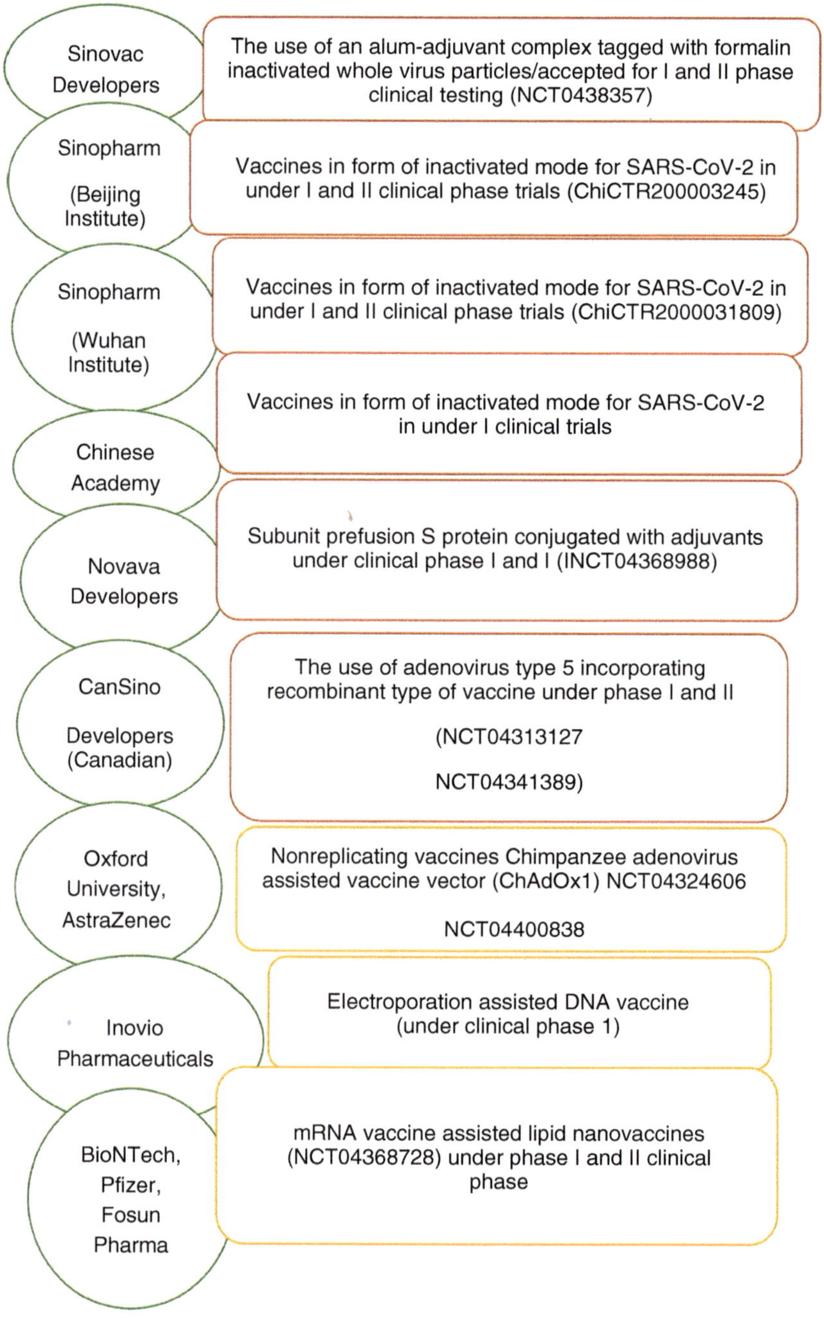

Figure 4.2 A breakthrough explaining the various COVID-19 vaccine initiatives by different companies and their ongoing clinical trials (nucleoside analogs to nanocoated advancement in vaccine process for COVID-19).

The modern implementation of mRNA vaccine is generally dependent on a lipid-based nanoparticle pathway. In addition to the mRNA-NANO integration, there are a wide range of cationic-based liposomes, polysaccharides, and dendrimers that have been incorporated for improving the efficacy, stability, and finally delivery process of mRNA-tagged vaccine [40, 44–46].

4.5.2 Nano-VLP Subunit Vaccines: A Stable and Ordered Vaccine Complex

The use of subunits for consideration in vaccine production can be alternatively used as nano-VLPs (nanovirus-like particles). These can be [47–49] produced through recombinant technology to incorporate ligands, modulators, and targeting sequences.

The self-assembly of nanoparticles and VLPs presents a highly specific, stable, and ordered nanovaccine formulation. However, the use of subunit nanovaccines has been processed by Medicago and iBio by utilization of *Nicotiana benthamiana* for producing VLPs incorporating the S protein and Adapt Vac on the other side producing VLPs from S2 protein using insect cell expression. These are to be processed further for clinical trials soon (Figure 4.3).

For utilization and presentation of multiple or multivalent antigen display for enhancing the immune response, these nanosubunits have been used as effective adjuvants, effective delivery vehicles, and sufficient nanocarrier platforms [50–56]. The vaccine used for influenza consists of an effective formulation incorporating liposomal nanoadjuvants with hemagglutinin as the influenza protein.

In addition to the advantage of displaying antigen multivalency, nanoparticles/nanovectors have an ability of efficient and accurate delivery of adjuvant–

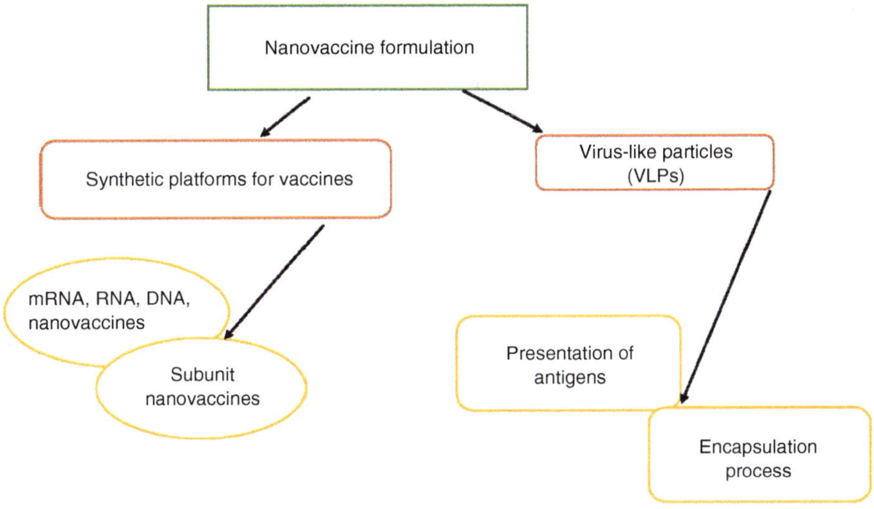

Figure 4.3 The different nanovaccine formulation constituting the different potential nanocandidates for COVID-19.

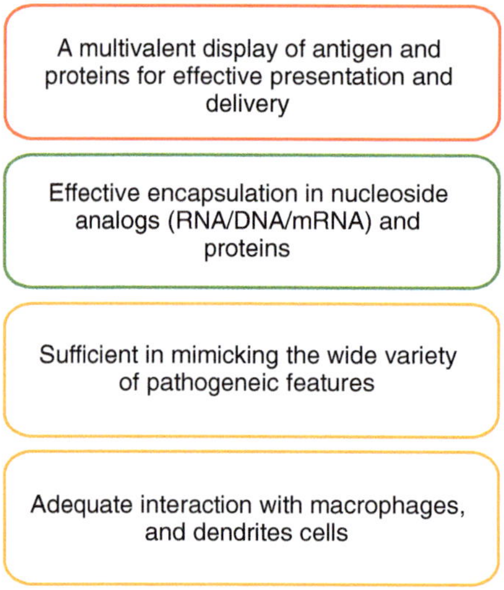

Figure 4.4 Advantages related to the use of nanovaccines for COVID-19.

antigen complex to the target organs, and preferential uptake by the concerned APCs (Antigen Presenting Cells), for acerbating the immune stimulus and leads to cross-antigen presentation by the t8 toxic cells tagged with MHC-1 pathway (Figure 4.3) [57].

4.5.3 Nanopeptide-Based Vaccines: "Hitchhiking Through Albumin"

These types of upcoming vaccine concept need an efficient working strategy of adjuvants for an effective delivery system. This purpose can be easily solved by the use of nanotechnology by including the advanced processes for targeting specific sites of lymph nodes or specific cellular and subcellular locations to address the specific diseases and their detection procedures (Figure 4.4).

A strategy called as "hitchhiking through albumin" assists a natural trafficking ability for many proteins and peptides such as albumin [58, 59].

A dual targeting HBV (hepatitis B vaccine) is being recently developed by using the inherent ability of the specific nanoparticles against dendritic cells and specified macrophages. This inherent ability of nanoparticles has resulted in clearance of virus from the different inhabited sites [60]. To ensure a fine APC presentation, cross presentation, and targeting of dendritic cells, various nanoparticles/nanocarriers are being used (polymeric assisted micelles-PEG-PE), which help in transformation of peptides (antigens) to facilitate delivery process efficiently.

4.6 Future Prospective: Resolving the Big Pandemics

The current scenario highlights a global health threat due to an ongoing pandemic situation, as the COVID-19 viral infection has crossed the barriers of high mortality to infection ratios surpassing the other viral entities. The physicians, researchers, and scientists have to make collaborative efforts to fight [61] against these pandemics and must work together to overcome the threat of SARS-CoV-2 and provide a strong base to block future pandemics. The fight against COVID-19 can only be sustained if research and technology gaps are filled with the potential methods. The main reason for the fast tracking and repurposing of various drugs and formulations during clinical and preclinical phase testing is the glitches observed in time management, as these drugs or vaccines take years to come into real play. Therefore, there is a need for "platform or lab-on-chip technologies," which establishes POC diagnosis and are easy to repurpose on demand. Nanotechnology thus offers a stable, cost-efficient, and time-effective tool to contribute toward understanding of their viral mechanism and finally to develop easy, quick nanovaccines to restrict the path of pandemics such as COVID-19. Nanovaccines/nanodevices/nanoformulations therefore hold a strong potential in terms of development of vaccines in the future scenario.

4.7 Conclusion

Nanotechnology has provided solutions in every possible sphere of science starting from agriculture to engineering and finally to therapeutics. The current situation of COVID-19 is waiting for the nanosolutions to combat its rapid spread. Nanotechnology is ready to provide its nanotools/nanovaccines/nanoformulation for easy drug development, delivery, administration, and its easy distribution. The ongoing research and studies have been working for collaborative efforts of medical field, research laboratories and industries. Further it can be easily observed that repurposing the nanomedicine formulations may enable quick immune responses, easy and specific targeting of macrophages, antigen-presenting cells, dendritic cells, phagocytic cells upon delivery through intravenous route, and pulmonary routes.

References

1 Waris, A., Khan, A.U., Ali, M. et al. (2020). COVID-19 outbreak: current scenario of Pakistan. *New Microbes New Infect.* 35: 100681.

2 Kaul, D. (2020). An overview of coronaviruses including the SARS-2 coronavirus – molecular biology, epidemiology and clinical implications. *Curr. Med. Res. Pract.* https://www.ncbi.nlm.nih.gov/pubmed/32363221.

3 Curley, M. and Thomas, N. (2004). Human security and public health in Southeast Asia: the SARS outbreak. *Aust. J. Int. Aff.* 58: 17–32.

4 Kumar, S. and Rathi, B. (2020). Coronavirus disease COVID-19: a new threat to public health. *Curr. Top. Med. Chem.* 20 (8): 599–600. https://www.ncbi.nlm.nih.gov/pubmed/32133964.

5 (a) Agarwal, H.K., Chhikara, B.S., Doncel, G.F., and Parang, K. (2017). Synthesis and anti-HIV activities of unsymmetrical long chain dicarboxylate esters of dinucleoside reverse transcriptase inhibitors. *Bioorg. Med. Chem. Lett.* 27: 1934–1937; (b) Gnanasangeetha, D. and Sarala Thambavani, D. (2013). One pot synthesis of zinc oxide nanoparticles via chemical and green method. *Res. J. Mater. Sci.* 2320: 6055.

6 Chhikara, B.S. (2020). Corona virus SARS-CoV-2 disease COVID-19: infection, prevention and clinical advances of the prospective chemical drug therapeutics. *Chem. Biol. Lett.* 7: 63–72. www.ncbi.nlm.nih.gov/pubmed/28351588.

7 Chotiwan, N., Brewster, C.D., Magalhaes, T. et al. (2017). Rapid and specific detection of Asian-and African-lineage Zika viruses. *Sci. Transl. Med.* 9: eaag0538. https://pubmed.ncbi.nlm.nih.gov/28469032.

8 Yang, W., Dang, X., Wang, Q. et al. (2020). Rapid detection of SARS-CoV-2 using reverse transcription RT-LAMP method. *medRxiv*. https://doi.org/10.1101/2020.03.02.20030130.

9 Lu, R., Zhao, X., Li, J. et al. (2020). Genomic characterisation and epidemiology of 2019 novel coronavirus: implications for virus origins and receptor binding. *Lancet* 395: 565–574.

10 Chan, W.C.W. (2020). Nano research for COVID-19. *ACS Nano* 14: 3719–3720. https://pubmed.ncbi.nlm.nih.gov/32227916/.

11 Law, S., Leung, A.W., and Xu, C. (2020). Severe acute respiratory syndrome (SARS) and coronavirus disease-2019 (COVID-19): from causes to preventions in Hong Kong. *Int. J. Infect. Dis.* https://pubmed.ncbi.nlm.nih.gov/32251790.

12 Yang, W., Dang, X., Wang, Q. et al. (2020). Rapid detection of SARS-CoV-2 using reverse transcription RT-LAMP method. *medRxiv*. https://doi.org/10.1101/2020.03.02.20030130.

13 Zhu, X., Wang, X., Han, L. et al. (2020). Reverse transcription loop-mediated isothermal amplification combined with nanoparticles-based biosensor for diagnosis of COVID-19. *medRxiv*. https://doi.org/10.1101/2020.03.17.20037796.

14 Yu, L., Wu, S., Hao, X. et al. (2020). Rapid colorimetric detection of COVID-19 coronavirus using a reverse transcriptional loop-mediated isothermal amplification (RT-LAMP) diagnostic plat-form: iLACO. *medRxiv*. https://doi.org/10.1101/2020.02.20.20025874.

15 Lamb, L.E., Bartolone, S.N., Ward, E., and Chancellor, M.B. (2020). Rapid detection of novel coronavirus (COVID-19) by reverse transcription-loop-mediated isothermal amplification. *medRxiv*. https://doi.org/10.1101/2020.02.19.20025155.

16 Xiang, J., Yan, M., Li, H. et al. (2020). Evaluation of enzyme-linked immunoassay and colloidal gold-immunochromatographic assay kit for detection of novel coronavirus (SARS-CoV-2) causing an outbreak of pneumonia (COVID-19). *medRxiv*. https://doi.org/10.1101/2020.02.27.20028787.

17 Udugama, B., Kadhiresan, P., Kozlowski, H.N. et al. (2020). Diagnosing COVID-19: the disease and tools for detection. *ACS Nano* 14 (4): 3822–3835. https://doi.org/10.1021/acsnano.0c02624. PMID: 32223179; PMCID: PMC7144809.

18 Wang, M., Fu, A., Hu, B. et al. (2020). Nanopore target sequencing for accurate and comprehensive detection of SARS-CoV-2 and other respiratory viruses. *medRxiv*. https://doi.org/10.1101/2020.03.04.20029538.

19 Liu, R., Fu, A., Deng, Z. et al. (2020). Promising methods for detection of novel coronavirus SARS-CoV-2. *Small* 16, 2002169.

20 Yu, L., Tong, Y., Shen, G. et al. (2020). Immunodepletion with hypoxemia: a potential high risk subtype of coronavirus disease 2019. *medRxiv*. https://doi.org/10.1101/2020.03.03.20030650.

21 Peeri, N.C., Shrestha, N., Rahman, M.S. et al. (2020). The SARS, MERS and novel coronavirus (COVID-19) epidemics, the newest and biggest global health threats: what lessons have we learned? *Int. J. Epidemiol.* https://www.ncbi.nlm.nih.gov/pubmed/32086938.

22 Ziaie, S., Koucheck, M., Miri, M. et al. (2020). Review of therapeutic agents for the treatment of COVID-19. *J. Cell. Mol. Anesth.* 5: 32–36.

23 Song, Z., Hu, Y., Zheng, S. et al. (2020). Hospital pharmacists' pharmaceutical care for hospitalized patients with COVID-19: recommendations and guidance from clinical experience. *Res. Social Adm. Pharm.* https://www.ncbi.nlm.nih.gov/pmc/articles/PMC7129111/.

24 Tobaiqy, M., Qashqary, M., Al-Dahery, S. et al. (2020). Therapeutic management of COVID-19 patients: a systematic review. *Infect. Prev. Pract.* 2: 100061.

25 Donders, F., Lonnée-Hoffmann, R., Tsiakalos, A. et al. (2020). COVID I. ISIDOG recommendations concerning COVID-19 and pregnancy. *Diagnostics* 10: 243. https://www.ncbi.nlm.nih.gov/pubmed/32338645.

26 Sanders, J.M., Monogue, M.L., Jodlowski, T.Z., and Cutrell, J.B. (2020). Pharmacologic treatments for coronavirus disease 2019 (COVID-19): a review. *JAMA*. https://www.ncbi.nlm.nih.gov/pubmed/32282022.

27 Wang, L.S., Wang, Y.R., Ye, D.W., and Liu, Q.Q. (2020). A review of the 2019 novel coronavirus (COVID-19) based on current evidence. *Int. J. Antimicrob. Agents* 55: 105948. https://www.ncbi.nlm.nih.gov/pmc/articles/PMC7156162/.

28 Kaura, K.K., Allahbadiab, G., and Singhc, M. (2020). A comprehensive review on epidemiology, aetiopathogenesis, diagnosis and treatment of the novel coronavirus syndrome – COVID-19. *Iberoam J. Med.* 2 (2): 110–123.

29 Dong, L., Hu, S., and Gao, J. (2020). Discovering drugs to treat coronavirus disease 2019 (COVID-19). *Drug Discov. Ther.* 14: 58–60. https://www.ncbi.nlm.nih.gov/pubmed/32147628.

30 Razzaque, M. (2020). COVID-19 pandemic: can maintaining optimal zinc balance enhance host resistance? *Tohoku J. Exp. Med.* 251 (3): 175–181. https://doi.org/10.1620/tjem.251.175. PMID: 32641644.

31 Eby, G.A. (1997). Zinc ion availability – the determinant of efficacy in zinc lozenge treatment of common colds. *J. Antimicrob. Chemother.* 40: 483–493. https://www.ncbi.nlm.nih.gov/pubmed/9372416.

32 Read, S.A., Obeid, S., Ahlenstiel, C., and Ahlenstiel, G. (2019). The role of zinc in antiviral immunity. *Adv. Nutr.* 10: 696–710. https://www.ncbi.nlm.nih.gov/pubmed/31305906.

33 Eby, I.I.I.G.A. (2010). Zinc lozenges as cure for the common cold – a review and hypothesis. *Med. Hypotheses* 74: 482–492. https://www.ncbi.nlm.nih.gov/pubmed/19906491.

34 Field, C.J., Johnson, I.R., and Schley, P.D. (2002). Nutrients and their role in host resistance to infection. *J. Leukocyte Biol.* 71: 16–32. https://pubmed.ncbi.nlm.nih.gov/11781377.

35 Prasad, A.S. (2007). Zinc: mechanisms of host defense. *J. Nutr.* 137: 1345–1349. https://www.ncbi.nlm.nih.gov/pubmed/17449604.

36 Overbeck, S., Uciechowski, P., Ackland, M.L. et al. (2008). Intracellular zinc homeostasis in leukocyte subsets is regulated by different expression of zinc exporters ZnT-1 to ZnT-9. *J. Leukocyte Biol.* 83: 368–380. https://www.ncbi.nlm.nih.gov/pubmed/17971500.

37 Lambert, S.A., Jolma, A., Campitelli, L.F. et al. (2018). The human transcription factors. *Cell* 172: 650665. https://www.ncbi.nlm.nih.gov/pubmed/30290144.

38 Wimalawansa, S.J. (2020). Global epidemic of coronavirus – COVID-19: what can we do to minimize risks. *Eur. J. Biomed. Pharm. Sci.* 7: 432–438.

39 Król, A., Pomastowski, P., Rafińska, K. et al. (2017). Zinc oxide nanoparticles: synthesis, antiseptic activity and toxicity mechanism. *Adv. Colloid Interface Sci.* 249: 37–52. https://www.ncbi.nlm.nih.gov/pubmed/28923702.

40 Pardi, N., Hogan, M.J., Porter, F.W., and Weissman, D. (2018). mRNA vaccines – a new era in vaccinology. *Nat. Rev. Drug Discovery* 17: 261–279.

41 Iavarone, C., O'hagan, D.T., Yu, D. et al. (2017). Mechanism of action of mRNA-based vaccines. *Expert Rev. Vaccines* 16: 871–881.

42 BIONTECH (2020). BioNTech and Pfizer announce regulatory approval from German authority Paul-Ehrlich-Institut to commence first clinical trial of COVID-19 vaccine candidates. https://investors.biontech.de/news-releases/news-release-details/biontech-and-pfizer-announce-regulatory-approval-german (accessed 02 March 2021).

43 Zeng, C., Hou, X., Yan, J. et al. (2020). Leveraging mRNAs sequences to express SARS-CoV-2 antigens in vivo. *bioRxiv*. https://www.biorxiv.org/content/10.1101/2020.04.01.019877v1.

44 ARCTURUS therapeutics (2020). Arcturus Therapeutics and Duke-NUS Medical School partner to develop a coronavirus (COVID-19) vaccine using STARR™ Technology. https://ir.arcturusrx.com/news-releases/news-release-details/arcturus-therapeutics-and-duke-nus-medical-school-partner (accessed 02 March 2021).

45 Takashima, Y., Osaki, M., Ishimaru, Y. et al. (2011). Artificial molecular clamp: a novel device for synthetic polymerases. *Angew. Chem. Int. Ed.* 50: 7524–7528.

46 Lim, M. et al. (2020). Engineered nanodelivery systems to improve DNA vaccine technologies. *Pharmaceutics* 12: 30.

47 Liu, H., Badruddoza, A.Z.M., Firdous, J. et al. (2017). Improvement of pharmacokinetic profile of TRAIL via trimer-tag enhances its antitumor activity in vivo. *Sci. Rep.* 7: 8953.

48 Hotez, P.J. and Bottazzi, M.E. (2020). Developing a low-cost and accessible COVID-19 vaccine for global health. https://www.preprints.org/manuscript/202003.0464/v1 (accessed 02 March 2021).

49 Kanekiyo, M., Bu, W., Joyce, M.G., et al. (2015). Rational design of an Epstein–Barr virus vaccine targeting the receptor-binding site. *Cell* 162: 1090–1100.

50 Kanekiyo, M., Joyce, M.G., Gillespie, R.A., et al. (2019). Mosaic nanoparticle display of diverse influenza virus hemagglutinins elicits broad B cell responses. *Nat. Immunol.* 20: 362–372.

51 Sharma, J., Shepardson, K., Johns, L.L., et al. (2020). A self-adjuvanted, modular, antigenic VLP for rapid response to influenza virus variability. *ACS Appl. Mater. Interfaces* 12: 18211–18224.

52 Brune, K.D. and Howarth, M. (2018). New routes and opportunities for modular construction of particulate vaccines: stick, click, and glue. *Front. Immunol.* 9: 1–15.

53 Ross, K., Senapati, S., Alley, J., et al. (2019). Single dose combination nanovaccine provides protection against influenza A virus in young and aged mice. *Biomater. Sci.* 7: 809–821.

54 Patterson, D.P., Rynda-Apple, A., Harmsen, A.L. et al. (2013). Biomimetic antigenic nanoparticles elicit controlled protective immune response to influenza. *ACS Nano* 7: 3036–3044.

55 Bachmann, M.F. and Jennings, G.T. (2010). Vaccine delivery: a matter of size, geometry, kinetics and molecular patterns. *Nat. Rev. Immunol.* 10: 787–796.

56 Herzog, C., Hartmann, K., Künzi, V., et al. (2009). Eleven years of Inflexal® V – a virosomal adjuvanted influenza vaccine. *Vaccine* 27: 4381–4387.

57 Wang, S., Qin, L., Yamankurt, G., et al. (2019). Rational vaccinology with spherical nucleic acids. *Proc. Natl. Acad. Sci. U.S.A.* 116: 10473–10481.

58 Bediz, B., Korkmaz, E., Khilwani, R., et al. (2014). Dissolvable microneedle arrays for intradermal delivery of biologics: fabrication and application. *Pharm. Res.* 31: 117–135.

59 Liu, H., Moynihan, K.D., Zheng, Y., et al. (2014). Structure-based programming of lymph-node targeting in molecular vaccines. *Nature* 507: 519–522.

60 Wang, W., Zhou, X., Bian, Y., et al. (2020). Dual-targeting nanoparticle vaccine elicits a therapeutic antibody response against chronic hepatitis B. *Nat. Nanotechnol.* 15: 406–416.

61 Parhi, R. (2019). Review of microneedle based transdermal drug delivery systems. *Int. J. Pharm. Sci. Nanotechnol.* 12: 4511–4524.

5

Biosensing Approach for SARS-CoV-2 Detection

Varun Rawat, Sonam, Diksha Gahlot, Kritika Nagpal, and Seema R. Pathak

Amity University Haryana, Amity School of Applied Sciences, Department of Chemistry, Biochemistry & Forensic Science, Gurugram, 122413, India

5.1 Introduction

The latest outbreak of the pandemic "severe acute respiratory syndrome coronavirus 2" christened as SARS-CoV-2 has triggered respiratory illnesses all over the world. Because no specific coronavirus disease-2019 (COVID-19) drug and vaccine is yet available, there is an urgent need for the use of biosensors so that the COVID-19 test can be carried out more accurately and productively. Most clinical laboratories studying the virus are currently using molecular diagnostic assays focused on quantitative reverse transcription polymerase chain reaction (RT-PCR). Polymerase chain reaction (PCR) requires amplification of the genetic material's minute traces, i.e. the virus' ribonucleic acid (RNA), leading to highly precise and responsive identification. PCR tests, however, are normally performed by highly trained staff in centralized diagnostic services and their findings can take from four hours to three days.

In addition, PCR testing is sensitive to the aforementioned major drawbacks:

(i) Sampling error: To take mucus from the ventilator system, nasopharyngeal swab is properly performed; however, this can give rise to false negatives as an ideal sampling moment is vulnerable to error.
(ii) In addition, as sample preparation is needed for PCR analysis (including cell lysis and nucleic acid purification), a limiting factor in rising COVID-19 testing is not just the number of tests but also the number of extraction kits [1].
(iii) PCR samples generally need specialized handling and transportation. During inadequate transport, the genetic material may be denatured, resulting in false negatives as well.
(iv) The standard of reagents used by various manufacturers of PCR kits can also affect the accuracy of the results.
(v) In COVID-19 patients with unidentified clinical symptoms, standard PCR methods can lack responsiveness, possibly giving rise to inaccurate negative results [2].

Detection and Analysis of SARS Coronavirus: Advanced Biosensors for Pandemic Viruses and Related Pathogens, First Edition. Edited by Chaudhery Mustansar Hussain and Sudheesh K. Shukla.

(vi) For recovered patients, even weeks after full recovery, the existence of genetic information of the dead SARS-CoV-2 can be reported by PCR testing, resulting in false-positive results (not particularly reinfections) [3].

The invention of highly sensitive and fast biosensing devices has therefore become increasingly relevant as these biosensors can have an enormous impact on the transformation of current analytical techniques into diagnostic strategies by transforming their sensing strategies. The recently developed rapid biosensor such as localized surface plasmon resonance (LSPR), field-effect transistor (FET), cell-based biosensing system, other biosensors, and finally some possible biomarkers will be mentioned here in this chapter as a biosensor application for detection and early diagnostics.

5.2 SARS-COVID-19 Structure and Genome

Coronaviruses (CoVs) are positive, single-stranded RNA viruses that refer to the coronavirinae subfamily, coronaviridae group, Nidovirales order. Four genotypes of CoVs are classified, that is alpha-coronavirus (αCoV), beta-coronavirus (βCoV), delta-coronavirus (δCoV), and gamma-coronavirus (γCoV). In bats and rats with αCoV and βCoV, the virus type is highly contagious, whereas δCoV and αCoV are present in bird species. Coronaviridae is a large group of viruses infecting animals and humans. The common forms of human coronaviruses include Middle East Respiratory Syndrome Coronavirus (MERS-CoV), SARS-CoV, and SARS-CoV-2 or COVID-19 [4, 5]. It appears that COVID-19 does not vary much in its clinical characteristics from SARS-CoV. The COVID-19 genome involves four main protein molecules: spike (S), membrane (M), envelope (E), and also nucleocapsid (N) [6]. However, SARS-CoV-2 is spread even more widely in the community [7, 8]. Coronaviruses, SARS-CoV-2, are a wrapped virus with approximately 60–140 nm in diameter, roughly spherical, or mildly pleomorphic [9].

The design of biosensor platforms for the detection of SARS-CoV-2 requires three important aspects:

(1) the purpose of recognition, e.g. viral RNA, viral proteins, or human immunoglobulins.
(2) the process of recognition, e.g. through nucleic acid samples, aptamers, antibodies, and receptors, where the interaction of antibody–antigen binding or receptor–ligand can be identified through conformative changes. One additional method of identification is enzymatic reactions, such as the detection of complex protease proteolytic cleavage.
(3) electrochemical, electrical, optical, surface plasmon resonance (SPR), fluorescent signals, and mechanical devices for signal amplification and transduction devices [10].

Samplers need to be integrated to treat samples directly from the field for environmental applications. For example, if the samples were to be taken from ambient

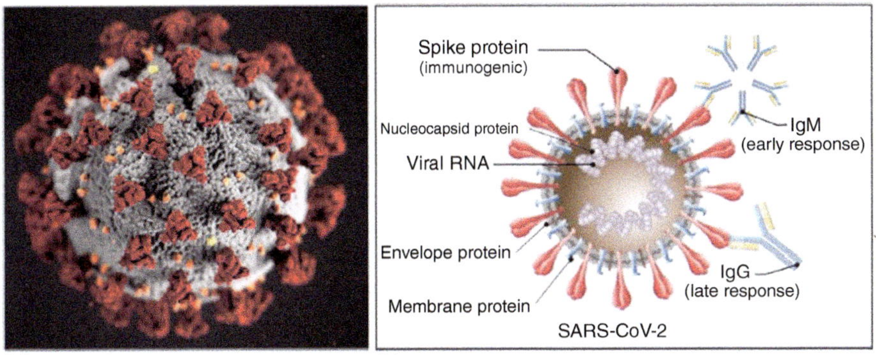

(a) (b)

Figure 5.1 Illustration showing SARS-CoV-2 with possible targeting sites. Source: Xu et al. [14].

air, then the kits were needed to be designed in a manner suitable for the condition [11]. The potential methods of targeting SARS-CoV-2 by various biomolecules are summarized in Figure 5.1a,b. In addition to the viral RNA, novel coronaviruses exhibit immunogenic spike proteins for the COVID-19 test; thus, the immune system can generate immunoglobulins to cause an immune response against the pathogen [12]. Relevantly, not only are these immunoglobulins useful for the detection of COVID-19 but also for their potential treatment [13]. Immunoglobulin M (IgM) antibodies are produced during the onset of the infectious disease (between 4 and 10 days), whereas immunoglobulin G (IgG) response is produced later (around 2 weeks) [14–16].

5.3 SARS-COVID-19 Sensors

5.3.1 Localized Surface Plasmon Resonance (LSPR) Sensor

SPR- and LSPR-based viral biosensors were referred earlier [15, 16]. In nucleic acid recognition along with virulent disease identification, these thermoplasmonic techniques are extremely useful. This SPR-based sensor has recently been reported for the detection of nucleocapsid antibodies specific to SARS-CoV-2 in undiluted human serum instead of oropharyngeal swabs. LSPR is an optical phenomenon produced when light waves are trapped in conductive nanoparticles that are small compared to the wavelength of light. To achieve coherent localized plasmon oscillation, the interaction of incident light and surface electrons in the conduction band is essential. The frequency of resonance is susceptible to local changes such as refractive index variation and molecular binding [17]. In the nanomolar range, anti-SARS-CoV-2 antibodies were detected by this peptide monolayer-coated SPR sensor and functionalized with the SARS-CoV-2 nucleocapsid. This bioassay is therefore quick, label-free and can diagnose specimens within 15 minutes of sample/sensor contact [18]. Dual-functional, plasmonic biosensor plasmonic photothermal (PPT), and LSPR were also investigated for the identification of the new

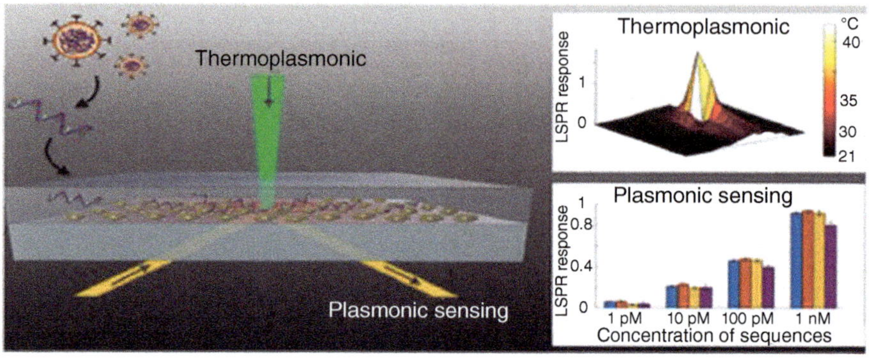

Figure 5.2 Representation of plasmonic sensing. Source: Qiu et al. [19].

pandemic (SARS-CoV-2). Functioning with complementary DNA (cDNA) receptors and integrating PPT effect and LSPR sensing techniques, 2D gold nanoislands (AuNIs) provide an alternative and promising solution for the detection by nucleic acid hybridization of clinical COVID-19. This dual-functional LSPR biosensor is highly sensitive to the selected SARS-CoV-2 sequences, with a detection limit of up to 0.22 pM. For the detection of various viral sequences, including RdRp-COVID, ORF1ab COVID, and E genes from SARS-CoV-2, a dual-functional plasmonic biosensor using the PPT effect and LSPR sensing transduction was used to detect the particular target in a multigen mixture. A stable heat source was generated by the converted PPT heat energy in the vicinity of gold to enhance the in situ hybridization of SARS-CoV-2 RdRp and its cDNA. Without the photothermal effect, the slope of the photothermal enhanced LSPR curve was larger than the device (Figure 5.2). The proposed sensor could differentiate between SARS-CoV viruses and SARS-CoV-2 viruses. For the RdRp-SARS series, a false-positive response signal was obtained without the assistance of the photothermal unit. The sensor showed an LOD of 0.22 pM [20].

5.3.2 Field Effect Transistor (FET)

Because of the electrostatic field, the FET transducer is based on modulating carrier mobility over a biased semiconductor. The FET gate surface is covered by a coating that can be modified by biomolecules for selective target detection [21]. The FET-based biosensing devices use a monoclonal antibody against the SARS-CoV-2 spike protein in the coating of the graphene sheets of the FET (Figure 5.3). Using antigen protein, cultured virus, and nasopharyngeal swab specimens from COVID-19 patients, they determined its sensitivity. A 1 fg/ml concentration (conc.) of SARS-CoV-2 spike protein in phosphate-buffered saline (PBS) and 100 fg/ml conc. could be detected by this FET biosensor unit [22]. Many electrochemical biosensors based on nanoparticle graphene FET have been decorated with SARS-CoV-2 spike S1 subunit protein (CSAb) or angiotensin-converting enzyme 2 (ACE2) antibodies to detect SARS-CoV-2 spike protein S1. The binding on the graphene surface of the S1 protein that has a slightly positive charge

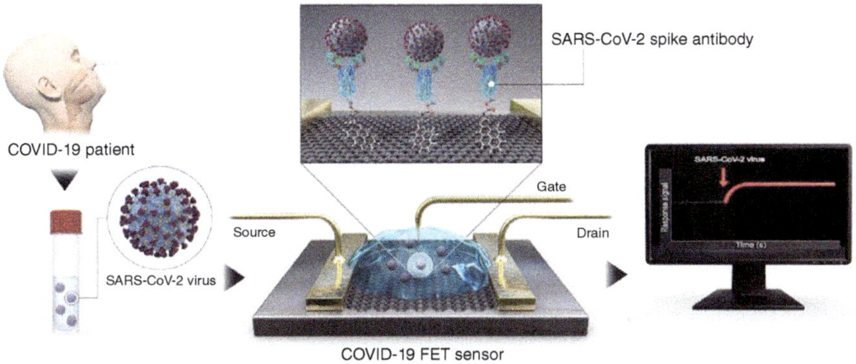

Figure 5.3 Diagram of COVID-19 FET-based biosensor operation. Source: Seo et al. [21].

with the CSAb/ACE2 receptors altered the conductance/resistance in graphene FET, which was deemed to be the source of detection. Because of the higher affinity of this antibody, CSAb-modified graphene FET exhibited better sensitivity. An LOD of 0.2 pM was shown by the proposed sensor [23]. Based on the changes in the channel surface potential and its effect on the electrical response, the FET system detected SARS-CoV-2. The S protein is an excellent antigen, as previously mentioned, because it is a major transmembrane protein of the virus and shows the diversity of amino acid sequences among coronaviruses. The biosensor was also able to differentiate between the SARS-CoV-2 antigen protein and the MERS-CoV protein, suggesting the platform's strong selectivity [24].

5.3.3 Cell-Based Potentiometric Biosensor

A membrane-engineered kidney cell modified by electro-insertion with the SARS-CoV-2 Spike S1 antibody was added to detect the SARS-CoV-2 S1 antigen. The membrane's potential is modified by the antibody's interaction with the target protein. Signal (volts) is received by binding the viral protein to membrane-bound antibodies with a detection limit of 1 fg/ml determined by the membrane potential difference. It can be used with a ready-to-use platform, including a smartphone/tablet powered portable reading device (Figure 5.4). The device was manufactured on eight-gold screen-printed electrodes protected by an eight-well polydimethylsiloxane (PDMS) layer. Suspension of the modified membrane was applied to the well of PDMS, accompanied by the addition of protein solution and potentiometer signal calculation [25].

5.3.4 eCovSens

Recently, Gandhi's research group fabricated an in-house built biosensor device. Using three electrode-electrochemical systems, they built this biosensor using disposable screen-printed carbon electrodes. In spiked saliva specimens, the eCovSens detection limits are 90 fM. Within 10–30 seconds, the as-fabricated eCovSens device

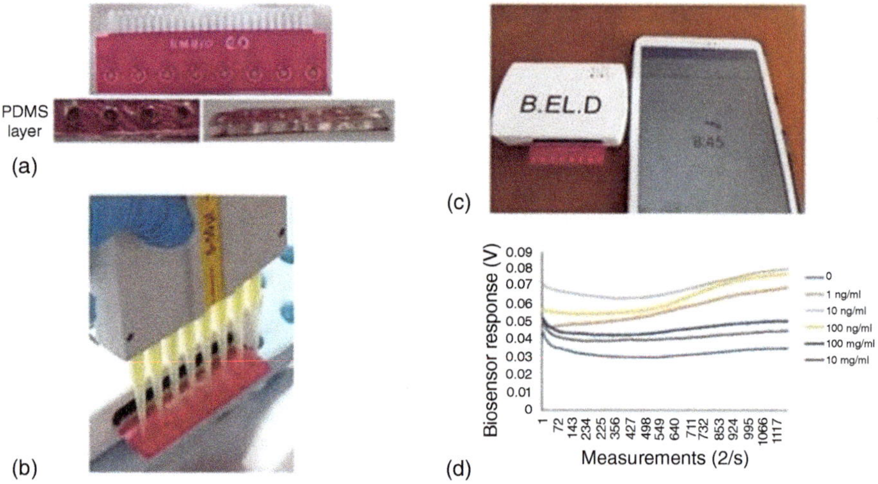

Figure 5.4 Schematics of the Vero/anti-S1 cell-based biosensor's assembly. (a) An eight-channel gold screen-printed electrode. (b, c) Process of sample application. (d) Visualization of signals. Source: Mavrikou e al. [23]. Licensed under CC BY 4.0.

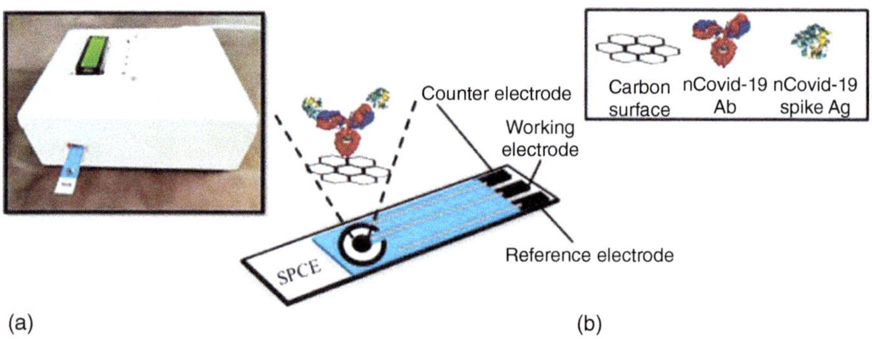

Figure 5.5 (a) Electrochemical eCovSens device. (b) Schematic of fabrication process of SPCE electrode. Source: Mahari et al. [24].

can detect COVID-19 Ag quickly. To detect traces of COVID-19 Ag in the saliva samples of infected individuals with improved selectivity and specificity, this platform can therefore be used as an alternative diagnostic system. The eCovSens unit is very inexpensive and compact, uses a very low voltage of 1.3–3 V, and can also be powered with a battery, so it can be used as a diagnostic point of care [26] (Figure 5.5).

5.3.5 CRISPR/Cas12

For SARS-CoV-2, a CRISPR (clustered regularly interspaced short palindromic repeats)/Cas12-based biosensor in combination with a lateral flow assay was recently developed to enable a test result in about 30 minutes (Figure 5.6) [27]. The so-called DETECTR conducts simultaneous reverse transcription and isothermal amplification after extraction of RNAs from patient samples using loop-mediated

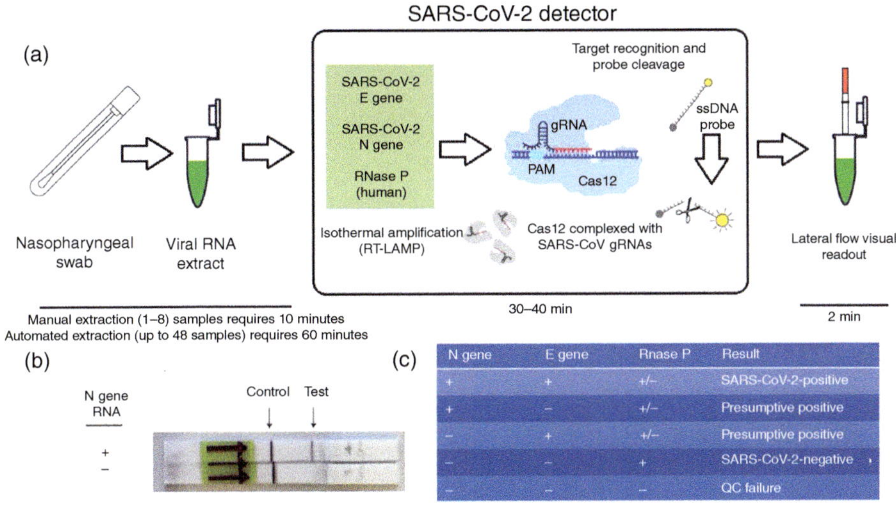

Figure 5.6 Schematic of SARS-CoV-2 DETECTR workflow. Source: Broughton et al. [25].

amplification (RT-LAMP) at 62 °C for 20 minutes, followed by Cas12 detection of predefined coronavirus sequences at 37 °C for 10 minutes, after which the cleavage of a reporter molecule confirms the detection of the virus as shown on a late coronavirus sequence. The lateral strip compatibility also allows this CRISPR-based technology to be used for point-of-care research away from the clinical diagnostics laboratory. Similarly, a rapid CRISPR-Cas12a fluorescent reporter assay coupled with one-step isothermal recombinase polymerase amplification (RPA) techniques for amplifying target regions from extracted viral RNAs was reported by Kishony and colleagues, with a sample-to-answer time of ~50 minutes, and an LOD of two copies per sample [28, 29]. In 96-well microtiter plates, this assay can be easily performed.

5.3.6 DNA Nanoscaffold Hybrid Chain Reaction (DNHCR)-Based Fluorescence Biosensor

Jiao et al. developed a DNA nanoscaffold hybrid chain reaction (DNHCR)-based method for detecting SARS-CoV-2 RNA. The DNA nanoscaffolds constructed by self-assembly of long DNA strands and self-quenching probes (H1) act as the sensing element in this biosensor. The hybridization of H1 and free H2 DNA probes along the nanoscaffold is then triggered by the target RNAs to illuminate the DNA nanostring, which represents the concentration of the virus. SARS-CoV-2 can be detected within 10 minutes and under mild conditions (15–35 °C) by this DNHCR biosensor, showing great potential for routine clinical diagnosis. Moreover, by implementing a pooling method to enable the simultaneous detection of dozens of samples [30], researchers have conceived new concepts to greatly improve the detection ability of quantitative reverse transcription polymerase chain reaction (qRT-PCR) testing. A study shows that a positive sample can be detected by group testing among 64 different samples with adequate sensitivity [31]. Such pooling approaches may therefore,

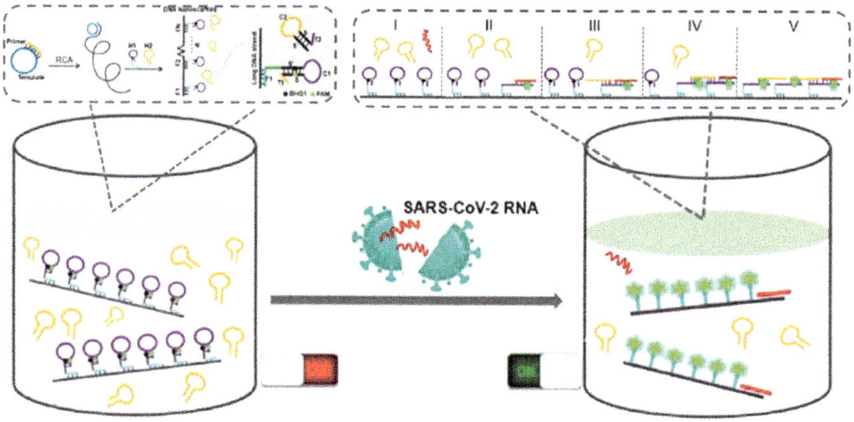

Figure 5.7 DNA nanoscaffold hybrid chain reaction (DNHCR)-based method for the detection of SARS-CoV-2 RNA. Source: Jiao et al. [27].

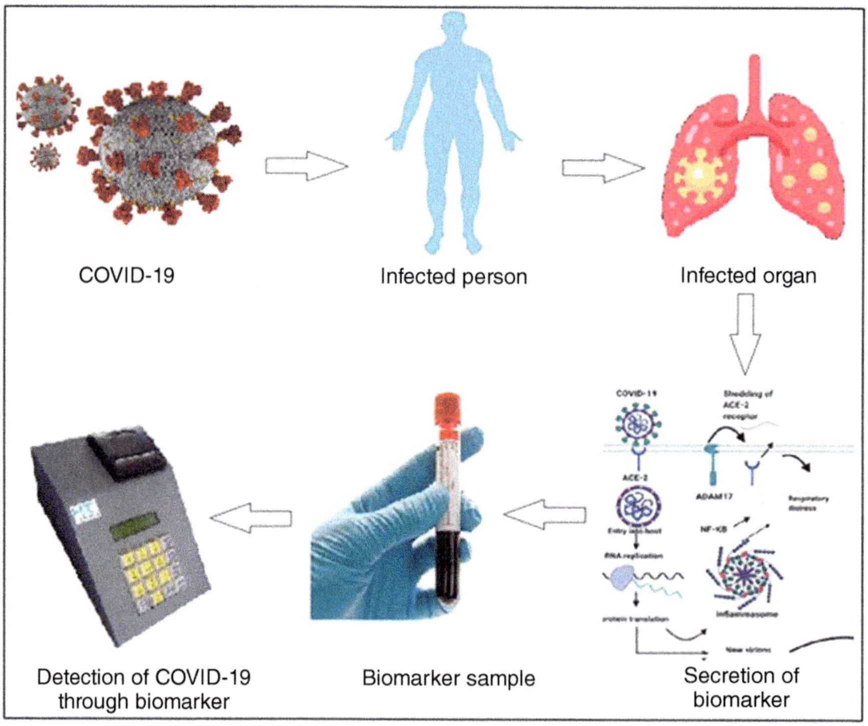

Figure 5.8 Schematic representation of biomarker-based biosensor. Source: Based on Xu et al. [32].

if properly scaled up, promote mass and large-scale research with less resource usage and faster time (Figure 5.7).

5.4 Biomarkers

Biomarkers are biomolecules specific to specific diseases that arise naturally, biomarkers may be a material that is inserted into an organism as a means of testing organ function or other health factors, such as CYFRA-21, a biomarker dependent on protein for oral cancer. A biomarker suggests a change in protein expression or state that is associated with the risk or progression of the disease or the susceptibility of the disease to treatment. Biomarkers can be of biological characteristics or molecules that can be found and measured in parts of the body, such as blood or tissue. In view of the urgent need for rapid detection of COVID-19,

Table 5.1 Potential biomarkers of SARS-CoV-2.

S.no.	Biomarker	Normal patient	Affective patient	Biological samples	References
1	Serum ferritin (ng/ml)	15–150	800.4 (452.9–1451.6)	Serum	[30]
2.	C-reactive protein (mg/l)	0–1	57.9 (20.9–103.2)	Plasma	[30]
3.	Interleukin-2R (U/ml)	223–710	757 (528.5–1136.3)	Serum	[30]
4.	IL-6 (pg/ml)	0–7	7.9	Blood	[31]
5.	D-dimer (μg/ml)	0–0.243	0.4	Plasma	[26]
6.	Serum amyloid A (SAA) (mg/l)	0–10	108.4	Serum/saliva	[26]
7	Procalcitonin (ng/ml)	0–0.5	<0.1	Blood	[32]
8.	Lymphocyte count (/l)	$0.8–4 \times 10^9$	$<0.6 \times 10^9$	Blood	[32]
9.	Cystatin C (mg/dl)	0.6–1.3	>1.1	Blood	[32]
10.	Serum direct bilirubin (μmol/l)	5.1–17	18–60	Blood	[33]
11.	Cholinesterase (kU/l)	8–18	1.5–8	Blood	[33]
12.	Lactate dehydrogenase (U/l)	140–280	300–600	Blood	[33]
13.	Creatinine (μmol/l)	60–120	30–80	Blood	[33]
14.	Urea (mmol/l)	2.5–7.1	4–22	Blood	[33]
15.	IL-10 (pg/ml)	0–9.1	>15	Blood	[34]

the sensor based on biomarkers will play a pivotal role as it will decrease the time to diagnose, be cost-effective, and also decrease the chance of virus transmission during diagnosis; we will look forward to integrating biosensor with microfluidics systems (Figure 5.8; Table 5.1).

5.5 Conclusion

Biosensors fulfill an important and useful role in the challenging situation such as COVID-19, where the healthcare facility providers are looking for some smart and innovative treatment or identifying a specific kind of devices for their patients. Traditional approaches such as PCR and sequencing are time consuming and can have clear false-positive outputs for individuals. However, the current challenges (such as accelerated mutations) and criteria (for mass populations) for the quicker and direct identification of viral pathogens may not be met by these approaches. Biosensors offer these opportunities to more conveniently and efficiently solve and confront the concerns and potential issues that have already been posed. In the near future, new technologies such as innovative platforms based on CRISPR-Cas, graphene FET, electrochemical biosensor, optical biosensor, and surface plasmon (SPR)-based biosensor could pave the way for fast, highly sensitive, and more promising biosensing cum diagnostic devices for COVID-19 and other unprecedented pandemics to be effective.

References

1 Satyanarayana, M. (2020). Shortage of RNA extraction kits hampers efforts to ramp up COVID-19 coronavirus testing [Internet]. *Chemical & Engineering News* (19 March). https://cen.acs.org/analytical-chemistry/diagnostics/Shortage-RNA-extraction-kits-hampers/98/web/2020/03.

2 An, J., Liao, X., Xiao, T. et al. (2020). Clinical characteristics of recovered COVID-19 patients with re-detectable positive RNA test. *Ann. Transl. Med.* https://doi.org/10.1101/2020.03.26.20044222.

3 Herald T. (2020). Tests in recovered patients found false positives, not reinfections, experts say. http://Koreaherald.com. http://www.koreaherald.com/view.php?ud=20200429000724 (accessed 22 October 2020).

4 Ali, M. (2020). The SARS-CoV-2 tears and ocular surface debate: what we know and what we need to know. *Indian J. Ophthalmol.* https://doi.org/10.4103/ijo.IJO_1881_20.

5 Malik, Y.A. (2020). Properties of coronavirus and SARS-CoV-2. *Malays. J. Pathol.* http://www.mjpath.org.my/2020/v42n1/properties-of-coronavirus.pdf.

6 Rota, P., Steven Oberste, M., Monroe, S.S. et al. (2020). Characterization of a novel coronavirus associated with severe acute respiratory syndrome. *Science* https://doi.org/10.1126/science.1085952.

7 Duan, L., Zheng, Q., Zhang, H. et al. (2015). The SARS-CoV-2 spike glycoprotein biosynthesis, structure, function, and antigenicity: implications for the design of spike-based vaccine immunogens. *Front. Immunol.* https://doi.org/10.3389/fimmu.2020.576622.

8 Petrosillo, N., Viceconte, G., Ergonul, O. et al. (2020). COVID-19, SARS and MERS: are they closely related? *Clin. Microbiol. Infect.* https://doi.org/10.1016/j.cmi.2020.03.026.

9 Taha, B., Al Mashhadany, Y., Mokhtar, M.H.H. et al. (2020). An analysis review of detection coronavirus disease 2019 (COVID-19) based on biosensor application. *Sensors* https://doi.org/10.3390/s20236764.

10 Krejcova, L., Michalek, P., Merlos Rodrigo, M. et al. (2015). Nanoscale virus biosensors: state of the art. *Nanobiosens. Dis. Diagn.* https://doi.org/10.2147/NDD.S56771.

11 Fronczek, C. and Yoon, J. (2015). Biosensors for monitoring airborne pathogens. *J. Lab. Automat.* https://doi.org/10.1177/2211068215580935.

12 Grifoni, A., Sidney, J., Zhang, Y. et al. (2020). A sequence homology and bioinformatic approach can predict candidate targets for immune responses to SARS-CoV-2. *Cell Host Microbe* https://doi.org/10.1016/j.chom.2020.03.002.

13 Chen, L., Xiong, J., Bao, L., and Shi, Y. (2020). Convalescent plasma as a potential therapy for COVID-19. *Lancet Infect. Dis.* https://doi.org/10.1016/S1473-3099(20)30141-9.

14 Xu, L., Li, D., Ramadan, S. et al. (2020). Facile biosensors for rapid detection of COVID-19. *Biosens. Bioelectron.* https://doi.org/10.1016/j.bios.2020.112673.

15 Park, T., Hyun, M., Lee, H. et al. (2009). A self-assembled fusion protein-based surface plasmon resonance biosensor for rapid diagnosis of severe acute respiratory syndrome. *Talanta* https://doi.org/10.1016/j.talanta.2009.03.051.

16 Lee, T., Ahn, J., Park, S. et al. (2018). Recent advances in AIV biosensors composed of nanobio hybrid material. *Micromachines* https://doi.org/10.3390/mi9120651.

17 Petryayeva, E. and Krull, U. (2011). Localized surface plasmon resonance: nanostructures, bioassays and biosensing – a review. *Anal. Chim. Acta* https://doi.org/10.1016/j.aca.2011.08.020.

18 Djaileb, A., Charron, B., Jodaylami, M. et al. (2020). A rapid and quantitative serum test for SARS-CoV-2 antibodies with portable surface plasmon resonance sensing. *ChemRxiv* https://doi.org/10.26434/chemrxiv.12118914.v1.

19 Qiu, G., Gai, Z., Tao, Y. et al. (2020). Dual-functional plasmonic photothermal biosensors for highly accurate severe acute respiratory syndrome coronavirus 2 detection. *ACS Nano* https://doi.org/10.1021/acsnano.0c02439.

20 Ahmad, R., Mahmoudi, T., Ahn, M., and Hahn, Y.-B. (2018). Recent advances in nanowires-based field-effect transistors for biological sensor applications. *Biosens. Bioelectron.* https://doi.org/10.1016/j.bios.2017.09.024.

21 Seo, G., Lee, G., Kim, M. et al. (2020). Rapid detection of COVID-19 causative virus (SARS-CoV-2) in human nasopharyngeal swab specimens using field-effect transistor-based biosensor. *ACS Nano* https://doi.org/10.1021/acsnano.0c02823.

22 Samson, R., Navale, G., and Dharne, M. (2020). Biosensors: frontiers in rapid detection of COVID-19. *3 Biotech* https://doi.org/10.1007/s13205-020-02369-0.

23 Mavrikou, S., Moschopoulou, G., Tsekouras, V., and Kintzios, S. (2020). Development of a portable, ultra-rapid and ultra-sensitive cell-based biosensor for the direct detection of the SARS-CoV-2 S1 spike protein antigen. *Sensors* https://doi.org/10.3390/s20113121.

24 Mahari, S., Roberts, A., Shahdeo, D., and Gandhi, S. (2020). eCovSens-ultrasensitive novel in-house built printed circuit board based electrochemical device for rapid detection of nCovid-19 antigen, a spike protein domain 1 of SARS-CoV-2. *bioRxiv* https://doi.org/10.1101/2020.04.24.059204.

25 Broughton, J., Deng, X., Yu, G. et al. (2020). CRISPR–Cas12-based detection of SARS-CoV-2. *Nat. Biotechnol.* https://doi.org/10.1038/s41587-020-0513-4.

26 Huang, C., Wang, Y., Li, X. et al. (2020). Clinical features of patients infected with 2019 novel coronavirus in Wuhan, China. *Lancet* https://doi.org/10.1016/S0140-6736(20)30183-5.

27 Jiao, J., Duan, C., Xue, L. et al. (2020). DNA nanoscaffold-based SARS-CoV-2 detection for COVID-19 diagnosis. *Biosens. Bioelectron.* https://doi.org/10.1016/j.bios.2020.112479.

28 Yelin, I., Aharony, N., Tamar, E. et al. (2020). Evaluation of COVID-19 RT-qPCR test in multi sample pools. *Clin. Infect. Dis.* https://doi.org/10.1101/2020.03.26.20039438.

29 Kaur, M., Tiwari, S., and Jain, R. (2020). Protein based biomarkers for non-invasive Covid-19 detection. *Sens. Bio-Sens. Res.* https://doi.org/10.1016/j.sbsr.2020.100362.

30 Qin, C., Zhou, L., Hu, Z. et al. (2020). Dysregulation of immune response in patients with COVID-19 in Wuhan, China. *Clin. Infect. Dis.* https://doi.org/10.1093/cid/ciaa248.

31 Wang, D., Hu, B., Hu, C. et al. (2020). Clinical characteristics of 138 hospitalized patients with 2019 novel coronavirus–infected pneumonia in Wuhan, China. *JAMA* https://doi.org/10.1001/jama.2020.1585.

32 Xu, Z., Shi, L., Wang, Y. et al. (2020). Pathological findings of COVID-19 associated with acute respiratory distress syndrome. *Lancet Respir. Med.* https://doi.org/10.1016/S2213-2600(20)30076-X.

33 Xiang, J., Wen, J., Yuan, X. et al. (2020). Potential biochemical markers to identify severe cases among COVID-19 patients. *medRxiv* https://doi.org/10.1101/2020.03.19.20034447.

34 Gong, J., Dong, H., Xia, Q. et al. (2020). Correlation analysis between disease severity and inflammation-related parameters in patients with COVID-19 pneumonia. *medRxiv* https://doi.org/10.1101/2020.02.25.20025643.

6

Role of Nanotechnology in Coronavirus Detection

AbdulGafar O. Tiamiyu[1], Bashir Adelodun[2,3], Hashim O. Bakare[1], Fidelis O. Ajibade[4,5,6], Kola Y. Kareem[2], Rahmat G. Ibrahim[7], Golden Odey[3], Madhumita Goala[8], and Jamiu A. Adeniran[9,10]

[1] *University of Ilorin, Department of Chemical Engineering, Ilorin PMB 1515, Nigeria*
[2] *University of Ilorin, Department of Agricultural and Biosystems Engineering, Ilorin, PMB 1515, Nigeria*
[3] *Kyungpook National University, Department of Agricultural Civil Engineering, Daegu 41566, Korea*
[4] *Federal University of Technology, Department of Civil and Environmental Engineering, Akure, PMB 704, Nigeria*
[5] *Chinese Academy of Sciences, Research Centre for Eco-Environmental Sciences, Key Laboratory of Environmental Biotechnology, Beijing 100085, PR China*
[6] *University of Chinese Academy of Sciences, Beijing, 100049, PR China*
[7] *Kwara State Ministry of Health, Ilorin, Kwara State, Nigeria*
[8] *Affiliated Assam University, Nehru College, Pailapool, Silchar, Cachar, 788098, Assam, India*
[9] *University of Ilorin, Department of Chemical Engineering, Environmental Engineering Research Laboratory, Ilorin, PMB 1515, Nigeria*
[10] *Peking University, Department of Atmospheric and Oceanic Sciences, Atmospheric Chemistry and Modelinq Group,, Beijinq China*

6.1 Introduction

The roles of nanoparticles (NPs) in the scientific and engineering world are immense and have generated great interest in closing the gap between bulk materials and atomic structures. These features make the nanoparticles unique in terms of large surface area from such a small bulk of the material and numerous uses [1]. The general application and wide usage of nanoparticles or nanomaterials have been attributed to features such as mechanical, catalytic, electrical, optical, and anti-microbial [2]. Nanomaterials are classified based on their size limits of 100 nm or 10^{-9} m or can be said to be within the range of 1–100 nm, also known as the nanoscale range [2–4].

The classifications of nanomaterials/nanoparticles are of two main types, which are natural nanoparticles and synthetic nanoparticles. The former may be formed because of natural actions such as decaying and chemical weathering, while the latter are artificial nanomaterials produced for particular purposes using chemical and physical research processes [2]. Synthesis of NPs depends on the application, and two major techniques top-down and bottom-up involve the reduction of the material to nanosize and gathering of materials from the bottom, respectively. However, regardless of the synthesis routes, the physical properties of the nanomaterials are

Detection and Analysis of SARS Coronavirus: Advanced Biosensors for Pandemic Viruses and Related Pathogens, First Edition. Edited by Chaudhery Mustansar Hussain and Sudheesh K. Shukla.

the same if properties such as microstructure, crystallinity, and chemical composition are similar [1]. In an effort to improve the shortcomings in the utilization of NP synthesis by a physical and chemical process, the green method using plant extracts had gained great attention from researchers, and successful synthesis of metallic NPs was recorded using *Aloe vera* plant extracts, *Capsicum annuum* leaf, geranium leaf, and neem leaf [5].

Having examined the various methods of synthesizing NPs, this chapter aims to review the existing virus (coronavirus disease 2019 [COVID-19]) detection means using nanotechnology. This paper critically looked into the application of different conventional technologies and nano-based ones while also examining the side effect of nanomaterials on human health.

6.2 Application of Nanomaterials

6.2.1 Silver Nanoparticles

Liu and Jiang [6] reported the vast application of silver nanoparticles in different fields such as healthcare, catalyst, electric, and optical applications, which have accounted for about 25% of the total nanomaterials in the market and are abundantly in commercial quantities. Silver nanoparticle synthesis, using the green method, was reported to have higher adsorption capacity than activated carbon in the removal of dibenzothiophene from oil [7]. This and many other applications of AgNPs such as anticoagulant, decolorization, and thrombolytic features of the nanoparticles have been ascertained [8]. Lateef et al. [9] investigated a biosynthesis (green synthesis) of AgNPs using both seed and shell of *Cola nitida* for antibacterial actions on drug-resistant strains of bacteria. While the seed shell was reported to have more activities than the seed extracts, the two biogenic nanoparticles were eco-friendly and proved effective for the antibacterial activities [9].

6.2.2 Gold Nanoparticles

Gold nanoparticles are another set of nanomaterials that have influenced the revolutionization of the nanotechnology industry, including many other nanomaterials, such as AgNPs and AuNPs, with important features that can find its applicability in all fields [10]. Gold nanoparticles fall under the category of metallic NPs and can be synthesized via different techniques, such as chemical, physical, and green (biological) methods. However, the biological methods of NP synthesis proved to be more economically viable because of low temperature, pH, pressure, high yield, and cost-effectiveness, thereby making it eco-friendly [10]. Kalimuthu et al. [10] and Lee et al. [11] reported AuNP synthesis via a green method using plants (leaves, fruits, rhizome, flowers, peel, seeds, bark, biomass, nuts, and latex), bacteria, enzymes, and fungi such as molds, yeast, and mushrooms.

The use of green chemistry for AuNP synthesis has gained the attention of the research community lately. Bacteria, enzymes, fungi, and plants act as both reducing and stabilizing agents against the physical and chemical synthesis of NPs that had brought upon the release of hazardous and toxic substances into the environment [11]. The multifaceted properties of AuNPs, including characterization, biological stability and compatibility, tunable surface, and modification, had increased the applications in biomedicine [12]; electronics such as single-electron switch, sensors, and transistors [13]; and catalysis, especially in the removal of volatile organic compounds (VOCs), toluene, hexane, nitrogen, sulfur, and acetaldehyde in water pollution treatment [14].

6.2.3 Carbon Nanotubes

Carbon nanotubes (CNTs) are a group of compounds having different features. These unique characteristics result from their formation, otherwise known as carbon nanotubes synthesis, and they are mostly in a tubular structure, i.e. the diameter is in nanometers and the length ranged from micrometers to centimeters [15]. CNTs exhibit special structural, electronic, and mechanical properties, making it readily available for an array of applications, including electron field emitters and gas breakdown sensors [16]. Other properties of CNTs include multifarious handedness, good thermal ability, large surface-to-volume ratio, and well-arranged carbon atoms, making it a suitable material in numerous solar cells applications [17].

The CNTs are often called a one-dimensional structure because of their high length-to-diameter ratio [17]. In addition to this, CNTs are classified into two different types: single-walled carbon nanotubes (SWCNTs) and multi-walled carbon nanotube (MWCNTs). The SWCNTs comprise a single sheet of graphene and have a tube diameter ranging from 1 to 5 nm, while the MWCNTs are made of 2–10 graphene sheets, making the MWCNTs possess an internal tube diameter of 10 nm and above [15, 17, 18]. The methods of synthesis of CNTs are of three types: laser ablation (LA) technique, chemical vapor deposition (CVD), and electric arc discharge (EAD). The three methods are applied based on the application of the CNTs, level of purity, quality, solubility, and mechanical properties [15]. Examples of these applications of CNTs include adsorbents for removing inorganic and organic pollutants in the environmental engineering aspect [18], hybrid solar cell application, biomedical imaging biosensors [17], and sensitive electrochemical sensors [15].

There are several other forms of metal, oxides of metal, dioxide, and polymer-coated nanoparticles such as Cu-NPs, CuO-NPs, Pt-NPs, Si-NPs, Ni-NPs, FeO-NPs, and TiO_2-NPs [19]. These nanoparticles are classified according to different applications such as excellent physiochemical properties (metallic NPs, polymeric NPs, semiconductor NPs, and ceramic NPs), chemical nature (organic and inorganic NPs), crystallinity (crystalline, polycrystalline, and amorphous NPs), origin (natural and man-made NPs), and magnetic properties (paramagnetic NPs such as CdS-NPs and Fe_3O_4-NPs and diamagnetic NPs such as TiO_2-NPs and $MgFe_2O_4$-NPs) [19]. These nanoparticles have numerous applications in engineering, medicine, and biomedical engineering. For instance, TiNPs were synthesized using the green methods (aqueous extract of *Falcaria vulgaris* leaves) for application in the wound healing process under in vitro and in vivo conditions, and the remedial effects were outstanding [20]. Other areas of applications of TiNPs are in cancer treatment and inactivation of antibiotic-resistant bacteria using photodynamic inactivation [21]. In a study conducted by Irshad et al. [22], TiNP synthesis via two plant extracts (*Trianthema portulacastrum* and *Chenopodium quinoa*) and a chemical (sol–gel) method exhibited excellent antifungal activity when applied to wheat rust. Nanoparticle and nanomaterial application in nanomedicine, including drug delivery, therapeutics, and biosensors, will be discussed further in Section 6.3.

6.3 Nanotechnology and Application in Medicine

Nanotechnology has given medicine a revolutionary rebranding through the exploitation of new materials with properties capable of giving new methods of combating human diseases. This new approach, known as nanomedicine, has further enhanced how scientists deal with disease processes at the molecular and cellular levels [1, 23]. Commercialization of medicine from the entry stage to the final stage does take about 10–15 years, with the huge cost running in millions of dollars while this problem is extended to personalized medicine, thereby making it look unrealistic [24]. However, with nanomedicine and the emerging sub-sections such as biobarriers, molecular imaging, early detection, and nanodiagnostics, the challenges in medical fields have got a light beam of hope. These are areas where the immense contribution of nanotechnology in the medical field has been felt and proven successful.

6.3.1 Biobarriers

Nanotechnology advancement has enabled the science and engineering community, especially in biomedical engineering and clinical medicine, where the process of transporting the right quantity of the drug to the right place and at an expected duration is now achievable with ease [23]. The problems associated with conventional drug delivery including the use of overdose to ensure active agents reached the intended sites, resulting in significant side effects or drug abuse. It was reported that only 1 out of 100 000 (0.000 01%) molecules of drug administered reached the pathological site [23]. The drug delivery challenges encountered due to bio-barriers have now been made easy by developing nanocarriers that allow delivery through the gastrointestinal tract, skin, and lung [25]. Luther et al. [26] reported the use of inorganic nanoparticles as nanocarriers, such as gold, silica, iron oxide, and lanthanide-based NPs, owing to their nanoscopic optical, magnetic, structural, and functional properties for drug delivery in combating diseases and improving the human health.

6.3.2 Molecular Imaging

In medical diagnosis, the information needed is mostly obtained from physical examination and history of the patient; meanwhile, multiple diagnostic imaging techniques are carried out to obtain some essential results [27]. Advancement of nanomedicine in the molecular imaging sections will help collect information about tissue anatomy, metabolism, physiology, early disease detection, and disease progression, which are expected to guide through the therapy and treatment sessions [23, 27]. The different types of imaging techniques available are ultrasound (US), magnetic resonance imaging (MRI), and computed tomography (CT) and are categorized under morphological imaging techniques. At the same time, the other category is functional or molecular imaging, which includes optical imaging (OI), single-photon emission computed tomography (SPECT), and positron emission tomography (PET) [24]. Rai et al. [27] listed out photothermal therapy (PTT), photodynamic therapy (PDT), and neuron capture therapy (NCT) as a theranostic modality technique for molecular imaging. However, these techniques have different basic physical principles with unique demerits in sensitivity and specificity to contrast agents, resolution, tissue penetration, tissue contrast, and quantitative estimation.

Molecular imaging of nanomedicine has enhanced the development of contrast agents that have the correct contrast-generating materials such as radioactive, paramagnetic, superparamagnetic, fluorescent or electron-dense, biocompatible coating, and targeting groups [28]. Nanoparticle molecular (NM) imaging helps in getting brighter and exact tissue imaging that can help detect disease at the early stages and, often, before disease exhibition [28]. NMs possess unique properties such as electronic, optical, structure, and magnetic properties, thereby making them suitable and most preferred for in vitro and in vivo utilization [24]. Examples of the NMs applied for biomedical applications are CNTs and graphene [25].

6.3.3 Early Detection

Improving personalized medicine through the application of nanotechnology will help in early detection probing, sophisticated drug delivery devices, nano-based injectable therapeutics, contrast agents, and tissue engineering. These will give the patients the freedom to self-care testing, as simple point-of-care (POC) testing give assess to an array of usage ranging from in vivo testing to glucose tests [25]. Meanwhile, efficient diagnostic methods must have excellent sensitivity and should support early detection of diseases so that a better treatment option can be sought for by a physician, as inadequate disease detection process or monitoring will result in the death of patients [24, 29].

There is a high expectation of nanotechnology as a field that will enhance cellular and subcellular diagnosis, with high-performance imaging processes and sensors [29]. Early detection methods of nanotechnology have been extended to quality changes in food products; therefore, it is important to develop nanosensors for food-borne pathogen detection as an alternative and improvement over the traditional method [30].

6.3.4 Nanodiagnostics

Advancement in diagnosis, which is brought about nanotechnology, is called nanodiagnostics, and this often occurs at the nanoscale level using a designed device [31]. Early diagnosis and excellent sensitivity can be achieved using nanodiagnostics or a combination of conventional diagnostics techniques and nanodiagnostics, thereby complementing the shortcomings of the traditional ones [31]. In addition, challenges with the traditional diagnostics include minimal portability, high cost, the possibility of cross-contamination, and the tendency to produce false-negative results [29]. Owing to the numerous advantages of nanodiagnostics over the conventional ones, the development needs to be stepped up because there is high expectation on developing clinically related biomarkers in a swift, sensitive, and cost-effective way [23, 29].

As an application of nanomedicine, disease diagnosis is critical in disease care and control using nanodiagnostics because early detection of disease will yield better clinical results and minimal therapeutic requirements [1]. The examples of the conventional methods for the fast diagnostics of infectious disease are DNA microarrays or DNA chips that are basically regarded as high-performance methods. The detection method is solely on oligonucleotides hybridized to pathogen genes for the diagnosis of infectious disease; however, usability in clinical settings is not feasible because of the problem of identifying pathogen-specific target genes among other genes and difficulties in the design of oligonucleotides that will magnify exact pathogen genes in complex polymerase chain reaction (PCR) without generic intensifications [32]. Nanotechnology in infectious diseases will make quick and actual detection possible using small volumes of samples from patients [32].

6.4 Biosensors for Infectious Disease Detection

Consistently, human health had received global threats in the form of epidemic or pandemic from infectious diseases such as influenza, SARS, Hendra, and Nipah, and there is an urgent need, as usual, to contain their spread and treat as required [33]. These infectious diseases need proper diagnostic tools to eliminate/reduce them to minimalize the chances of their outbreak [33]. Pathogens are also known infectious agents that can cause disease. They are mostly found as foodborne, airborne, and waterborne, with numerous modes of infections that have caused over 15 million deaths per year [34, 35]. The examples of these pathogens include protozoans, bacteria, fungi, and infectious agents (viruses and prions). However, the common pathogens include viruses such as norovirus, influenza virus, and bacteria, such as *Staphylococcus aureus* and *Escherichia coli* [35].

Researchers have always focused on developing highly effective and sensitive techniques for detecting pathogenic microorganisms. This forms the basis of monitoring and early detection campaign in global health research. These techniques are required as concise tools for analyzing and controlling pathogenic microorganisms without laborious cultivation, enrichment, and pure culture isolation [36]. Cultivation of viruses is the basis of every virus detection campaign. It is a process of keeping, growing, and studying the virus's ability to infect other cells. It is indeed an exact and delicate laboratory procedure, usually taking days to get results [37]. Electrochemical methods, immunology-based methods, PCR, and enzyme-linked immunosorbent assay (ELISA) are the current laboratory techniques; however, stringent conditions such as high-grade and costly reagents, special laboratory setup/conditions, well-trained lab workers, and highly accurate and precise instruments make the procedures not economically viable [37]. Therefore, it is as a matter of urgency to develop sensitive, precise, fast, and low-budget diagnostic tools to detect infected individuals to enable proper isolation, quarantine, and treatment [38].

6.4.1 Biosensors

Because of high demand and urgent need, biosensors had received a great deal of attention in technological advancement, especially in healthcare systems (molecular diagnostics and pathogen detection), environmental monitoring, and food industries. The market demand for biosensors has been predicted to be around US$28 billion with an annual growth rate (compound) of 8.4% by 2022 and more than US$70 billion by 2030–2035 [33, 35, 38, 39]. Biosensor systems have three main modules: a bioreceptor, a transducer, and a digital output detector [39]. They are based on the electrochemical reaction processes that are known to have the basis of biological component and nature of processes such as a biocatalytic agent (enzymes), an immunological agent (antibody), and a nucleic acid component (DNA) [33].

Biosensors are classified based on their technology usage and the difference in designs incorporated in the methods of detection with the analytical devices. The two main types are biosensors with labels and the ones without labels [36]. Label-free biosensors have low analysis cost, portable, highly sensitive, short analysis time, reduced usage of organic solvents, and can detect small molecules [36].

6.4.2 Nano-Based Biosensors

Nanotechnology brings a fresh and enhanced approach to all fields, including biosensor development, because the quest to develop a rapid, portable, multiplex, and the low-cost device is demanding and important [40]. El Moutaouakil et al. [37] posited that incorporating nanomaterials into biosensors will enhance sensitivity, specificity, integration of nanoscale measurement, instantaneous detection, and novel label-free sensing techniques owing to their unique physical, mechanical, electrical, and optical properties. An example of such nanomaterial with the best fit is graphene, giving a graphene-based biosensor. Nano-based biosensors will give new functionality such as biocompatibility, microfluidic manifolds, and compactness (device packaging) [41].

6.5 Coronavirus Detection

SARS-CoV-2 with the code name COVID-19 is a novel coronavirus disease that emerged as a global threat to human and the environment in 2019, and its rapidly evolving transmission and survival through the different environment is a significant issue for researchers and policymakers [40, 42–46]. As of 30 September 2020, SARS-CoV-2 has spread to over 235 countries and territories with over 33 million confirmed cases and more than 1 million deaths [47], with the number expected to be on the rise while lasting solution through vaccine development and rapid testing devices are being developed.

6.5.1 Biosensors for COVID-19 Detection

Early detection of viruses provides prevention of viral spread and give room for proper planning in control measures and vaccine development [48]. In view of this, there are conventional means of virus detection/diagnosis, which will be discussed in this section. Table 6.1 presents the list of selected viral diagnostic techniques; however, these methods are reported to contain some limitations, which include low accuracy and sensitivity, time-consuming in sample preparation and purification, sophisticated instrument, laboratory setup, and maintenance, required highly trained and skilled personnel and not usable for quick on-site analysis [49]. Therefore, there is a need to develop a viral biosensor that can solve the challenges posed

Table 6.1 Viral diagnostic methods.

S. no.	Diagnostic tests	Types
1.	Nucleic acid detection and amplification	PCR, RT-PCR, qPCR
2.	Isothermal amplification technologies	NASBA, LAMP, helicase-dependent amplification (HAD), RCA, NEAR, SDA, TMA
3.	Immunoassays	Fluorescent antibody (FA) staining, hemagglutination inhibition, immuno-perox staining, EIA/ELISA (FPIA, MEIA, chemiluminescence immunoassay [CLIA])
4.	DNA sequencing	Sanger sequencers, next-generation sequencers, DNA microarrays
5.	Mass spectrometric methods	MALDI-TOF
6.	Direct visualization of viruses	Electron microscopy
7.	Microelectronics and microfluidics-based techniques	Lab-on-a-chip (LOC) technologies, Point of care (POC) testing, surface plasmon resonance (SPR) technique

PCR, polymerase chain reaction; qPCR, quantitative polymerase chain reaction; RT-PCR, real-time polymerase chain reaction; NASBA, nucleic acid sequence-based amplification; LAMP, loop-mediated isothermal amplification; HDA, helicase-dependent amplification; RCA, rolling circle amplification; NEAR, nicking enzyme amplification reaction; SDA, strand displacement amplification, ™A transcription-mediated amplification; EIA/ELISA, enzyme immunoassay/enzyme-linked immunosorbent assay; ESI, electrospray ionization; FPIA, fluorescence polarization immunoassay; MEIA, micro-particle enzyme immunoassay; MALDI-TOF, matrix-assisted laser desorption ionization time-of-flight.
Source: Modified from Samson et al. [49].

Table 6.2 Biosensor applications and limitations.

S. no.	Types		Application	Limitation
1.	Electrochemical biosensor	Potentiometric, amperometric, cyclic and impedimetric	MERS-CoV and SARS-CoV-2	Required isolation and filtration before detection using a saliva sample
2.	Electronic biosensor	Field-effect transistor (FET)	SARS-CoV-2	Incubation time is longer
3.	Physical biosensor	Piezoelectric and magnetic sensors	COVID-19	Lack of accuracy, time-consuming, not highly sensitive
4.	Optical biosensors	Photon crystal fiber (PCF), fiber optics, and nanolaser	COVID-19, SARS, MERS	Isolation and purification of sample is time consuming, longer incubation time
5	Thermal biosensors		MERS, SARS-CoV	

Source: Samson et al. [49] and Taha et al. [39].

by conventional diagnostic assays while also being economically viable, portable, and efficient.

Design specificity of biosensors for SARS-CoV-2 required three essential aspects, which are recognition of targets (including viral ribonucleic acid [RNA], proteins, or human immunoglobulins), the method of recognition (such as detection of antibody–antigen binding through nucleic acid, aptamers, antibodies, receptors), and signal intensification and transduction system (such as surface plasmon resonance, electrochemical, electrical, optical, and mechanical systems) [50]. Based on categorization, biosensors are of different types, techniques, analytical methods, and limitations. The various types of biosensors, viral detection/diagnosis, and limitations are summarized in Table 6.2.

6.5.2 Nano-Based Biosensors for Coronavirus Detection

One of the primary means of pathogen detections is the Real-Time Reverse Transcription-Polymerase Chain Reaction (RT-PCR), which is currently being used for the detection of SARS-CoV-2 according to some laboratory procedures [4, 50–52]. It is also important to note that COVID-19 is spreading fast around the world, even more than the earlier SARS and Middle East Respiratory Syndrome (MERS). There are currently three types of diagnostic methods for SARS-CoV-2 (COVID-19), which include chest CT scan with clinical symptoms, reverse transcription-polymerase chain reaction (RT-qPCR)-based RNA detections, lateral flow immunochromatographic strip (LFICS), chemiluminescence assay, and ELISA-based antibody diagnosis. Unfortunately, these methods have many demerits, such as affordability, time-consuming, high false-negative results, differentiating viruses or specificity, and not suitable for asymptomatic case detection [38].

Chauhan et al. [48] reported about 47 different testing techniques, including nucleic acid and protein-based tests, that are available and currently being utilized despite the challenges in optimization and viral load change. However, because of the problems associated with this diagnosis methods, it is very urgent to research and develop POC devices that will test rapidly, be cost-effective, and allow self-tests. Nanotechnology and nanoscience-based approaches must be utilized to help combat coronavirus (COVID-19) and also other emerging viral outbreaks in the future, in different ways, including vaccines and drug development, highly sensitive POC devices, nano-based filters for personal protective equipment (such as face masks), and nanocoating and disinfection [53].

A number of nanomaterials are qualified in this regard for biosensors and have outstanding features such as stability, biocompatibility, excellent surface chemistry, high surface energy, and strong intensification effect on signals, and example of the NPs are metallic nanoparticles, graphene, CNTs, photonic crystals (PCs), and nano/microgels [36, 50]. Alphandéry [54] highlighted some nano-detection methods developed to serve as an alternative to the existing standard methods such as PCR; the methods include fluorescence diagnosis of green fluorescent protein (GFP) with gold NPs, a calorimetric assay using AuNPs, Au nanoislands functionalized with specific DNA strands binding to SARS-CoV-2 nucleic acid, use of nanotraps for capturing CoV, and use of carbon electrodes containing AuNPs.

Several studies were conducted on the combination of field-effect transistor (FET) coated with graphene sheets. This is very useful as detection of COVID-19 protein was observed in different media such as clinical samples, phosphate-buffered saline, and culture medium [54]. Nasrollahzadeh et al. [4] reported the application of hybridized nanoparticles using 2*D gold nanoislands with excellent detection rate and in different media and the application of the FET graphene sheet biosensor, giving top-notch results in SARS-CoV-2 detection. The benefit of using the nanomaterials with the biosensors in this regard can decrease the time of analysis and enhance sensitivity, thereby opening the way for a more robust detection device development with higher performance for COVID-19 and other future viral infections [55]. Consequently, a method of NP-based biosensor was reported to have been successful; however, it could not meet up with the rapid and efficient diagnosis need as reported by [56]. In this technique, the sensor used a single-step reverse transcription loop-mediated isothermal amplification (RT-LAMP) with nanoparticle-based assay (NBS), additional isothermal heat at 63 °C was needed for 40 minutes, while it took about 1 hour to take the sample and analyze results. This calls for more intense research on balancing quality diagnosis and rapid detection time and rate as nanotechnology is a proven aspect of research and development to combat viral infections (Figure 6.1).

6.6 Emerging Concerns on COVID-19

Persistence of SARS-CoV-2 in different environments is a growing concern for public health and safety. There are several reports on the transmission routes of

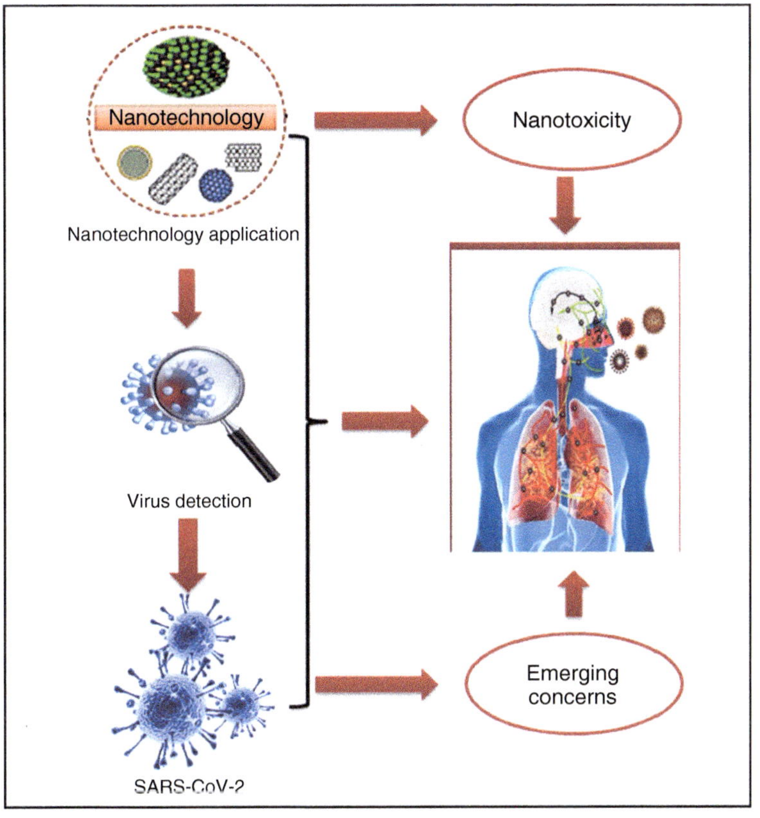

Figure 6.1 Flow chart demonstrating the application of nanotechnology to tackle the emerging concerns of SARS-CoV-2.

SARS-CoV-2 through the fecal–oral routes [40, 55, 57–59]. However, there is no study yet to substantiate this speculation, which demands critical analysis considering that the most vulnerable populations who dwell most on using non-potable water such as streams, rivers, and even underground water like open well are prone to an infection by a waterborne pathogenic virus [34, 60, 61].

6.6.1 Nanotechnology in COVID-19 Contaminated Water

Overtime, nanoparticles have been applied in different engineering fields, including environmental engineering, mostly water and wastewater treatments. Landmark achievements were achieved, and certainly, NP development in virus detection in water needs to be developed and utilized as required. This could be another focus for researchers in understanding the prevalence of SARS-CoV-2 and other viral infections in water and the development of a detection device that is nano-based.

6.7 Nanotoxicity

With so much focus on the advancement of nanoscience in improving nanomedicine and nano-biomedical engineering, the adversative result of nanotechnology is sometimes treated separately by many researchers. The same unique properties of nanoparticles, such as small sizes, large surface area, and shapes, make its applicability enormous and needed to determine its effectiveness in causing harm to the human body. Nanomaterials may pose a danger to the cells if there is a heightened generation of oxidative stress and inflammatory mediators in different tissues; therefore, perfused organs such as the liver, spleen, lungs, heart, and kidney receive quite a quantifiable amount of any material that finds its way into the body via absorption or injection. Hence, nanomaterials can cause nephrotoxicity, hepatotoxicity, cardiotoxicity, immunotoxicity, and genotoxicity [62].

Based on their size, shape, and surface area of nanomaterials, the smaller the size, the easier it is to get them to stick to the surface by translocation, hence modifying the cellular digestion by interacting with subcellular organelles [63]. Patel and Nanda [63] reported a comparative study of CuO nanorods and CuO nanospheres, and the result showed that the nanorods with higher surface area released more ions, hence resulting to more toxicity. CNT handling has been reported to have the same handling concerns as with asbestos, while highly purified ones have short-term toxicity and may be tagged as biocompatible [64]. Nanomaterials used in drug delivery accumulate in the liver where an excessive immune response may bring about permanent impairment; the accumulation of silica or quartz dust may also cause pulmonary fibrosis and silicosis [64].

6.8 Conclusion

Coronavirus (COVID-19) is currently ravaging the world, thereby bringing it to a standstill. The debilitating effect was so intense that countries that were severely affected had to embrace lockdown as the only possible transmission control measure because vaccines and rapid diagnosis development were and are still evolving. Traditional diagnosis/detection techniques have proved helpful but were met with various challenges of time constraints, false-negative results, specificity, and compatibility. Nanotechnology, as an ever-growing field in medicine, has played a leading role in advancing viral infection treatment, detection, and vaccine development. The application of nano-based biosensors has helped in the rapid testing of infected persons, and research is ongoing in developing more easy-to-use POC devices that can allow self-diagnosis irrespective of time, location, and type of virus.

As a double-edged sword, nanotechnology application, as widely embraced and utilized, should be treated as a potential threat to human health because of the emerging concerns of nanomaterials accumulation in sensory organs of the body. Therefore, a detailed risk assessment of NMs should be developed, and this should contain assessments such as hazard identifications, characterizations (with a focus on critical organs), dose-response, exposure assessments in different conditions, and characterization of risks. This approach will help to determine the viability of existing NPs and may give rise to developing less toxic ones, which will help in the nanomedicine and biomedical engineering field.

References

1 Soni, S., Salhotra, A., and Suar, M. (2014). *Handbook of Research on Diverse Applications of Nanotechnology in Biomedicine, Chemistry, and Engineering*. IGI Global.

2 Hussain, C.M. (2020). The ELSI handbook of nanotechnology: risk, safety, ELSI and commercialization. In: *The ELSI Handbook of Nanotechnology* (ed. C.M. Hussain), 165–194. Hoboken, NJ: Wiley.

3 Uskoković, V. (2020). Why have nanotechnologies been underutilized in the global uprising against the coronavirus pandemic? *Nanomedicine (Lond.)* 15 (17): 1719–1734.

4 Nasrollahzadeh, M., Sajjadi, M., Soufi, G.J. et al. (2020). Nanomaterials and nanotechnology-associated innovations against viral infections with a focus on coronaviruses. *Nanomaterials* 10 (6): 1072.

5 Elizondo, N., Segovia, P., Coello, V. et al. (2012). Green synthesis and characterizations of silver and gold nanoparticles. In: *Green Chemistry – Environmentally Benign Approaches* (eds. M. Kidwai and N.K. Mishra), 139–156. InTech.

6 Liu, J. and Jiang, G. (2015). *Silver Nanoparticles in the Environment.* Springer-Verlag Berlin Heidelberg.

7 Olajire, A.A., Abidemi, J.J., Lateef, A., and Benson, N.U. (2017). Adsorptive desulphurization of model oil by Ag nanoparticles-modified activated carbon prepared from brewer's spent grains. *J. Environ. Chem. Eng.* 5 (1): 147–159.

8 Lateef, A., Akande, M.A., Azeez, M.A. et al. (2016). Phytosynthesis of silver nanoparticles (AgNPs) using miracle fruit plant (*Synsepalum dulcificum*) for antimicrobial, catalytic, anticoagulant, and thrombolytic applications. *Nanotechnol. Rev.* 5 (6): 507–520.

9 Lateef, A., Azeez, M.A., Asafa, T.B. et al. (2015). *Cola nitida*-mediated biogenic synthesis of silver nanoparticles using seed and seed shell extracts and evaluation of antibacterial activities. *Bionanoscience* 5 (4): 196–205.

10 Kalimuthu, K., Cha, B.S., Kim, S., and Park, K.S. (2020). Eco-friendly synthesis and biomedical applications of gold nanoparticles: a review. *Microchem. J.* 152 (September 2019): 104296.

11 Lee, K.X., Shameli, K., Yew, Y.P. et al. (2020). Recent developments in the facile bio-synthesis of gold nanoparticles (AuNPs) and their biomedical applications. *Int. J. Nanomed.* 15: 275–300.

12 Badeggi, U.M., Ismail, E., Adeloye, A.O. et al. (2020). Green synthesis of gold nanoparticles capped with procyanidins from leucosidea sericea as potential antidiabetic and antioxidant agents. *Biomolecules* 10 (3): 452.

13 Jafarizad, A., Safaee, K., Vahid, B. et al. (2019). Synthesis and characterization of gold nanoparticles using *Hypericum perforatum* and nettle aqueous extracts:

a comparison with Turkevich method. *Environ. Prog. Sustainable Energy* 38 (2): 508–517.

14 Yunus, I.S., Harwin, Kurniawan, A. et al. (2012). Nanotechnologies in water and air pollution treatment. *Environ. Technol. Rev.* 1 (1): 136–148.

15 Porto, L.S., Silva, D.N., de Oliveira, A.E.F. et al. (2020). Carbon nanomaterials: synthesis and applications to development of electrochemical sensors in determination of drugs and compounds of clinical interest. *Rev. Anal. Chem.* 38 (3): 1–16.

16 Ngô, C. and Van de Voorde, M. (2014). *Nanotechnology in a Nutshell.* Atlantis Press.

17 Kumar, U., Sikarwar, S., Sonker, R.K., and Yadav, B.C. (2016). Carbon nanotube: synthesis and application in solar cell. *J. Inorg. Organomet. Polym. Mater.* 26 (6): 1231–1242.

18 Haghi, A.K. and Zaikov, G.E. (2012). *Handbook of Research on Nanomaterials, Nanochemistry and Smart Materials.* Nova Biomedical.

19 Salem, S.S. and Fouda, A. (2020). Green synthesis of metallic nanoparticles and their prospective biotechnological applications: an overview. *Biol. Trace Elem. Res.* 199: 344–370.

20 Ahmeda, A., Abbasi, N., Ghaneialvar, H. et al. (2020). Application of titanium nanoparticles containing natural compounds in cutaneous wound healing. *Appl. Organomet. Chem.* 34 (4): 1–13.

21 Ziental, D., Czarczynska-Goslinska, B., Mlynarczyk, D.T. et al. (2020). Titanium dioxide nanoparticles: prospects and applications in medicine. *Nanomaterials* 10 (2): 387.

22 Irshad, M.A., Nawaz, R., Zia ur Rehman, M. et al. (2020). Synthesis and characterization of titanium dioxide nanoparticles by chemical and green methods and their antifungal activities against wheat rust. *Chemosphere* 258: 127352.

23 Vizirianakis, I.S. (2013). *Handbook of Personalized Medicine: Advances in Nanotechnology, Drug Delivery and Therapy.* Jenny Stanford Publishing.

24 Zhou, Z. and Lu, Z.R. (2017). Molecular imaging of the tumor microenvironment. *Adv. Drug Delivery Rev.* 113: 24–48.

25 Fadeel, B. (2015). *Handbook of Safety Assessment of Nanomaterials: From Toxicological Testing to Personalized Medicine.* Jenny Stanford Publishing.

26 Luther, D.C., Huang, R., Jeon, T. et al. (2020). Delivery of drugs, proteins, and nucleic acids using inorganic nanoparticles. *Adv. Drug Delivery Rev.* 156: 188–213.

27 Rai, D.B., Gupta, N., Pooja, D., and Kulhari, H. (2020). *Dendrimers for Diagnostic Applications.* Elsevier Inc.

28 Nazar, H. (2015). The use of nanotechnology in disease diagnosis and molecular imaging. *Pharm. J.* 290: 1–9.

29 Rai, N.K., Singh, S.K., Tiwari, M., and Sharma, L.K. (2020). Implications of nanotechnology in cancer diagnostics and therapeutics. In: *Functional Bionanomaterials. Nanotechnology in the Life Sciences* (eds. D. Thangadurai, J. Sangeetha and R. Prasad), 271–291. Cham: Springer International Publishing.

30 Kumar, H., Kuča, K., Bhatia, S.K. et al. (2020). Applications of nanotechnology in biosensor-based detection of foodborne pathogens. *Sensors (Switzerland)* 20 (7): 1–19.

31 Khan, F.A. (2020). Major nano-based products: nanomedicine, nanosensors, and nanodiagnostics. In: *Applications of Nanomaterials in Human Health* (ed. F.A. Khan), 211–228. Singapore: Springer.

32 Hammou, R.A., Benhassou, M., and Ennaji, M.M. (2020). Application of nanodiagnostics in viral infectious diseases. In: *Emerging and Reemerging Viral Pathogens* (ed. M.M. Ennaji), 179–195. Elsevier.

33 Singh, S., Kumar, V., Dhanjal, D.S., and Datta, S. (2020). Chapter 14, Biological biosensors for monitoring and diagnosis. In: *Microbial Biotechnology: Basic Research and Applications*, Environmental and Microbial Biotechnology (eds. J. Singh, A. Vyas, S. Wang and R. Prasad), 317–335. Springer Nature.

34 Adelodun, B., Ajibade, F.O., Ighalo, J.O. et al. (2021). Assessment of socioeconomic inequality based on virus-contaminated water usage in developing countries: a review. *Environ. Res.* 192 (September 2020): 110309.

35 Cesewski, E. and Johnson, B.N. (2020). Electrochemical biosensors for pathogen detection. *Biosens. Bioelectron.* 159 (April): 112214.

36 Andryukov, B.G., Besednova, N.N., Romashko, R.V. et al. (2020). Label-free biosensors for laboratory-based diagnostics of infections: current achievements and new trends. *Biosensors* 10 (2): 11.

37 El Moutaouakil, A., Poovathy, S., Belmoubarik, M., and Peng, W.K. (2020). Review: graphene-based biosensor for viral detection. *J. Japan Prosthodont. Soc.* 28 (5): 972–973.

38 Cui, F. and Zhou, H.S. (2020). Diagnostic methods and potential portable biosensors for coronavirus disease 2019. *Biosens. Bioelectron.* 165: 112349.

39 Taha, B., Arsad, N., and Al Mashhadany, Y. (2020). *An Analysis Review, Detection Coronavirus Disease 2019 (COVID-19) Based on Biosensor Application.* MDPI. Preprints.

40 Sharifi, M., Avadi, M.R., Attar, F. et al. (2019). Cancer diagnosis using nanomaterials based electrochemical nanobiosensors. *Biosens. Bioelectron.* 126 (July 2018): 773–784.

41 Mujawar, M.A., Gohel, H., Bhardwaj, S.K. et al. (2020). Nano-enabled biosensing systems for intelligent healthcare: towards COVID-19 management. *Mater. Today Chem.* 17: 100306.

42 Adelodun, B., Ajibade, F.O., Ibrahim, R.G. et al. (2020). Snowballing transmission of COVID-19 (SARS-CoV-2) through wastewater: any sustainable preventive measures to curtail the scourge in low-income countries? *Sci. Total Environ.* 742: 140680.

43 Chhikara, B.S., Rathi, B., Singh, J., and Poonam, F.N.U. (2020). Corona virus SARS-CoV-2 disease COVID-19: infection, prevention and clinical advances of the prospective chemical drug therapeutics: a review on corona virus disease COVID-19, epidemiology, prevention, and anticipated therapeutic advances. *Chem. Biol. Lett.* 7 (1): 63–72.

44 Heller, L., Mota, C.R., and Greco, D.B. (2020). COVID-19 faecal-oral transmission: are we asking the right questions? *Sci. Total Environ.* 729: 138919.
45 Orive, G., Lertxundi, U., and Barcelo, D. (2020). Early SARS-CoV-2 outbreak detection by sewage-based epidemiology. *Sci. Total Environ.* 732: 139298.
46 Sharma, V.K., Jinadatha, C., and Lichtfouse, E. (2020). Environmental chemistry is most relevant to study coronavirus pandemics. *Environ. Chem. Lett.* 18 (4): 993–996.
47 WHO (2020) Coronavirus disease (COVID-19): numbers at a glance. World Health Organization. Coronavirus disease (COVID-19) (who.int) (accessed 1 October 2020).
48 Chauhan, D.S., Prasad, R., Srivastava, R. et al. (2020). Comprehensive review on current interventions, diagnostics, and nanotechnology perspectives against SARS-CoV-2. *Bioconjugate Chem.* 31 (9): 2021–2045.
49 Samson, R., Navale, G.R., and Dharne, M.S. (2020). Biosensors: frontiers in rapid detection of COVID-19. *3 Biotech* 10 (9): 1–9.
50 Liang, K.H., Chang, T.J., Wang, M.L. et al. (2020). Novel biosensor platforms for the detection of coronavirus infection and severe acute respiratory syndrome coronavirus 2. *J. Chin. Med. Assoc.* 83 (8): 701–703.
51 Palestino, G., García-Silva, I., González-Ortega, O., and Rosales-Mendoza, S. (2020). Can nanotechnology help in the fight against COVID-19? *Expert Rev. Anti-infect. Ther.* 18 (9): 849–864.
52 Seo, G., Lee, G., Kim, M.J. et al. (2020). Rapid detection of COVID-19 causative virus (SARS-CoV-2) in human nasopharyngeal swab specimens using field-effect transistor-based biosensor. *ACS Nano* 14 (4): 5135–5142.
53 Weiss, C., Carriere, M., Fusco, L. et al. (2020). Toward nanotechnology-enabled approaches against the COVID-19 pandemic. *ACS Nano* 14 (6): 6383–6406.
54 Alphandéry, E. (2020). The potential of various nanotechnologies for coronavirus diagnosis/treatment highlighted through a literature analysis. *Bioconjugate Chem.* 31 (8): 1873–1882.
55 Nikaeen, G., Abbaszadeh, S., and Yousefinejad, S. (2020). Application of nanomaterials in treatment, anti-infection and detection of coronaviruses. *Nanomedicine (Lond.)* 15 (15): 1501–1512.
56 Sivasankarapillai, V.S., Pillai, A.M., Rahdar, A. et al. (2020). On facing the SARS-CoV-2 (COVID-19) with combination of nanomaterials and medicine: possible strategies and first challenges. *Nanomaterials* 10 (5): 1–23.
57 Al-Gheethi, A., Noman, E., Al-Maqtari, Q., Almoheer, R., Mohamed, R., Hamdan, R. (2020) *Survival and Disinfection of SARS-CoV-2 in Environment and Contaminated Surface*. Preprints.
58 Arslan, M., Xu, B., and Gamal El-Din, M. (2020). Transmission of SARS-CoV-2 via fecal-oral and aerosols–borne routes: environmental dynamics and implications for wastewater management in underprivileged societies. *Sci. Total Environ.* 743: 140709.
59 Paleologos, E.K., O'Kelly, B.C., Tang, C.-S. et al. (2020). Post COVID-19 water and wastewater management to protect public health and geoenvironment. *Environ. Geotech.*: 1–5.

60 Adelodun, B., Ogunshina, M.S., Ajibade, F.O. et al. (2020). Kinetic and prediction modeling studies of organic pollutants removal from municipal wastewater using *Moringa oleifera* biomass as a coagulant. *Water* 12 (7): 2052.

61 Adelodun, B., Ajibade, F.O., Ogunshina, M.S., and Choi, K.-S. (2019). Dosage and settling time course optimization of *Moringa oleifera* in municipal wastewater treatment using response surface methodology. *Desalin. Water Treat.* 167: 45–56.

62 Chakravarty, M. and Vora, A. (2020). Nanotechnology-based antiviral therapeutics. *Drug Delivery Transl. Res.*: 1–40.

63 Patel, S. and Nanda, R. (2015). Nanotechnology in healthcare: applications and challenges. *Med. Chem. (Los Angeles)* 05 (12): 528–533.

64 Riehemann, K., Schneider, S.W., Luger, T.A. et al. (2009). Nanomedicine – challenge and perspectives. *Angew. Chem. Int. Ed.* 48 (5): 872–897.

Part III

Biosensors for Analysis of SARS-CoV-2/COVID-19

7

Sensor Development for Coronavirus

Ranjita D. Tandel[1], Nagappa L. Teradal[2], and Sudheesh K. Shukla[3]

[1]*Gokhale Centenary College, Department of Chemistry, Ankola, Karnataka 581 314, India*
[2]*GE Society's, J.S.S. Arts Science and Commerce College, Department of Chemistry, Gokak, Karnataka 591 307, India*
[3]*Department of Biomedical Engineering, School of Biological Engineering and Life Science, Shobhit Institute of Engineering & Technology (Deemed-to-be University), Modipuram, Meerut 250110, India*

7.1 Introduction

Severe acute respiratory syndrome (SARS) is an infectious disease that was first detected in China and has caused serious infection causing death in a great proportion of patients. The novel coronavirus was first identified in December 2019. Coronavirus (CoV) disease is a newly emerging human infectious disease associated with respiratory distress. The disease was first named as 2019 novel coronavirus (COVID-19) and later renamed as severe acute respiratory syndrome coronavirus 2 (SARS-CoV-2) [1]. The current outbreak of SARS-CoV-2 has presented epidemiologists around the globe with an unprecedented challenge. The World Health Organization (WHO) has classified COVID outbreak as a pandemic as it was transforming human to human rapidly [2]. SARS-CoV-2 is highly reported because the virus is easily transmitted through respiratory droplets, air droplets, and by contact. The incubation period for CoV-2 ranges from 2 to 14 days, and the asymptomatic spread occurs before the onset of symptoms [3]. It is believed that it has high viral load at the initial stage of the infection, which further led to pandemic. CoVs are developed in the cellular cytoplasm of host, promoting cell destruction. In addition, the virus was detected on the contacted surface of plastic and stainless steel up to 72 hours [4].

Coronavirus belongs to the family Coronaviridae, the genome size of CoV is approximately 26–32 Kb, and it is the largest genome known of RNA virus [5]. Figure 7.1 represents the three-dimensional structure and cross-sectional view of SARS-CoV-2. The size of the virus ranges from 60 to 140 nm in diameter, having cube-shaped spike proteins. On the basis of phylogeny, Coronavirinae

Detection and Analysis of SARS Coronavirus: Advanced Biosensors for Pandemic Viruses and Related Pathogens,
First Edition. Edited by Chaudhery Mustansar Hussain and Sudheesh K. Shukla.

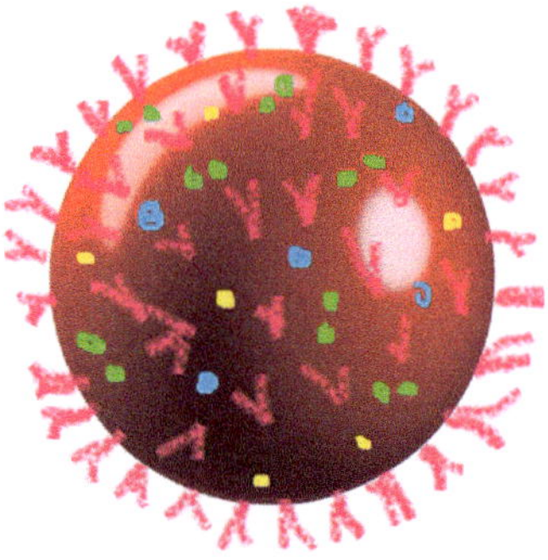

Figure 7.1 The three-dimensional structure of SARS-CoV-2.

consist of four genera: alpha coronavirus (α-CoV), beta coronavirus (β-CoV), gamma coronavirus (γ-CoV), and delta coronavirus (δ-CoV). The α- and β-CoV cause respiratory problems in humans, whereas the γ- and δ-CoV infect birds [6]. Till today, seven CoVs have been found to be human pandemic. Among them, the human coronavirus (HCoV) – HKU, HCoV-NL63, HCoV-OC43, and HCoV-229E, causes respiratory disease [7–9]. The β-coronavirus, SARS-CoV-2, has a single positive-strand RNA genome, whereas CoV genomes encode four proteins: envelope (E), nucleocapsid (N), spike (S), and matrix (M). The human CoVs, viz., SARS-CoV and Middle East respiratory syndrome coronavirus (MERS-CoV), emerged in the year 2002 and 2012 [10, 11]. The causes of the two major outbreaks belong to CoVs including SARS-CoV (epidemic in 2002) and MERS-CoV (epidemic since 2012). The latest outbreak of β-CoVs subfamily has been carried out by SARS-CoV-2 since December 2019. WHO is concerned about the high-risk health systems as there is no specific vaccine to treat SARS-CoV-2.

Symptoms of SARS-CoV-2 include fever, loss of air, cough, sputum production, headache, cold, and, in severe cases, severe respiratory problems including respiratory insufficiency and associated comorbidities [12]. The risk factors for severe illness include age and underlying medical comorbidities such as diabetes, chronic respiratory disease, and cardiovascular disease [13]. As there is no specific vaccine or drugs are not yet available for COVID-19, early diagnosis and management are crucial for the outbreak. CoVs cause upper respiratory tract illness in both humans and animals.

The infected individuals were diagnosed by real-time (RT) reverse transcription polymerase chain reaction (RT-PCR). As the number of infected individuals increased day by day, it led to shortage of laboratory-based testing capacity and reagents. Thus, rapid and easy-to-use devices are in need for the detection of infection. Early identification and detection of CoV-infected patients and prevention of transmission of the SARS-CoV-2 outbreak are critical. The much need of the hour is its diagnosis. In view of this, the present chapter describes the developments in the fabrication of sensing devices, biosensors, chemical sensors, and electrochemical sensors for the detection and identification of SARS-CoV-2.

A biosensor is an analytical device that converts biological reactions or interactions into measurable signals. The device consists of a biological component (bioreceptor) and a physiochemical detector (transducer). The biological materials, viz., enzymes, tissues, antibodies, microorganisms, cell receptors, etc., are immobilized on a physicochemical transducer and allowed to interact with the target analyte, producing useful measurable signals that may be electrochemical, optical, thermometric, or piezoelectric [14]. Depending on the interaction of the target molecule to be detected, different types of bioreceptors used ligands such as nucleic acids, biomimetic materials, antibodies/antigens, etc. Biosensors have found advanced applications in medical diagnostics and pharmaceutical, food, environmental, agriculture, and many other industries. Biosensors give more sensitive, specific, and reproducible results. Biosensors have been increasingly applied in clinical assays because of their portability and point-of-care testing. These biosensors can detect the target analyte in very low quantities and are considered to be a powerful tool to diagnose diseases at initial stage [15]. Since past decades, several biosensing methods have been used to detect proteins, viruses, cancer biomarkers, and other analyte samples. Biosensors have the advantages of being sensitive, selective, portable, and disposable as compared to other traditional analytical methods. Designing a biosensor requires fundamental knowledge on the chemical structure of the target analyte and concentration of the analyte to be measured. Biosensors are classified into three main groups based on the type of energies they produce upon receiving the signal from sensing element: (i) electrochemical, (ii) optical, and (iii) piezoelectric biosensors.

Electrochemical biosensors measure the electrochemical changes that occur on the transducer surface interaction or reaction with the analyte. Based on electrical changes, it can be an amperometric (change in current is measured at a fixed voltage), potentiometric (change in voltage is measured), and impedimetric (change in impedance is measured). The advantages of electrochemical biosensors are that they are cost-effective, sensitive, and easy to operate [16]. **Optical biosensors** provide multiplexed detection within a single device. These sensors measure the optical properties and characteristics of the transducer surface during the interaction of the target analyte with a recognition element. The changes in the properties of light correlate with the population of the target analyte in the given sample. Based on the transduction mechanism, optical biosensors can be categorized into absorbance, fluorescence, surface plasma resonance, and chemiluminescence biosensors. These biosensors are particularly applied for detection of bacterial pathogens and viruses and for studying the kinetics of antigen–antibody and DNA interactions [17, 18]. **Piezoelectric sensors**, also called mass sensors, work based on the interaction of the target analyte with the sensing element, which is usually a vibrating piezoelectric quartz crystal. The resonance frequency of the piezoelectric crystal changes upon the binding of the analyte molecules. The change in resonance frequency leads to alter an oscillating voltage that is further measured [19].

Nanotechnology plays a vital role in clinical applications, which is used in the development of sensing or biosensing devices for the detection of microorganisms. During the development of biosensing devices, incorporation of nanomaterials makes them to improve their sensitivity, relative selectivity, biocompatibility, and resistance against nonspecific adsorption. The application of nanomaterials including metal nanoparticles, carbon nanomaterials (carbon nanotubes, graphene/graphene oxide, and quantum dots [QDs]), nanosheets, and metal–organic frameworks could significantly improve the performance of the biosensors. The high performance of these nanomaterial-based biosensing devices could be due to a huge active surface-to-volume ratio, ease of surface passivation, and excellent electron transfer capabilities of smart nanomaterials [20]. Carbon nanomaterials, specifically graphene-based materials, have been explored widely in the design of biosensors because graphene is biocompatible. Graphene, an allotrope of carbon, is a two-dimensional honeycomb crystal structure. The excellent properties of graphene, viz., high conductivity, huge active surface area, good electron mobility at room temperature, and superior optical and electrochemical properties, make it possible to be used as a biosensing nanomaterial [21]. The biocompatibility and ease of surface modification enable its wide utility in biomedical field [22]. Graphene-metal nanoparticle composites are also applied to construct biosensors for various applications. In a similar way, metal-based nanoparticles, namely, gold, silver, zinc, and platinum, and others possess interesting properties such as huge surface area, biocompatibility, ease of surface functionalization, and surface plasmon resonance (SPR) phenomena [23]. In addition, they are good conductors of electricity and shows good optical and electrical properties. The superior properties of the metal-based nanoparticles have been used in the development of new biosensing devices for detection of CoVs. The other most commonly used nanomaterials for biosensing applications are QDs [24]. QDs are semiconductor crystals with unique confinement effects having nanometer size. The use of QDs for the development of optical biosensors is based on their photoluminescence properties and plasmon resonance phenomena. They have broad excitation and a narrow size emission band width [25].

In this context, biosensors based on nanomaterials have been demonstrated to fulfill the desired diagnosis techniques. Biosensors can be used to determine the virus mediated infections such as nucleic acid, antibody, antigen dependent and aptamer [26]. Thus, owing to the advantages of physicochemical properties of nanomaterials, specifically electrical, optical, and magnetic, attempts have been made to report nanomaterial-enabled biosensing techniques for the diagnosis of CoVs.

As an example, Layqah and Eissa have developed an immunosensor based on an array of disposable carbon electrodes (DEP) modified with gold nanoparticles (AuNPs) for the detection of MERS-CoV [27]. In the present report, authors have used recombinant spike protein S1 as a biomarker for the selective recognition of MERS-CoV. The pictorial representation of biosensor development is illustrated in Figure 7.2.

Initially, a disposable carbon electrode is modified with gold nanoparticles and then the surface is functionalized with a biomarker, recombinant spike protein S1, through chemical linkage. The biosensor works on an indirect assay that is

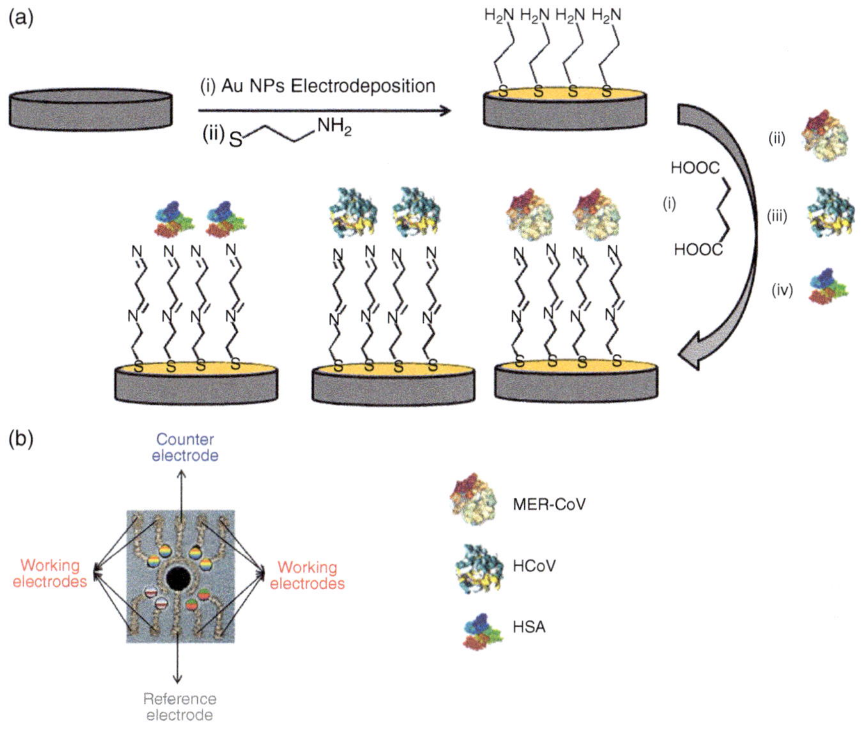

Figure 7.2 The fabrication of immunosensor. (a) Initially, a disposable carbon electrode is modified with gold nanoparticles and then chemically linked by the biorecognition protein for selective sensing of SARS-CoV-2. (b) Immunosensor array for SARS-CoV-2 detection [27].

the competition between free virus in the sample and immobilized MERS-CoV protein for a fixed concentration of added antibody to the sample. For the determination of MERS-CoV, the demonstrated biosensor showed a linear voltammetric response in the concentration range of 0.001–100 ng/ml. The sensitive and specific immunosensor was successfully used to detect MERS-CoV and human CoV proteins in spiked nasal samples. The results of the study showed good recovery percentages [27]. In an another report, Mahari fabricated an electrochemical biosensor through the use of fluorine-doped tin oxide electrode modified with gold nanoparticles (FTO/AuNPs) and then FTO/AuNPs immobilized with SARS-CoV-2 monoclonal antibody (SARS-CoV-2Ab) for the specific detection of SARS-CoV-2 in spiked saliva samples [28]. The output response of the proposed biosensing device is the change in the electrical conductivity upon the binding of SARS-CoV-2. In device fabrication, AuNPs acted as a bridge between SARS-CoV-2Ab and the FTO substrate. Under optimized conditions, the developed immunosensor is able to detect SARS-CoV-2 in the concentration range of 1 fM to 1 μM. The analytical utility of the proposed device is established in determining SARS-CoV-2 in spiked saliva samples with a detection limit of 90 fM [28]. Yet in another work, Hsu et al.

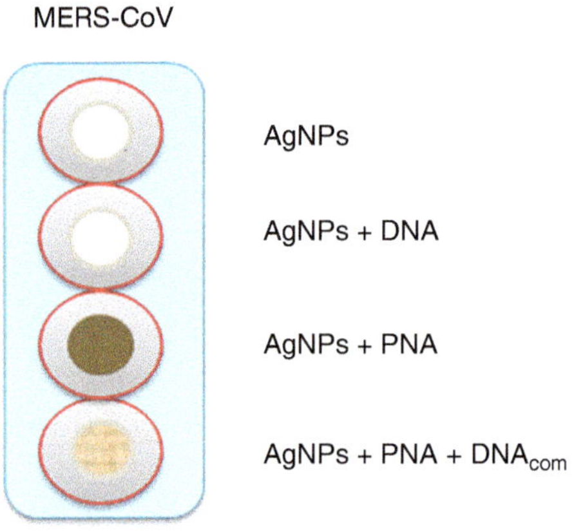

Figure 7.3 The colorimetric paper-based biosensor. Visual color change observed upon the detection of MERS-CoV in the presence of DNA$_{com}$. The top layer of sensor contains detection zones and four control zones. Each zone contained AgNPs with single PNA probe to provide selectivity for DNA. The base layer has four wax-defined channels extending outward from a sample inlet [31].

explored the applicability of an aluminum gallium nitride high electron mobility transistor (AlGaN/GaN HEMT)-based biosensor for the effective diagnostic strategy for CoV [29]. In the present work, the N protein of SARS-CoV-2 was targeted. Initially, the AlGaN/GaN layer is prepared through a metal–organic chemical vapor deposition method and then the layer is immobilized with DNA (double stranded), which is isolated from the SARS-CoV-2 genome. Upon binding of the N protein of SARS-CoV-2, a RT current response is generated, which is directly proportional to the concentration of SARS-CoV-2. The demonstrated biosensor is able to detect the SARS-CoV N protein with a minimum concentration of 0.003 nM [29]. Martínez-Paredes et al. designed a biosensor using thiolated oligonucleotides onto gold nanoparticles modified screen printed carbon electrodes (SPCEs) (genosensor) for the detection of SARS-CoV-2 [30]. The genosensor response was found to be linear in the range of 2.5–50 pmol/l with a limit of detection of 2.5 pmol/l [30].

A paper-based colorimetric biosensor is based on pyrrolidinyl peptide nucleic acid (PNA)-induced AgNP aggregation for the detection of Middle East respiratory syndrome, MERS-CoV, reported by Teengam et al. [31]. Figure 7.3 depicts the graphical representation of the paper-based colorimetric biosensor. In the present work, authors have used PNA as a probe alternative to DNA and RNA because of its chemical and biological stability, ease of synthesis, and efficiency in hybridization with the complementary DNA (cDNA) strands. The PNA probe contains a single positive charge from the lysine at the C-terminus and it causes aggregation of citrate anion-stabilized silver nanoparticles (AgNPs) in the absence of cDNA. The sensing mechanism of a colorimetric biosensor, upon target analyte binding, there is a formation of the anionic DNA–PNA duplex, results in dispersion of the AgNPs as a result of electrostatic repulsion, which gave rise to a visible color change. The proposed colorimetric biosensor showed a detection limit of 100 nM for MERS-CoV [31].

Huang et al. developed a localized surface plasmon-coupled fluorescence (LSPCF) fiber-optic biosensor that combines sandwich immunoassay with a LSP technique for the detection of SARS-CoV nucleocapsid (N protein). In this report, authors have fabricated two monoclonal antibodies against SARS-CoV N protein tagged with GST (glutathione S-transferase). The presented biosensor showed lowest detection sensitivity of 1 pg/ml for GST-N protein human serum. Experimental results showed a linear response between the fluorescence signal and the concentration of GST-N protein in the range of 0.1 pg to 1 ng/ml. The demonstrated technology could be used for the clinical diagnosis of SARS-CoV-2 infection at early stages. In comparison with other conventional enzyme-linked immunosorbent assay, the detection limit of the proposed biosensor for the GST-N protein is improved to 10^4 folds [32].

A new biosensor has been developed, Mavrikou et al., through the use of an engineered biorecognition element (mammalian cells bearing the human chimeric spike S1 antibody) for the detection of S1 spike protein expressed on the surface of SARS-CoV-2 [33]. In the present work, the binding of the protein to the antibodies results in a selective detection and considerable change in the cellular bioelectric properties measured by bioelectric recognition assay. The device showed actual availability of SARS-CoV-2 via change in the membrane potential and other electric properties of the cells in a highly sensitive, speed, and reproducible mode. The biosensor provided a detection limit of 1 fg/ml and a linear response range was between 10 fg and 1 μg/ml. The biosensor was also configured as a ready-to-use platform, including a portable read-out device operated through a tablet or a smartphone. The demonstrated biosensor was potentially applied for the mass screening of SARS-CoV-2 surface antigens without prior sample processing [33].

An SPR-based biosensor was developed by Lei et al. for specific detection of nine common respiratory virus, including influenza A and influenza B, hemagglutinin type 1 and neuraminidase type 1 (H1N1), respiratory syncytial virus (RSV), parainfluenza virus 1–3 (PIV1, 2, 3), adenovirus, and SARS coronavirus [34]. In the report, the SPR-based biosensor was developed by immobilizing respiratory virus-specific oligonucleotides in an SPR chip. Thin chips were prepared by vacuum evaporation and of 2.5 nm chromium and 47 nm gold onto SF10 glass wafers. Further, before printing, activation of carboxyl group was done by immersing it into an aqueous mixture of 0.4 M EDC (1-ethyl-3-(3-dimethylaminopropyl)carbodiimide) and 0.1 NHS (*N*-hydroxysuccinimide). A 10 μM specific respiratory virus probe was printed onto the activated surface. Further, to increase the sensitivity of biosensors, they have used biotin to label the PCR primer. The presented result confirms that the probe displayed specific hybridization properties, which complement their target DNA in PCR products. Furthermore, biotin labeling leads to enhancement of the detection limits [34].

Roh and Jo reported a new QD-conjugated RNA aptamer with high sensitivity for the detection of SARS-CoV-2 N protein using an on-chip system. Using an optical QD-based RNA aptamer chip, SARS-CoV-2 N protein was detected as low as 0.1 pg/ml. Detection of SARS-CoV-2 N protein was analyzed by measuring the fluorescence intensity using confocal microscopy. Results indicated that the QD-supported RNA aptamer conjugates showed high fluorescence signals on the chip. However, with the addition of SARS-CoV-2 N protein, the signal intensity was found to increase gradually. Hence, authors have concluded that the QD-conjugated RNA aptamer specifically targets the SARS-CoV N protein on an immobilized chip. The results of the report suggested that the QD-conjugated RNA aptamer can recognize the SARS-CoV-2 N protein with a detection limit of 0.1 pg/ml [35].

By considering the huge public health issue of SARS-CoV-2, Zhu et al. have devised a multiplex reverse transcription loop-mediated isothermal amplification (mRT-LAMP) coupled with a nanoparticle-based lateral flow biosensor (LFB) assay (mRT-LAMP-LFB) for diagnosing SARS-CoV-2 [36]. The primer sets of LAMP have ORF1ab (opening reading frame), and the N (nucleoprotein) genes of SARS-CoV-2 were simultaneously amplified in a single tube reaction and detected. The use of LFB provided a rapid, convenient, and easily interpretable readout of COVID-19 mRT-LAMP results. The limit of detection for SARS-CoV-2 mRT-LAMP-LFB assay was 12 copies each for ORF1ab (open reading frame 1ab) and N-plasmid. The sensitivity of SARS-CoV-2 was found to be 100% (33/33 oropharynx swab samples collected from SARS-CoV-2 patients). From the presented results, it is concluded that the analytical sensitivity and specificity of SARS-CoV-2 mRT-LAMP-LFB method was sensitive, reliable, and feasible for the diagnosis of SARS-CoV-2 infection. Hence, SARS-CoV-2 mRT-LAMP-LFB method is a valuable diagnostic tool for the detection of SARS-CoV-2 infections in clinical and point-of-care field settings. The total diagnostic test was completed within one hour from sample collection to result interpretation [36].

In order to detect the protein induced from SARS-CoV-2 replication at very low concentrations, the coupling of a paramagnetic nanoparticle with an aptamer-PQC (piezoelectric quartz crystal)-based biosensor was developed by Dharmatov [37]. The PQC sensor unit provided a promising detector system as it operated by a direct interaction between the target analyte and the PQC sensor surface. From the flow injection analysis, the presented biosensor showed a limit of detection of 3.5 ng/ml. The aptamer-coated crystal exhibits linearity for SARS-CoV-2 helicase in the concentration range of 0.05–1 μg/ml [37].

Park et al. developed a biosensor by the application of SPR for the diagnosis of SARS-CoV-2 using genetically engineered protein fusing gold binding polypeptides (GBPs) to SARS-CoV-2 surface antigen (SCVme) [38]. The biosensor was fabricated by the following strategy: firstly, mobilization of SCVme onto the SPR gold sensor chip is performed. Further, the GBP domain of the fusion protein acts as an anchoring component on the gold surface, and SCV is a recognition element for anti-SCVme antibody, the target analyte. The fabricated biosensor offers advantages such as a stable and specific platform for binding activity with anti-SCVme. AFM-coupled SPR imaging demonstrated that anti-SCVme was specifically bound to the fusion protein on the gold surface, indicating that the appropriate orientation of the fusion protein by GBP resulted in efficient capture of the anti-SCVme antibody. The packing density of the fusion protein on the SPR chip has a detection limit of 10 g/ml. However, the fusion protein-coated SPR chip showed a lower detection limit of 200 ng/ml anti-SCVme. Further, authors concluded that GBP fusion proteins with other functional groups may open an

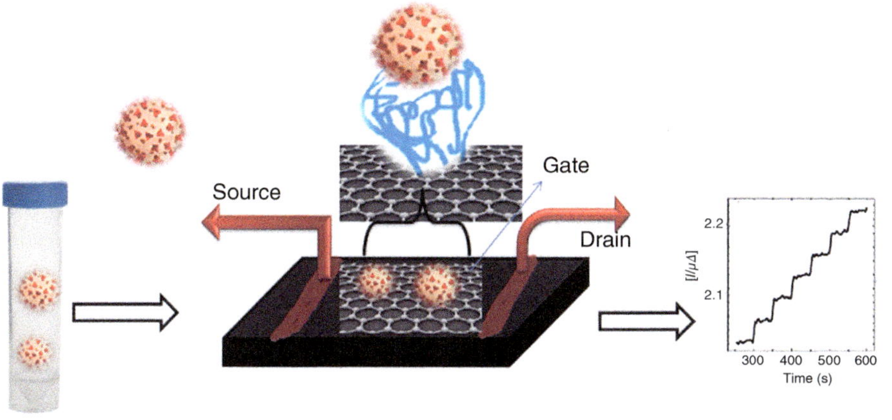

Figure 7.4 Schematic representation of FET-based biosensor and its operation. The biosensor is fabricated by the conjugation of SARS-CoV-2 spike antibody and graphene through 1-pyrenebutyric acid *N*-hydroxysuccinimide ester (PBASE). The surface-induced current is measured upon the binding of SARS-CoV-2.

opportunity for the development of various gold substrate-based biosensor systems. The fusion protein provides a simple and effective method for construction of SPR sensing platforms, permitting sensitive and selective detection of anti-SCVme antibody [38].

Field effect transistor (FET), changes in the surface potential or current induced by the binding of analyte target molecules, based biosensor, proposed by Seo et al., for the detection of SARS-CoV-2 in clinical samples [39]. Figure 7.4 represents the FET-based biosensor fabrication and its operation. The biosensor was prepared by conjugating SARS-CoV-2 spike antibody to a graphene sheet through 1-pyrenebutyric acid *N*-hydroxysuccinimide ester (PBASE), which acts as an active area for sensing. The FET was covered with a phosphate buffer solution of pH 7.4 as an electrolyte to maintain the gating effect. The proposed biosensor was characterized by AFM, HR-TEM, Raman, and XPS techniques. The demonstrated biosensor device detected SARS-CoV-2 spike protein at a low concentration of 1 fg/ml in phosphate buffer solution and at 100 fg/ml in spiked clinical samples. In addition, the FET biosensor successfully detected SARS-CoV-2 in a culture medium and clinical samples with a limit of detection of 1.6×10^1 pfu/ml and 2.42×10^2 copies/ml, respectively. Hence, the reported FET biosensor provides simple, rapid, and highly responsive detection of SARS-CoV-2 in clinical samples [39].

Ishikawa et al. reported the application of antibody mimic proteins (AMPs) in the field of nanobiosensors [40]. The polypeptide AMPs bind to target analytes with high affinity and specificity. The In_2O_3 nanowire-based biosensor was configured with an AMP (fibronectin, Fn) to detect a biomarker for SARS-CoV-2 nucleocapsid protein (N). Bovine serum albumin (BSA) was used for passivation of the free surface area to get rid of nonspecific binding interactions of proteins. The device characteristics were measured by utilizing a liquid gate electrode. The In_2O_3 nanowire FET device exhibited an excellent transistor behavior in the 0.01 phosphate buffer solution. The linear plot of source or drain current vs. source or drain voltage (I_{ds}–V_{ds}) suggested a good contact between the nanowires and electrodes. The reported biosensor was capable of detecting the N protein at subnanomolar concentrations in the presence of 44 μM BSA. Further, the binding constant of AMP to Fn was also determined from the concentration dependence of the response of this biosensor. Authors have concluded that the N protein was detected at sensitivity comparable to current immunological detection methods but the results obtained within shorter time. The present report also demonstrated the potential for nanobiosensors to be used as an accurate, convenient, and rapid tool to measure complex systems [40].

An alternative promising analytical tool for the SARS-CoV-2 diagnosis was developed based on a dual-functional plasmonic biosensor combining the plasmonic photothermal (PPT) effect and localized surface plasmon resonance (LSPR) transduction [41]. The sensitive detection of SARS-CoV-2 is through nucleic acid hybridization, which was done through two-dimensional gold nanoislands (AuNIs) functionalized with cDNA receptors. By using two different angles of incidence, the plasmonic resonance of PPT and LSPR was excited at two different wavelengths, which further enhanced the sensing ability, sensitivity, and reliability. In the present work, a dual-functional LSPR biosensor exhibits a high sensitivity toward the selected SARS-CoV-2 sequences with a detection limit of 0.22 pM. The reported results confirmed that the in situ PPT enhancement on the AuNI chip dramatically improved the hybridization kinetics and the specificity of nucleic acid detection. The proposed biosensor provided a reliable platform to improve the diagnostic accuracy in clinical sample tests and relieve the pressure on PCR-based tests [41].

Hong et al. reported a one-step single-tube-accelerated RT quantitative reverse transcription loop-mediated amplification (LAMP) assay for the detection of SARS-CoV-2 replicase gene [42]. The RT-LAMP evaluation showed a sensitivity and specificity of 100% and 87%, respectively, in comparison with RT-PCR evaluation. The present assay involves a simple procedure, and the results were obtained in less than 1 hour (as early as 11 minutes). Therefore, RT-LAMP has advantages of simple operation, rapid amplification, and easy detection [42].

A selective colorimetric biosensor was developed by Moitra based on AuNPs capped with thiol-modified antisense oligonucleotides (ASOs) specific for N-gene of SARS-CoV-2 [43]. The demonstrated biosensor, specific for the N-gene of SARS-CoV-2, is used to detect the SARS-CoV-2 infections within 10 minutes. The detection process involved a process where AuNPs-ASOs were agglomerated in the presence of SARS-CoV-2 RNA, which led to shift (40 nm) in the UV absorbance spectrum owing to the SPR effect. Further, the addition of RNase H cleaves the RNA strand from the composite hybrid RNA and Au-ASOmix, which leads to a visually detectable precipitate. It is also reported that the selectivity of the colorimetric biosensor toward the MERS-CoV-2 is in the dynamic range of 0.2–3 ng/μl with a detection limit of 0.18 ng/μl for SARS-CoV-2 [43]. In another report, Kumar developed a similar colorimetric biosensor based on an optical sensing platform to detect the RNA-dependent RNA polymerase

(RdRp) gene of SARS-CoV-2 [44]. In the assay, the formation of an oligo-probe target hybrid led to salt-induced aggregation and changes in color from pink to blue. The limit of detection for SARS-CoV-2 RNA was 0.5 ng and capable of detecting SARS-CoV-2 infection in human nasopharyngeal samples in less than 30 minutes [44].

Chen reported a sensitive and simple analytical tool, lateral flow immunoassay (LFIA), for the detection of SARS-CoV-2 IgG in human serum [45]. In the present work, the biosensor was fabricated using lanthanide-doped polystyrene nanoparticles (prepared by miniemulsion polymerization method). The biosensor platform was modified using mouse anti-human IgG and rabbit IgG on the lanthanide surface, which acts as a fluorescent probe. A nitrocellulose membrane was used to immobilize a recombinant nucleocapsid phosphor-protein of SARS-CoV-2, which is responsible to confine the specific IgG. Then, investigations revealed that LIFA was able to detect SARS-CoV-2 IgG in human serum samples in 10 minutes. Further, the validation for the clinical application of LIFA was done through the RT-PCR technique. It was noticed that the results obtained through LIFA were the same as those obtained through the RT-PCR technique [45].

A microcantilever-based biosensor was developed by Velanki and Ji for the detection of feline SARS-CoV-2 [46]. In the report, the microcantilever was modified with Feline infectious peritonitis (FIP) type I antiviral, and it was used for the detection of FIP type I virus. If the sample contains FIP type I virus, the microcantilever bends upon the recognition of the FIP type I virus by the antiserum on the surface of the cantilever. The detection limit of the sensor was found to be 0.1 μg/ml [46].

A graphene-based field effect transistor (GFET) biosensor was developed by Rockx et al. to detect SARS-CoV-2 in clinical samples [47]. In the present study, authors have targeted spike protein as an antigen because of its high immunogenicity. Using a coupling agent, PBASE, the SARS-CoV-2 specific spike antibody, was conjugated to the graphene sheets. The S protein was specifically bonded to the spike antibody-functionalized GFET, which resulted in a change in the channel surface potential, which in turn produced a real-time electrical response that was dose dependent. The developed S antibody-functionalized GFET was able to detect SARS-CoV-2, S protein for a low concentration of about 1 fg/ml in phosphate buffer solution [47].

Later in 2020, Mavrikou et al. demonstrated a label-free cell-based bioassay for the detection of SARS-CoV-2 spike protein [33]. In the present work, the electro-insertion of SARS-CoV-2 spike S1 antibodies into the membrane-engineered mammalian kidney cells, the membrane potential was altered, thus generating a potentiometric signal on the gold screen-printed electrode. The obtained potentiometric signal was proportional to the concentration of the S1 antibodies. The reported cell-based bioassay was able to reach a detection limit of 1 fg/ml for S1 protein detection [33].

The detection of genomic DNA of SARS-CoV-2 was done by Costa-Garcia and group [48]. In the report, they have made use of strong thiol–gold interaction to immobilize SH-DNA probes on a sputtered gold film. Upon hybridization with the biotinylated 30-mer SARS-CoV-2 sequence, conjugation with alkaline phosphate (AP)-labeled streptavidin occurred, thus promoting the conversion of the enzymatic substrate 3-indoxy phosphate (3-IP) into the electroactive indigo carmine (IC). The voltammetric signal was then measured and correlated with the presence and amount of the target analyte. The selectivity of the developed biosensor was assessed by using a three-base mismatch DNA target as the negative control. Because of the nature of the SARS-CoV-2 virus to undergo potential mutation in order to adapt to the environment, the same group showed that the developed biosensor was also able to distinguish among the complementary target and sequences carrying one and two base mismatches, respectively [48]. Yet in an another report, Diaz-Gonzalez et al. have used the same approach for the immobilization of the biotinylated probe through electrostatic interactions on SPCEs modified with positively charged polylysine [49].

Mark et al. has developed a commercial electrochemical biosensor, ePlex® SARS-CoV-2, for the qualitative detection of SARS-419 CoV-2 in nasopharyngeal swab specimens. The detection process involves the extraction of viral RNA from the swab samples and reverse-transcribed to cDNA before PCR amplification. Next, a complementary ferrocene-labeled signaling probe was combined with the amplified target DNA to form a target DNA/signaling probe complex. Upon hybridization with the specific capture probes immobilized onto the surfaces of the gold electrode microarray, the electrochemical signal from the ferrocene label was correlated with the presence of the target DNA. This study showed the clinical performance of the ePlex SARS-CoV-2 test to be comparable to the approved RT-PCR, with less susceptibility to contamination and reduction in the sample preparation time required [50].

7.2 Conclusions

The pandemic of corona virus has motivated many researchers to make efforts toward the development of an advanced approach that includes high efficiency and cable of responding to the present demand of early diagnosis to manage the spread of virus. Because there is no vaccine available for the treatment of SARS-CoV-2, therefore, management of this pandemic is only possible through monitoring, prevention, and early detection of SARS-CoV-2 infection. However, this virus is more complex, and hence, it is necessary to invent a rapid, sensitive, and low-cost technique for the early diagnosis of this infection. The sensors have shown their potential toward the diagnosis and sensing of viral infections and hence may perhaps fulfill the current demand for early diagnosis of SARS-CoV-2 cases. A number of biosensing methods including impedimetric, amperometric, and colorimetric mediated methods have been established. The reported method could be used to detect SARS-CoV-2 within a short period of time. Although fewer studies on the development of sensors for SARS-CoV-2 detection have been reported, the biosensing techniques offer an alternative approach to PCR-based testing for SARS-CoV-2 infection.

7.3 Future Perspectives

Compared with the traditional PCR methods, sensors have a number of advantages, including easier modification, simple synthesis, smaller size, and high accuracy for the viral detection. Although biosensors and sensors for the detection of SARS-CoV-2 have many advantages, they also suffer from some limitations: (i) biological complexity of samples, (ii) preparation and screening of a suitable antigen antibody, and (iii) improvement of selectivity and sensitivity of sensor. In future research, the application of electrochemical sensors and biosensors by the use of various stable nanomaterials could be studied in detail, and modification of electrode substrates with advanced nanomaterials, QDs, and carbon dots could be utilized to functionalize their surface to achieve selectivity. The affinity of sensor to specific biological targets could be explored to generate sensitive response.

References

1 Singhal, T. (2020). A review of coronavirus disease-2019 (COVID-19). *Indian J. Pediatr.* 87 (4): 281–286.

2 WHO (2020). Coronavirus disease 2019 (COVID-19): situation report – 52. https://www.who.int/docs/default-source/coronaviruse/situation-reports/20200312-sitrep-52-covid-19.pdf?sfvrsn=e2bfc9c0_4 (accessed 05 March 2021).

3 Cascella, M., Rajnik, M., Cuomo, A. ct al. (2020). Features, evaluation and treatment of coronavirus (COVID-19). In: *StatPearls*. Treasure Island, FL. PMID: 32150360.

4 van Doremalen, N., Bushmaker, T., Morris, D.H. et al. (2020). Aerosol and surface stability of SARS-CoV-2 as compared with SARS-CoV-1. *N. Engl. J. Med.* 382: 1564–1567.

5 Su, S., Wong, G., Shi, W. et al. (2016). Epidemiology, genetic recombination, and pathogenesis of coronaviruses. *Trends Microbiol.* 24 (6): 490–502.

6 Chan, J.F.-W., Kok, K.-H., Zhu, Z. et al. (2020). Genomic characterization of the 2019 novel human-pathogenic coronavirus isolated from a patient with atypical pneumonia after visiting Wuhan. *Emerging Microbes Infect.* 9: 221–236.

7 Bemtgen, X., Krüger, K., Supady, A. et al. (2020). First successful treatment of coronavirus disease 2019 induced refractory cardiogenic plus vasoplegic shock by combination of percutaneous ventricular assist device and extracorporeal membrane oxygenation: a case report. *ASAIO J.* 66: 607–609.

8 Lv, D.F., Ying, Q.M., Weng, Y.S. et al. (2020). Dynamic change process of target genes by RT-PCR testing of SARS-CoV-2 during the course of a coronavirus disease 2019 patient. *Clin. Chim. Acta* 506: 172–175.

9 Xu, Y.H., Dong, J.H., An, W.M. et al. (2020). Clinical and computed tomographic imaging features of novel coronavirus pneumonia caused by SARS-CoV-2. *J. Infect.* 80: 394–400.

10 Li, G., Fan, Y., Lai, Y. et al. (2020). Coronavirus infections and immune responses. *J. Med. Virol.* 92: 424–432.

11 de Wit, E., van Doremalen, N., Falzarano, D. et al. (2016). SARS and MERS: recent insights into emerging coronaviruses. *Nat. Rev. Microbiol.* 14: 523–534.

12 Pan, L., Mu, M., and Yang, P. (2020). Clinical characteristics of COVID-19 patients with digestive symptoms in Hubei, China: a descriptive, cross-sectional, multicenter study. *Am. J. Gastroenterol.* 115: 766–773.

13 CDC (2020). Interim clinical guidance for management of patients with confirmed coronavirus disease (COVID-19).

14 Nurunnabi, M., Cho, K.J., Choi, J.S. et al. (2010). Targeted near-IR QDs-loaded micelles for cancer therapy and imaging. *Biomaterials* 31 (20): 5436–5444.

15 Giepmans, B.N.G., Adams, S.R., Ellisman, M.H. et al. (2006). The fluorescent toolbox for assessing protein location and function. *Science* 312 (5771): 217–224.

16 Di Tocco, A., Robledo, S.N., Osuna, Y. et al. (2018). Development of an electrochemical biosensor for the determination of triglycerides in serum samples based on a lipase/magnetite-chitosan/copper oxide nanoparticles/multiwalled carbon nanotubes/pectin composite. *Talanta* 190: 30–37.

17 Watts, H.J., Lowe, C.R., and Pollard-Knight, D.V. (1994). Optical biosensor for monitoring microbial cells. *Anal. Chem.* 66 (15): 2465–2468.

18 Syam, R., Davis, K., Pratheesh, M. et al. (2012). Biosensors: a novel approach for pathogen detection. *Vet. Scan.* 7 (1): 14–18.

19 Jandas, P.J., Luo, J., Quan, A. et al. (2020). Highly selective and label-free Love-mode surface acoustic wave biosensor for carcinoembryonic antigen detection using a self-assembled monolayer bioreceptor. *Appl. Surf. Sci.* 518: 146061.

20 Wang, J., Wu, Y., Zhou, P. et al. (2018). A novel fluorescent aptasensor for ultra sensitive and selective detection of acetamiprid pesticide based on the inner filter effect between gold nanoparticles and carbon dots. *Analyst* 143: 5151–5160.

21 Pelle, F.D. and Compagnone, D. (2018). Nanomaterial-based sensing and biosensing of phenolic compounds and related antioxidant capacity in food. *Sensors* 18 (2): 462.

22 Singh, V., Joung, D., Zhai, L. et al. (2011). Graphene based materials: past, present and future. *Prog. Mater Sci.* 56: 1178–1271.

23 Zhan, T., Yang, J., Guo, L. et al. (2019). Targets regulated formation of boron nitride quantum dots–gold nanoparticles nanocomposites for ultrasensitive detection of acetylcholinesterase activity and its inhibitors. *Sens. Actuators, B* 279: 61–68.

24 Ko, J., Jeong, B.G., Chang, J.H. et al. (2020). Chemically resistant and thermally stable quantum dots prepared by shell encapsulation with cross-linkable block copolymer ligands. *NPG Asia Mater.* https://doi.org/10.1038/s41427-020-0200-4.

25 Chen, N., He, Y., Su, Y. et al. (2012). The cytotoxicity of cadmium based quantum dots. *Biomaterials* 33: 1238–1244.

26 Ozer, T., Geiss, B.J., and Henry, C.S. (2020). Review-chemical and biological sensors for viral detection. *J. Electrochem. Soc.* 167 (3): 037523.

27 Layqah, L.A. and Eissa, S. (2019). An electrochemical immunosensor for the corona virus associated with the Middle East respiratory syndrome using an array of gold nanoparticle-modified carbon electrodes. *Microchim. Acta*. https://doi.org/10.1007/s00604-019-3345-5.

28 Mahari, S. (2020). CovSens-ultrasensitive novel in-house built printed circuit board based electrochemical device for rapid detection of nCOVID-19. *bioRxiv*. https://doi.org/10.1101/2020.04.24.059204.

29 Hsu, Y.-R., Lee, G.-Y., Chyi, J.-I. et al. (2013). Detection of severe acute respiratory syndrome (SARS) coronavirus nucleocapsid protein using AlGaN/GaN high electron mobility transistors. *ECS Trans.* 50: 239–243.

30 Martínez-Paredes, G., González-García, M.B., and Costa-García, A. (2009). Genosensor for SARS virus detection based on gold nanostructured screen-printed carbon electrodes. *Electroanalysis* 21 (3–5): 379–385.

31 Teengam, P., Siangproh, W., Tuantranont, A. et al. (2017). Multiplex paper-based colorimetric DNA sensor using pyrrolidinyl peptide nucleic acid-induced AgNPs aggregation for detecting MERS-CoV, MTB, and HPV oligonucleotide. *Anal. Chem.* 89: 5428–5435.

32 Huang, J.C., Chang, Y.-F., Chen, K.-H. et al. (2009). Detection of severe acute respiratory syndrome (SARS) coronavirus nucleocapsid protein in human serum using a localized surface plasmon coupled fluorescence fiber-optic biosensor. *Biosens. Bioelectron.* 25: 320–325.

33 Mavrikou, S., Moschopoulou, G., Tsekouras, V. et al. (2020). Development of a portable, ultra-rapid and ultra-sensitive cell-based biosensor for the direct detection of the SARS-CoV-2 S1 spike protein antigen. *Sensors* 20 (1): 3121.

34 Shi, L., Sun, Q., He, J. et al. (2015). Development of SPR biosensor for simultaneous detection of multiplex respiratory viruses. *Bio-Med. Mater. Eng.* 26 (1): S2207–S2216.

35 Roh, C. and Jo, S.K. (2011). Quantitative and sensitive detection of SARS coronavirus nucleocapsid protein using quantum dots-conjugated RNA aptamer on chip. *J. Chem. Technol. Biotechnol.* 86 (12): 1475–1479.

36 Zhu, X., Wang, X., Han, L. et al. (2020). Multiplex reverse transcription loop-mediated isothermal amplification combined with nanoparticle-based lateral flow biosensor for the diagnosis of COVID-19. *Biosens. Bioelectron.* 166: 112437.

37 Dharmatov, R.B. (2018). Piezoelectric quartz crystal aptamer biosensor for detection and quantification of SARS CoV helicase protein. 17th International Meeting on Chemical Sensors. https://doi.org/10.5162/IMCS2018/BS5.3

38 Park, T.J., Hyun, M.S., Lee, H.J. et al. (2009). A self-assembled fusion protein-based surface plasmon resonance biosensor for rapid diagnosis of severe acute respiratory syndrome. *Talanta* 79 (20): 295–301.

39 Seo, G., Lee, G., Kim, M.J. et al. (2020). Rapid detection of COVID-19 causative virus (SARS-CoV-2) in human nasopharyngeal swab specimens using field-effect transistor-based biosensor. *ACS Nano* 14 (4): 5135–5142.

40 Ishikawa, F.N., Chang, H.-K., Curreli, M. et al. (2009). Label-free, electrical detection of the SARS virus N-protein with nanowire biosensors utilizing antibody mimics as capture probes. *ACS Nano* 3 (5): 1219–1224.

41 Qiu, G., Gai, Z., Tao, Y. et al. (2020). Dual-functional plasmonic photothermal biosensors for highly accurate severe acute respiratory syndrome coronavirus 2 detection. *ACS Nano* 14 (5): 5268–5277.

42 Hong, T.C., Mai, Q.L., Cuong, D.V. et al. (2004). Development and evaluation of a novel loop-mediated isothermal amplification method for rapid detection of severe acute respiratory syndrome coronavirus. *J. Clin. Microbiol.* 42: 1956–1961.

43 Moitra, P. (2020). Selective naked-eye detection of SARS-CoV-2 mediated by N gene targeted antisense oligonucleotide capped plasmonic nanoparticles. *ACS Nano*. https://doi.org/10.1021/acsnano.0c03822.

44 Kumar, V. (2020). Development of RNA-based assay for rapid detection of SARS-CoV-2 in clinical samples. *bioRxiv*. https://doi.org/10.1101/2020.06.30.172833.

45 Chen, Z. (2020). Rapid and sensitive detection of anti-SARS-CoV-2 IgG, using lanthanide-doped nanoparticles-based lateral flow immunoassay. *Anal. Chem.* https://doi.org/10.1021/acs.analchem.0c00784.

46 Velanki, S. and Ji, H.F. (2006). Detection of feline coronavirus using microcantilever sensors. *Meas. Sci. Technol.* 17: 2964–2968.

47 Rockx, B., Kuiken, T., Herfst, S. et al. (2020). Comparative pathogenesis of COVID-19, MERS, and SARS in a nonhuman primate model. *Science* 368: 1012–1015.

48 Abad-Valle, P., Fernandez-Abedul, M.T., and Costa-Garcia, A. (2007). DNA single-base mismatch study with an electrochemical enzymatic genosensor. *Biosens. Bioelectron.* 22: 1642–1650.

49 Diaz-Gonzalez, M., de la Escosura-Muniz, A., Gonzalez-Garcia, M.B. et al. (2008). DNA hybridization biosensors using polylysine modified SPCEs. *Biosens. Bioelectron.* 23: 1340–1346.

50 GenMark Diagnostics, Inc. (2020). ePlex® SARS-CoV-2 test assay manual. https://www.fda.gov/media/136282/download (accessed 04 March 2021).

8

Chemical Sensor for the Diagnosis of Coronavirus

Gyandshwar K. Rao, Ashish K. Sengar, and Seema R. Pathak

Amity University Haryana, Amity School of Applied Sciences, Department of Chemistry, Biochemistry & Forensic Science, Gurugram 122413, India

8.1 Introduction

Coronavirus disease 2019 (COVID-19), an outbreak of the coronavirus in China, has caused global concern and led to a worldwide pandemic. More than 41 million people have been infected and around 1.12 million people deceased due to COVID-19 infection till 20th October, 2020, in more than 215 countries, according to a report published by the World Health Organization (WHO) (https://covid19.who.int) [1]. The symptoms associated with COVID-19 appear usually in two to five days. Initial symptoms of COVID-19 resemble mostly those of common cold and flu and are mainly fever, restlessness, body ache, and cough. In due course of time, the severity of COVID-19 may cause pneumonia, breathing difficulty, infection in the lower respiratory tract, diarrhea, etc. Moreover, many infected patients have been reported to be asymptomatic as well. In spite of significant efforts, only a few claims have been made to date for the development of vaccines to treat this virus. In such a scenario, the best strategy is to identify the infected person in the early stage followed by isolation and treatment. Several detection methods have been developed to identify COVID-19-infected individuals. In one of the detection methods, targeted antibodies are used. This method is not suitable as antibodies are observed in about 10 days from the day of infection and can create a severe condition for the patients. The second method involves the use of chemical sensors. These chemical sensors have great efficiency, are inexpensive, are much selective toward the target, and have a high demand for diagnosis of COVID-19 patients.

Generally, the molecular method reverse transcription-polymerase chain reaction (RT-PCR) is frequently used to detect a particular virus in the sample [2]. According to the latest version of "WHO interim guidance for laboratory testing for COVID-19 in humans," there are several molecular assays that have been discussed to detect the virus. The gene targets that are selected by different countries for RT-PCR molecular assays are genetically similar. RT-PCR is currently the most sensitive method of viral RNA detection by rapidly making many copies of a specific

Detection and Analysis of SARS Coronavirus: Advanced Biosensors for Pandemic Viruses and Related Pathogens, First Edition. Edited by Chaudhery Mustansar Hussain and Sudheesh K. Shukla.

sequence. A recent study showed that the sensitivity of severe acute respiratory syndrome coronavirus 2 (SARS-CoV-2) has reached 3.7 RNA copies on detecting the RNA-dependent RNA polymerase (RdRp) sequence [2]. However, RT-PCR also has some drawbacks and can also fail for various reasons, like its amplification of spurious nucleic acid contaminations. Many false-negative results have been reported using this RT-PCR assay [3]. Moreover, if the sample is from an upper respiratory tract, then the positive rate was found to be 30–40% for laboratory-confirmed cases.

Molecular tests that are used to detect COVID-19 currently are considered as most appropriate for SARS-CoV-2. However, they require a swab sample and the laboratory procedure involves is time-consuming. Such tests required special laboratory facilities that are available at limited test centers. This required sample transport to centers increases the load on the laboratory, which delays the test report and increases the burden on the healthcare system. Local community transmission allowed the spread of COVID-19, which means that knowing the source of infection is a highly difficult task [4]. Hence fast, less-expensive, and easy-to-use methods are required in order to figure out the infection at very earlier stages, even before the symptoms are observed, to decrease the transmission and fatality rates. Thus, the present chapter is focusing on the design and synthesis of various chemical sensors for the diagnosis of COVID-19. It also includes a critical analysis of the interaction of coronavirus with such chemical sensors. In the present chapter, chemical sensor used for the detection of COVID-19 has been classified into the Sections 8.2–8.8.

8.2 Multiplexed Nanomaterial-Based Sensor

A nanomaterial-based sensor array with multiplexed capabilities for fast detection and monitoring of COVID-19 has been developed. Such sensors are made of different gold nanoparticles linked to organic ligands and create a diverse sensing layer capable of swelling or shrinking upon exposure to volatile organic compounds (VOCs). The change in electric resistance was observed because of swelling or shrinking in the layers. The presence of inorganic nanomaterials along with the organic film element provides sites for the adsorption of VOCs, which is responsible for electrical conductivity in these layers [5, 6]. VOCs on exposure either diffuse to the sensing layer or fall on the sensing surface and react with the organic segment or the functional groups of the capping agent. This interaction is responsible for the volume change (swelling/shrinkage) in the nanomaterial film [5]. This phenomenon is responsible for an increase/decrease in conductivity [5]. When the nanomaterial layer is exposed to VOCs, this brings a swift charge transfer to/from the inorganic nanomaterial and produces variations in the measured conductivity even when there are no steric changes occur within the layer. This array is important because of its flexibility and possibility, as it is can be used to identify a wide range of chemical patterns in different conditions. The rationale behind this approach relies on findings showing that viral agents and/or their microenvironment emit

VOCs [7, 8] that can reach the exhaled breath. If the emergence of VOCs could occur in the early stages of the infection [9, 10], immediate detection of COVID-19 is possible. The feasibility of this approach as a pre-screening diagnostic system was examined via an observational case–control study at the origin of the COVID-19 outbreak, Wuhan, China, during March 2020.

Using an array based on the principle discussed above, gold nanoparticles have been developed and combined with electronic circuitry and an advanced apparatus that collects an exhaled breath sample from the distance of a few (1–2) centimeters and blowing into it for two to three seconds (Figure 8.1). There is a built-in sensor that indicates when the test is complete. As the breath is passed through the array, a mixture of VOCs present in the breath of COVID-19 patient reacts with the sensors

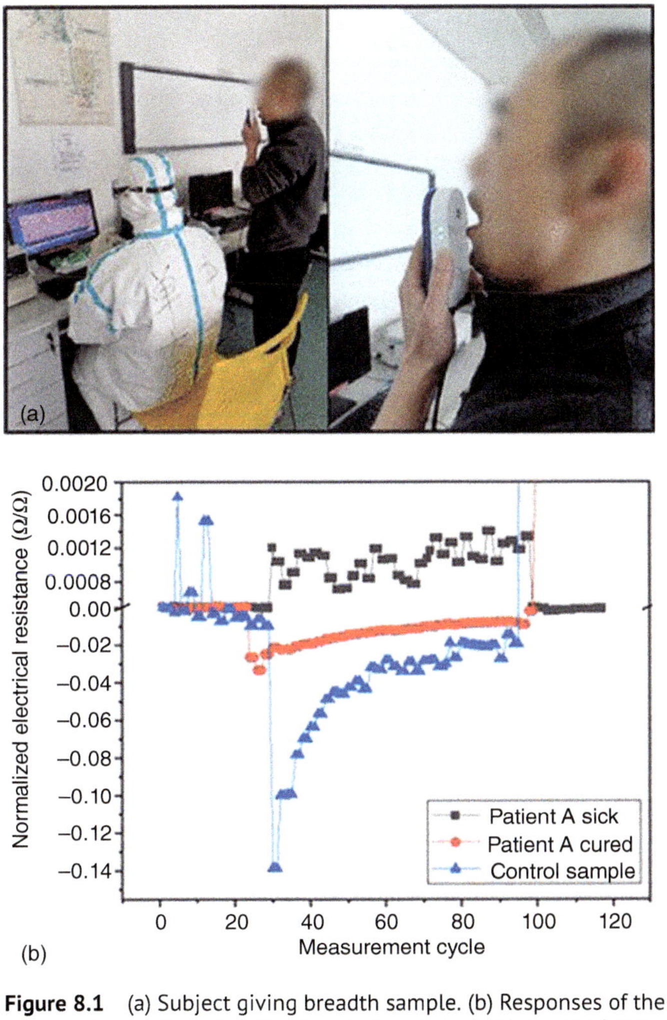

Figure 8.1 (a) Subject giving breadth sample. (b) Responses of the different subject samples (currently sick patient, patient cured, and the control sample). Source: Shan et al. [11].

and these emit a set of electrical resistance signals as a function of time. If the results are not satisfactory, then the test is repeated.

8.3 Nanomaterial-Mediated Paper-Based Sensors

Easy storage, transportability, simplicity of the paper-based analytical devices (PADs) made them a critical point-of-care (POC) tools for the detection of pathogens. These PAD scans can be easily used in the areas that have limited resources, in emergencies, as well as at home for medical diagnostics. One of the important features of these PADs is that they do not need any external devices for diagnosis [12, 13]. A variety of methods such as electrochemistry, conductivity, fluorescence, and colorimetry have been used to read out the signals in paper-based sensors [14]. The colorimetric paper-based biosensors are used to monitor the presence of specific pathogen by looking at the color change and can be distinguished with the naked eye. Such sensors have high demand and researchers are making continuous efforts for the fabrication of such sensors [15]. Increasing the surface area of nanoparticles is a common strategy to increase the number of receptors conjugated on the surface [16]. The physiochemical properties such as catalytic activity, conductivity, magnetic property, etc., are dependent on the surface-to-volume ratio. A high surface-to-volume ratio can easily enhance such properties. Such an approach has been utilized to enhance signal readout for colorimetric pathogen detection [17]. The intrinsic hydrophilic property of the porous structure of paper facilitates the flow of fluid by capillary action without requiring any external pump [13]. Generally, PADs have been classified into three types: dipstick assay, lateral flow assay (LFA), and microfluidic paper-based analytical device (μPAD) (Figure 8.2). LFA assays have several advantages such as fast response, cost-effectiveness, single-step operation, high sensitivity, user-friendliness, and sufficient selectivity and used to detect pathogens. Whiteside's group used a photolithography technique to generate microfluidic patterns in 2007 on the paper substrate [19]. This method involves the creation of hydrophilic channels on the paper using hydrophobic polymeric materials [20, 21]. 2D patterned papers have been constructed to 3D configuration

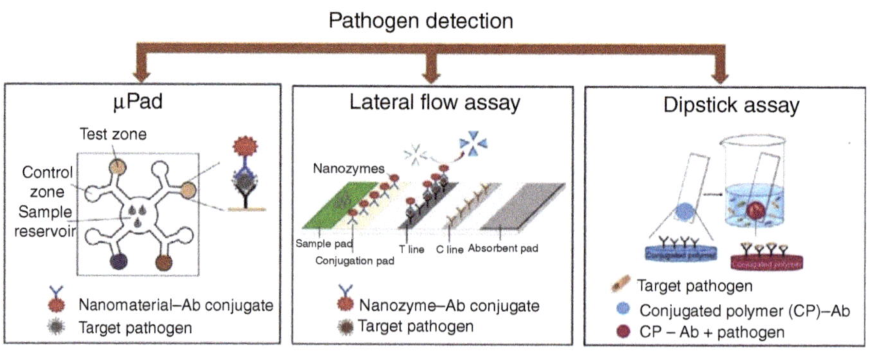

Figure 8.2 Classification of PADs. Source: Based on Hu et al. [18].

by using the folding process by Lim et al. [22]. Furthermore, printing on both sides of the paper resulted in the formation of 3D printed wax [23].

8.4 Molecularly Imprinted Polymer-Based Technology

Molecularly imprinted polymer (MIP)-based sensors have been used to detect SARS-CoV-2 because of their selectivity and sensitivity. Such techniques are very useful in the field of sensing because of their excellent ruggedness, thermal, chemical, and mechanical stabilities. MIPs are synthetic receptors having recognition sites and their shape and orientation match the target molecule [24]. MIPs are synthesized with molecular recognition cavities, which have specific selectivity for a particular template molecule [25, 26]. Theaflavin (TF), catechin (CAT), epigallocatechin-3-gallate (EGCG), and caffeine (CAF) have been successfully detected by sensors based on this technique [27–30]. Zika virus, Ebola virus, or Norovirus has been recognized by selecting their DNA or virus-specific aptamers as recognition element [31–35]. The detection of SARS-CoV-2 (coronavirus) is also possible using this method by selecting an appropriate polymer and using a specific aptamer of coronavirus as the recognition element. Polymers such as acrylamide, acrylic acid, methyl acrylate, methyl methacrylate, and ethyl acrylate can be selected to improve the sensing properties. These polymers can be functionalized on graphene or conducting metal oxides to provide better conductivity. The best performing polymer-conducting material nanocomposite can be selected based on repeated experiments. The process of imprinting and subsequent removal of the template is shown in Figure 8.3. The resultant sensor is embedded with an electronic circuitry fitted with an alarm such that the user can be immediately informed of his health status.

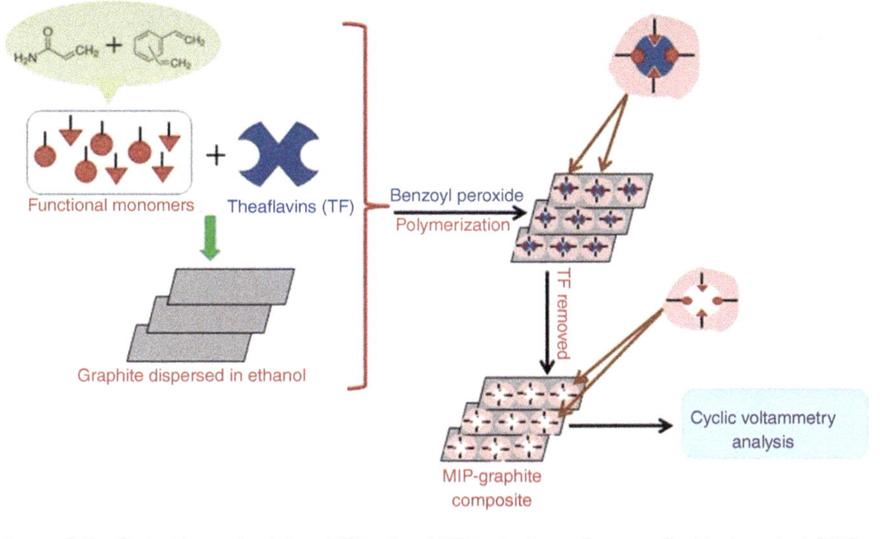

Figure 8.3 Detection principle of TF using MIP technique. Source: Chatterjee et al. [27].

8.5 Dual-Functional Plasmonic Photothermal Sensors for SARS-CoV-2 Detection

Biosensors play a vital role in monitoring and early detection of the coronavirus. They are ideal in providing accurate diagnosis that will further help in early treatment. When the different biosensing techniques are studied, localized surface plasmon resonance (LSPR) is a perfect alternative over the other methods. Biosensing applies to different types of analytes [36, 37]. In LSPR, there is a strong photon-driven coherent oscillation of the surface conduction electrons, which are modulated when coupling occurs at the surface of the plasmonic materials [38]. LSPR system shows high sensitivity to local variation, including the refractive index change and molecular binding. This process is ideal for real-time and label-free detection of micro- and nanoscale analytes [39]. surface plasmon resonance (SPR) has been used in recent research to investigate the biophysical properties of SARS-CoV-2 spike protein. It has been observed that the SARS-CoV-2 spike glycoprotein bounds angiotensin-converting enzyme 2 (ACE2) with a much higher affinity than SARS-CoV spike protein. In addition, several SARS-CoV receptor-binding domains (RBDs)-specific monoclonal antibodies did not have appreciable binding to the spike protein of SARS-CoV-2. The unique property of nucleic acids to be predictable and particular hybridization of complementary bases makes it useful in clinical diagnosis [40]. Hence, the LSPR technique is a promising technique for SARS-CoV-2 detection and COVID-19 diagnosis. The novel SARS-CoV-2 virus is a positive-sense, single-stranded RNA virus. The DNA–RNA hybridization has been widely used in RT-PCR as well as various biomedical sensors. The criteria for hybridization are based on nucleic acid strand melting [41, 42]. The two complementary strands can specifically hybridize with each other when the temperature is slightly lower than their melting temperature. It is important to note that the plasmonic nanoparticles normally exhibit large optical cross sections and the absorbed light can be non-radiatively relaxed, resulting in a significant heating energy [43, 44]. The converted plasmonic photothermal (PPT) heat energy, also known as the thermoplastic effect and is highly localized near the nanoparticles, which can be used as a source for controllable and uniform thermal processing. A dual-functional LSPR biosensor has been developed by Jing and coworkers. It has been used for SARS-CoV-2 detection and is based on plasmonic sensing transduction by combining the photothermal effect. The nanoabsorbers (AuNIs) have been found to generate local PPT heat and showed highly sensitive and accurate SARS-CoV-2 detection [45].

8.6 Zirconium Quantum Dot-Based Chemical Sensors

The group II–VI elements of the periodic table show exceptional optical and electrical behavior because of the quantum confinements of electrons and large surface area. Quantum dots (QDs), an example of nanoscale materials containing elements from the above-mentioned group, possess several significant characteristics. Strong

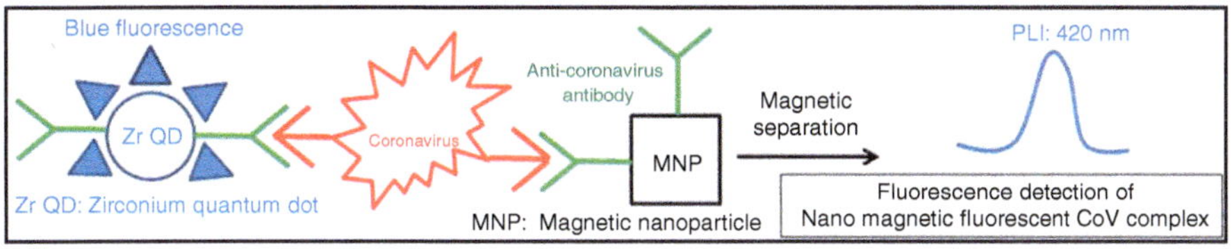

Figure 8.4 Fluorescent Zr QDs and magnetic nanoparticles are conjugated with antibodies that specifically bind to CoV. In the presence of CoV, a magnetic fluorescent complex is formed, which is isolated magnetically and detected by fluorescence measurements. Source: Alphandéry et al. [47].

fluorescence emission and optical stability, broad-range excitation wavelength, and better quantum yield (QY) are a few that are included in the promising characteristics. The introduction of chiral QDs' with optical and emission property would play a significant role in the development of several sensing applications. The chirality of nanomaterials has recently been deeply studied, and it can be considered as an alternative to natural chiral molecules. Zirconium, a group IV transition metal element, has been applied in various technological applications. Recently, chiral Zr QDs have been prepared by a one-step conversion of Zr NPs with the assistance of an autoclave. The synthesized Zr QDs show blue fluorescence emission as well as demonstrated aqueous dispersibility and was further applied in the biosensing of infectious bronchitis virus (IBV). Firstly, Zr QDs and magneto-plasmonic nanoparticles (MP NPs) are prepared, and then antibodies are conjugated with nanomaterials. Zr QDs and MP NPs will stay apart from each other since no attraction between them at this point. Antibody-conjugated Zr QDs and MP NPs will come closer and will make a magnetoplasmonic fluorescent nanohybrids structure while the target analyte is added. Then, an external magnet can be used to separate nanostructured magnetoplasmonic fluorescent, and the photoluminescence (PL) intensity of nanohybrids can be used to measure the analyte's concentration. The fluorescence and absorbance spectra of Zr QDs can be observed. Quantum yield and fluorescence intensity of Zr QDs is observed, which further confirm the antigen present. There is still much to be learned about the nature of Zr QDs, such as how excitonic properties arise on it, its size-tunable optical properties, surface modification with different biological molecules, and the control of the interaction between Zr QDs and plasmonic metals [46] (Figure 8.4).

8.7 Calixarene-Functionalized Graphene Oxide-Based Sensors

Electrochemical biosensors have been developed, which stand as an alternative and reliable solution to clinical diagnosis because of their tremendous properties and advantages, such as high sensitivity, low cost, user-friendliness, and robustness. High specificity and sensitivity of this sensor are the two main properties that helped it in gaining a lot of attention for a detailed study [48]. Capture probe

(CP), target sequence, label probe (LP), and auxiliary probe (AP) are the major components of this type of sensor [49]. Host–guest recognition has emerged tremendously and gained much attention in the fabrication of electrochemical biosensors. Host–guest recognition motifs are specific and biorthogonal, and they can form stable host–guest inclusion to increase the enrichment capability of guest molecules because of its own more rigid and well-defined cavity. Calixarenes such as CX8 show excellent supramolecular recognition and is also capable to be the electrochemical mediators of methylene blue and toluidine blue (TB) [50–53]. A supersandwich-type electrochemical biosensor based on *p*-sulfocalix[8]arene (SCX8)-functionalized graphene (SCX8-reduced graphene oxide [RGO]) is developed to enrich TB for SARS-CoV-2 RNA detection. Plug-and-play method is used to achieve the sensitive, accurate, and rapid detection of SARS-CoV-2 samples from different clinical specimens without RNA amplification (Figure 8.5).

8.8 AlGaN/GaN High Electron Mobility Transistor-Based Sensors

The development of an efficient and molecule labeling free binding assay with low cost and high sensitivity has been a subject of rigorous research. Recently, field-effect transistor (FET) devices have been shown its capability of biomaterial sensing and also used to measure the dissociation constants of ligand–receptor interactions, such as protein–protein interaction and aptamer–protein interaction without fluorescent labeling. Compared to electrophoresis mobility shift assay (EMSA), the FET sensor-based binding assay does not need to extract the free nucleic acids from nucleotide–protein mixture as in gel electrophoresis. In gel electrophoresis, the size of the nucleotide–protein complex was sometimes larger than the pole size of the gel [55], resulting in the complex being stuck in gel, which obstructs users to measure the dissociation constants in their system by EMSA. The cost of manufacturing nanowire-based FET is still high, so this kind of device could not be efficiently applied to support biological research immediately. Developing a planar semiconductor-based sensor for biological binding affinity investigation could decrease the cost for the detection of biomarkers. An AlGaN/GaN-HEMTs (high electron mobility transistors)-based sensor to detect the SARS-CoV C-terminal domains (CTDs) protein has been developed. The size of this device was in the micrometer scale, and it would be suitable for mass production due to the mature technology for semiconductor fabrication. AlGaN/GaN-HEMTs are useful in detecting biomolecule binding such as antibody–antigen, ligand–receptor, etc. [56–58]. The AlGaN/GaN HEMTs are also capable of detecting nucleic acid–protein interaction with high sensitivity. Figures 8.6a & b show the schematics of a DNA-immobilized AlGaN/GaN HEMT sensor and the plan-view microphotograph of the device, respectively. The AlGaN/GaN layers are developed by metal–organic chemical vapor deposition. The structure of the AlGaN/GaN layer consists of a 3 μm-thick undoped GaN buffer, 150 Å-thick undoped $Al_{0.25}Ga_{0.75}N$, and 10 Å-thick undoped GaN cap layer. The Mesa isolation is performed with an inductively

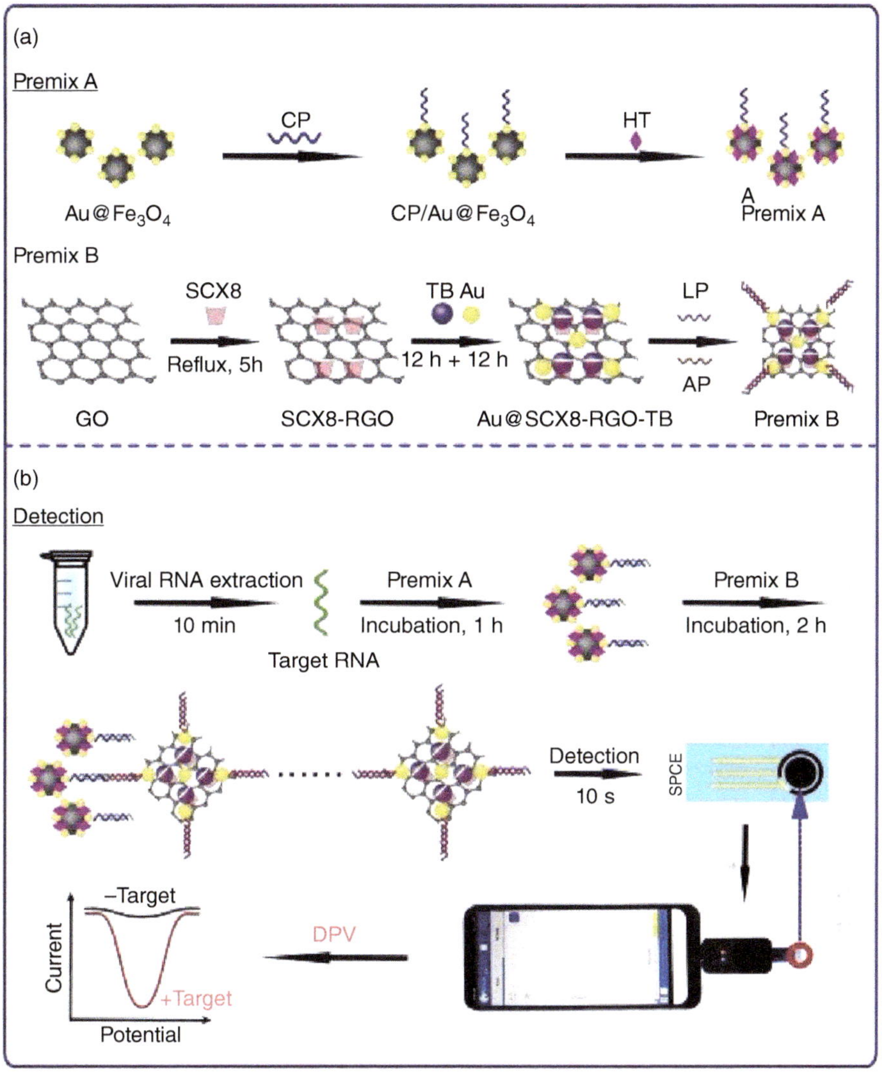

Figure 8.5 Representation of detection of SARS CoV-2. (a) Preparation of premix A and B and (b) process of electrochemical detection using a smartphone. Source: Zhao et al. [54].

Figure 8.6 (a) Schematic of the AlGaN/GaN HEMT sensor. (b) Plan view of a microphotograph of a completed device. Source: Hsu et al. [59].

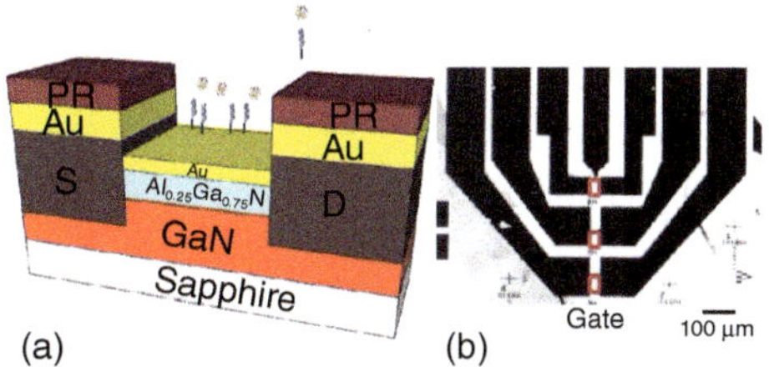

coupled plasma (ICP) etching with Cl_2/BCl_3/Ar-based discharges at −90 V dc self-bias, ICP power of 300 W. 10 × 50 μm^2 Ohmic contacts separated with gaps of 5 μm consist of e-beam deposits Ti/Al/Pt/Au patterned by lift-off, and annealing at 850 °C, for 45 seconds under flowing N_2. The dsDNA is the sensing element and immobilized on gate regions. Initially, the phosphate-buffered saline is dropped on the sensor. There is no net current change noticed until the target protein is added. A clear current change can be observed just as the system reaches a steady state. Real-time current monitoring can span the range of protein concentrations and shows saturation as the concentration of SARS-CoV CTD. AlGaN/GaN HEMTs can detect the nucleic acid-binding protein at allowing the detection limit. The AlGaN/GaN HEMTs are demonstrated to be an efficient tool for biological study and can potentially extract biochemical information such as dissociation constants and chemical dynamics.

8.9 Conclusion

We are seeing that what coronavirus has done to the world, causing so much chaos taking the life of people across the whole world. As the symptoms of the COVID-19 take few days to appear, it is more scary as an infected person can transmit the virus to many others causing the disease to them. So, we need methods that can help in early detection of the coronavirus. RT-PCR is currently the most sensitive method of viral RNA detection by rapidly making many copies of a specific sequence. However, there are limitations of this method, so we need chemical sensors for early detection of virus. These sensors are accurate, inexpensive, and easy to use. Looking at this, here we have discussed various chemical sensors such as nanomaterial-based sensors, polymer-based sensors, electrochemical sensors, etc., to detect coronavirus. Their working sensitivity, accuracy, and cost effectiveness have been discussed in detail. We have elaborated about the components used, fabrication of sensors, and all essential points and data related to these sensors.

References

1 WHO Coronavirus Disease (COVID-19) Dashboard. https://covid19.who.int (accessed 25 October 2020)

2 Corman, V.M., Landt, O., Kaiser, M. et al. (2020). Detection of 2019 novel coronavirus (2019-nCoV) by real-time RT-PCR. *Eurosurveillance* 25 (3).

3 Xie, X., Zhong, Z., Zhao, W. et al. (2020). Chest CT for typical 2019-nCoV pneumonia: relationship to negative RT-PCR testing. *Radiology* 296: 200343.

4 Bhatraju, P.K., Ghassemieh, B.J., Nichols, M. et al. (2020). COVID-19 in critically ill patients in the Seattle region case series. *N. Engl. J. Med.* 382: 2012–2022.

5 Broza, Y.Y. and Haick, H. (2013). Nanomaterial-based sensors for detection of disease by volatile organic compounds. *Nanomedicine* 8: 785–806.

6 Tisch, U. and Haick, H. (2010). Nanomaterials for cross-reactive sensor arrays. *MRS Bull.* 35: 797–803.

7 Broza, Y.Y., Mochalski, P., Ruzsanyi, V. et al. (2015). Hybrid volatolomics and disease detection. *Angew. Chem. Int. Ed.* 54: 11036–11048.

8 Lamote, K., Janssens, E., Schillebeeckx, E. et al. (2020). The scent of COVID-19: viral (semi-)volatiles as fast diagnostic biomarkers. *J. Breath Res.* 14: 042001.

9 Broza, Y.Y., Zhou, X., Yuan, M. et al. (2019). Disease detection with molecular biomarkers: from chemistry of body fluids to nature-inspired chemical sensors. *Chem. Rev.* 119: 11761–11817.

10 Traxler, S., Bischoff, A.C., Saß, R. et al. (2018). VOC breath profile in spontaneously breathing awake swine during influenza. A infection. *Sci. Rep.* 8: 14857.

11 Shan, B., Broza, Y.Y., Li, W. et al. (2020). Multiplexed nanomaterial-based sensor array for detection of COVID-19 in exhaled breath. *ACS Nano* 4 (9): 12125–12132.

12 Liu, L., Yang, D., Liu, G. et al. (2019). Signal amplification strategies for paper-based analytical devices. *Biosens. Bioelectron.* 136: 60–75.

13 Mahato, K., Srivastava, A., Chandra, P. et al. (2017). Paper based diagnostics for personalized health care: emerging technologies and commercial aspects. *Biosens. Bioelectron.* 96: 246–259.

14 Kant, K., Shahbazi, M.A., Dave, V.P. et al. (2018). Microfluidic devices for sample preparation and rapid detection of foodborne pathogens. *Biotechnol. Adv.* 36: 1003–1024.

15 Hristov, D.R., Rodriguez-Quijada, C., Gomez-Marquez, J. et al. (2019). Designing paper-based immunoassays for biomedical applications. *Sensors* 19: 554.

16 Fang, Y. and Ramasamy, R.P. (2015). Current and prospective methods for plant disease detection. *Biosensors* 5: 537–561.

17 Majdinasab, M., Hayat, A., Marty, J.L. et al. (2018). Aptamer-based assays and aptasensors for detection of pathogenic bacteria in food samples. *TrAC, Trends Anal. Chem.* 107: 60–77.

18 Hu, J., Wang, S., Wang, L. et al. (2014). Advances in paper-based point-of-care diagnostics. *Biosens. Bioelectron.* 54: 585–597.

19 Whitesides, G.M. (2013). Cool, or simple and cheap? Why not both? *Lab Chip* 13: 11–13.

20 Bhattacharya, S., Kumar, S., Agarwal, A.K. et al. (2019). *Paper Microfluidics Theory and Applications*. Springer Nature.

21 Nasseri, B., Soleimani, N., Rabiee, N. et al. (2018). Point-of-care microfluidic devices for pathogen detection. *Biosens. Bioelectron.* 117: 112–128.

22 Lim, H., Jafry, A.T., Lee, J. et al. (2019). Fabrication, flow control, and applications of microfluidic paper-based analytical devices. *Molecules* 24: 2869.

23 Li, X. and Liu, X. (2014). Fabrication of three-dimensional microfluidic channels in a single layer of cellulose paper. *Microfluid. Nanofluid.* 16: 819–827.

24 Yang, Y., Liu, X., Guo, M. et al. (2011). Molecularly imprinted polymer on carbon microsphere surfaces for adsorbing dibenzothiophene. *Colloids Surf., A* 377: 379–385.
25 Cheng, Y. et al. (2016). Preparation of a molecularly imprinted fluorescent chemosensor using quinoline modified vinyl-β-cyclodextrin and acrylamide as monomers for the selective recognition of spermidine. *Anal. Methods* 8: 5838–5842.
26 Motaharian, A. et al. (2016). Molecularly imprinted polymer nanoparticles-based electrochemical sensor for determination of diazinon pesticide in well water and apple fruit samples. *Anal. Bioanal. Chem.* 408: 6769–6779.
27 Chatterjee, T.N. et al. (2017). Detection of theaflavins in black tea using a molecular imprinted polyacrylamide-graphite nanocomposite electrode. *Sens. Actuators, B* 246: 840–847.
28 Chatterjee, T.N. et al. (2018). Molecular imprinted polymer-based electrode for sensing catechin (+C) in green tea. *IEEE Sens. J.* 18: 2236–2244.
29 Chatterjee, T.N. et al. (2019). Development of a nickel hydroxide nanopetal decorated molecularly imprinted polymer-based electrode for sensitive detection of epigallocatechin-3-gallate in green tea. *Sens. Actuators, B* 283: 69–78.
30 Das, D. et al. (2020). Titanium oxide nanocubes embedded molecularly imprinted polymer-based electrode for selective detection of caffeine in green tea. *IEEE Sens. J.* 20: 6240–6247.
31 Afsahi, S. et al. (2018). Novel graphene-based biosensor for early detection of Zika virus infection. *Biosens. Bioelectron.* 100: 85–88.
32 Kaushik, A. et al. (2018). A sensitive electrochemical immunosensor for label-free detection of Zika-virus protein. *Sci. Rep.* 8: 9700–9705.
33 Saylan, Y. et al. (2019). An alternative medical diagnosis method: biosensors for virus detection. *Biosensors* 9: 1–22.
34 Song, J. et al. (2016). Instrument-free point-of-care molecular detection of Zika virus. *Anal. Chem.* 88: 7289–7294.
35 Weerathunge, P. et al. (2019). Ultrasensitive colorimetric detection of murine norovirus using nanozyme aptasensor. *Anal. Chem.* 91: 3270–3276.
36 Masson, J.F. et al. (2017). Surface plasmon resonance clinical biosensors for medical diagnostics. *ACS Sens.* 2: 16–30.
37 Soler, M., Huertas, C.S., Lechuga, L.M. et al. (2019). Label-free plasmonic biosensors for point-of-care diagnostics: a review. *Expert Rev. Mol. Diagn.* 19: 71–81.
38 Willets, K.A. and Van Duyne, R.P. (2007). Localized surface plasmon resonance spectroscopy and sensing. *Annu. Rev. Phys. Chem.* 58: 267–297.
39 Qiu, G., Thakur, A., Xu, C. et al. (2019). Detection of glioma-derived exosomes with the biotinylated antibody-functionalized titanium nitride plasmonic biosensor. *Adv. Funct. Mater.* 29: 1806761.
40 Zhang, D.Y., Chen, S.X., Yin, P. et al. (2012). Optimizing the specificity of nucleic acid hybridization. *Nat. Chem.* 4: 208.
41 Stehr, J., Hrelescu, C., Sperling, R.A. et al. (2008). Gold nanostoves for microsecond DNA melting analysis. *Nano Lett.* 8: 619–623.

42 Zhang, J.X., Fang, J.Z., Duan, W. et al. (2018). Predicting DNA hybridization kinetics from sequence. *Nat. Chem.* 10: 91–98.

43 Jauffred, L., Samadi, A., Klingberg, H. et al. (2019). Plasmonic heating of nanostructures. *Chem. Rev.* 119: 8087–8130.

44 Lee, J.H., Cheglakov, Z., Yi, J. et al. (2017). Plasmonic photothermal gold bipyramid nanoreactors for ultrafast real-time bioassays. *J. Am. Chem. Soc.* 139: 8054–8057.

45 Qiu, G., Gai, Z., Tao, Y. et al. (2020). Dual-functional plasmonic photothermal biosensors for highly accurate severe acute respiratory syndrome coronavirus 2 detection. *ACS Nano* 14 (5): 5268–5277.

46 Ahmed, S.R., Kang, S.W., Oh, S. et al. (2018). Chiral zirconium quantum dots: a new class of nanocrystals for optical detection of coronavirus. *Heliyon* 4 (8): e00766.

47 Alphandéry, E. et al. (2020). The potential of various nanotechnologies for coronavirus diagnosis/treatment highlighted through a literature analysis. *Bioconjugate Chem.* 31 (8): 1873–1882.

48 Zou, X.L., Xiao, Y., Plaxco, K.W. et al. (2009). High specificity, electrochemical sandwich assays based on single aptamer sequences and suitable for the direct detection of small-molecule targets in blood and other complex matrices. *J. Am. Chem. Soc.* 131: 6944–6945.

49 Chen, X., Lin, Y.H., Li, J. et al. (2011). A simple and ultrasensitive electrochemical DNA biosensor based on DNA concatamers. *Chem. Commun.* 47: 12116–12118.

50 Dsouza, R.N., Pischel, U., and Nau, W.M. (2011). Fluorescent dyes and their supramolecular host/guest complexes with macrocycles in aqueous solution. *Chem. Rev.* 111: 7941–7980.

51 Yang, I., Zhao, H., Li, Y.C. et al. (2015). Fluorescent detection of tadalafil based on competitive host-guest interaction using *p*-sulfonated calix[6]arene functionalized grapheme. *ACS Appl. Mater. Interfaces* 7: 26557–26565.

52 Yang, L., Xie, X.G., Cai, L. et al. (2016). *p*-Sulfonated calix[8]arene functionalized grapheme as a "turn on" fluorescent sensing platform for aconitine determination. *Biosens. Bioelectron.* 82: 146–154.

53 Zhao, H., Yang, L., Li, Y.C. et al. (2017). A comparison study of macrocyclic hosts functionalized reduced graphene oxide for electrochemical recognition of tadalafil. *Biosens. Bioelectron.* 89: 361–369.

54 Zhao, H., Liu, F., Xie, W. et al. (2021). Ultrasensitive supersandwich-type electrochemical sensor for SARS-CoV-2 from the infected COVID-19 patients using a smartphone. *Sens. Actuators, B* 327: 128899.

55 Chen, C.Y., Chang, C.K., Chang, Y.W. et al. (2007). Structure of the SARS coronavirus nucleocapsid protein RNA-binding dimerization domain suggests a mechanism for helical packaging of viral RNA. *J. Mol. Biol.* 368 (4): 1075–1086.

56 Alur, S., Thapa, R., Gnaprakasa, T. et al. (2011). DNA hybridization sensor based on AlGaN/GaN HEMT. *Phys. Status Solidi C* 8: 2483–2485.

57 Pearton, S.J., Ren, F., Wang, Y.L. et al. (2010). Recent advances in wide bandgap semiconductor biological and gas sensors. *Prog. Mater. Sci.* 55 (1): 1–59.

58 Schwarz, S.U., Linkohr, S., Lorenz, P. et al. (2011). DNA-sensor based on AlGaN/GaN high electron mobility transistor. *Phys. Status Solidi A* 208 (7): 1626–1629.

59 Hsu, Y.R. et al. (2013). Detection of severe acute respiratory syndrome (SARS) coronavirus nucleocapsid protein using AlGaN/GaN high electron mobility transistors. *ECS Trans.* 50: 239.

9

Lab on a Paper-Based Device for Coronavirus Biosensing

Lucas Felipe de Lima[1], Ariana de Souza Moraes[2], Paulo de Tarso Garcia[3], and William Reis de Araujo[1]

[1] *Department of Analytical Chemistry, Institute of Chemistry, State University of Campinas (UNICAMP), Rua Josué de Castro, 126, Cidade Universitária, P.O. Box 6154, 13083-970, Campinas, São Paulo, Brazil*
[2] *Department of Physics, Chemistry, and Mathematics, Federal University of São Carlos (UFSCar), Rodovia João Leme dos Santos (SP-264), Km 110, 18052-780, Sorocaba, São Paulo, Brazil*
[3] *Institute of Exact Sciences, Faculty of Chemistry, Federal University of South and Southeast of Pará (UNIFESSPA), Avenida dos Ipês, s/n, Cidade Universitária, 68507-590, Marabá, Pará, Brazil*

9.1 Paper-Based Technology as Point-of-care Testing Devices

Paper is a type of material made from renewable resources, which has been used intensively for almost 2000 years in diverse daily activities such as drawing, writing, printing, packing, etc. In general, these processes aim to store information/facts. Based on the same idea, the paper was used in chemistry as a testing platform, where the cellulosic material allows the storage of chemicals and biological materials to perform portable assays. Paper platforms have proven their potentiality and versatility for the development of chemical sensors over the past decade [1]. Paper has many advantages for the manufacture of portable sensors for analyses directly in the field, such as low cost, large abundance, lightweight and flexibility to transport, disposability, biocompatibility, and many others [2]. The structure and physical characteristics of the cellulose enable a natural and spontaneous fluidic transport of samples and reagents. This attractive feature enables the creation of paper-based analytical devices (PADs) capable of pumping, mixing, separating, and preconcentrating solutions without any extra equipment or power source, allowing the complete integration of a routine laboratory process into a simple paper device (lab-on-paper).

Since the early 1980s, paper-based assays have been widely used for biomedical, environmental, and food-quality investigations. The use of paper for specific lateral flow (LF) test strips as a point-of-care (POC) diagnostic technique became an undeniable analytical tool [3]. The constant global health threats make us join efforts and tools of diverse disciplines to tackle the health adversities to guarantee the quality of life of our community. Development of low cost, portable, and easily operable tests enable the widespread deployment, large-scale testing, and population-level

Detection and Analysis of SARS Coronavirus: Advanced Biosensors for Pandemic Viruses and Related Pathogens, First Edition. Edited by Chaudhery Mustansar Hussain and Sudheesh K. Shukla.

surveillance in the case of an emergency like the current coronavirus pandemic outbreak, where the presence of reliable and decentralized tests provide fast decision-making about the start of medication protocol and need of quarantine, for example.

Lateral flow immunoassay (LFIA) is the most widely used and commercially available form of PAD. A simple and well-known representative of this kind of technology is the widespread pregnancy test. Lateral flow assays remain a foundation of modern medicine used to detect a wide range of pathogens and biomarkers. They typically utilize labeled antibodies to capture and detect a biomolecule through a colorimetric or fluorescent readout and are also used in other fields such as environmental monitoring and law enforcement [4]. The success of LFIAs is because they consist of a variety of advantageous parameters, including cost efficiency, portability, simplicity of use, and speed, which are not completely found in other conventional detection methods (enzyme-linked immunosorbent assay – ELISA, polymerase chain reaction – PCR, and cell culture). The overall format of LFIA uses the same rationale as ELISA, where an immobilized capture antibody or antigen is bound onto a solid-phase nitrocellulose (NC) membrane instead of a plastic well. The advantage here is the fact that the membrane enables a one-step assay, unlike that found in the multiple-step ELISA [3].

Figure 9.1a shows a chronological representation of the paper history in science from the oldest examples such as the use of litmus paper-like pH indicator in the early seventeenth century until 1937 when the first microfluidic structure was described by Yagoda [5], which employed paraffin to create hydrophilic microfluidic channels onto paper. In the same way, Müller and Clegg [6] optimized the original idea developed by Yagoda and collaborators and employed a paraffin-patterned paper microfluidic channel for improving paper chromatography in 1949. Since then, few studies have been reported using paper substrates for (bio)chemical tests. In 2017, Whitesides's group resumed the use of paper and its microfluidic properties as an alternative and low-cost material for the development of POC diagnostics introducing a new approach to pattern the paper [7]. Subsequently, there has been observed an exponential increase in research for the development of new methods to pattern the paper, exploring microfluidic properties and their applications for point-of-need analysis.

In the past years, PADs have demonstrated great potential in the (bio)chemical field because of some characteristics such as highly porous, flexibility, and wettability, which are a reflection of the flat structure of cellulose randomly arranged along its entire length. Besides, some characteristics such as affordability, portability, and disposability are also being explored to facilitate the application of the paper devices on the spot analysis [8].

The use of paper-based substrates is an interesting approach to develop portable methods because the nature of cellulosic material provides the porosity and physicochemical properties to allow passive microfluidic transport and diverse biochemical functionalization. These characteristics provide the essentials to achieve the ambitious goal of integrating all functions for point-of-need analysis, such as sample pretreatment, reagent delivery, mixing, separation, and detection in a miniaturized

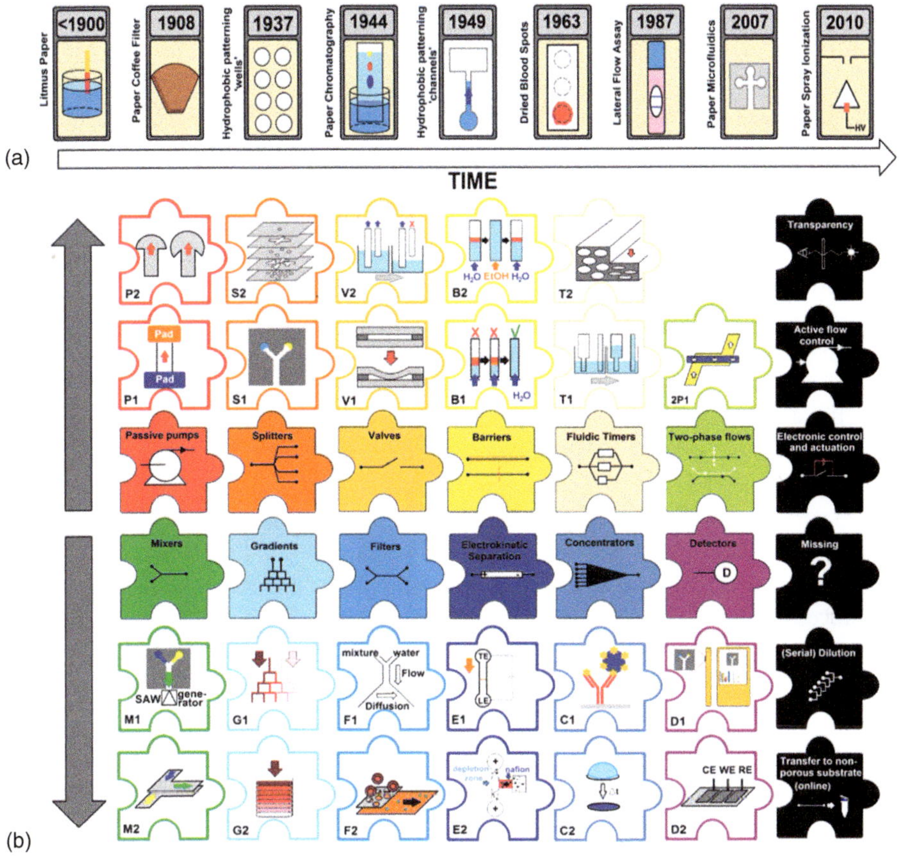

Figure 9.1 (a) The most important milestones of paper history in science. (b) Schematic illustration of paper microfluidic with their mean categories. The two central rows correspond to tool categories with color-coded examples in adjacent pieces (above or below). Black pieces represent tools that are still missing from the toolbox. Pumps: absorbent pad (P1) and fan (P2). Splitters: 2D (S1) and 3D (S2). Valves: mechanical (V1) and fluidic disconnects (V2). Barriers: dissolvable (B1) and selectively permeable (B2). Fluidic timers: geometrical (T1) and compressed (T2). Two-phase flows: countercurrent (2P1). Mixers: surface acoustic wave function (M1) and geometrical (M2). Gradient: branched (G1) and stacked (G2). Filters: Y-filter (F1) and membrane (F2). Electrokinetic separation: isotachophoresis (E1) and ion concentration polarization (E2). Concentrators: affinity-based (C1) and bulk (C2). Detectors: colorimetric/smartphone (D1) and electrochemical (D2). Source: Salentijn et al. [8].

device [9]. Thus, the called lab-on-paper technology has attracted great attention and relevance in diverse academic and industrial fields because of the simplicity of fabrication, accessibility of materials used, and diverse relevant applications. To facilitate the total integration of analytical steps on PADs, many tools have been developed over the years. Figure 9.1b illustrates the main tools used to develop those devices by a toolbox color-coded, where parts in black are remaining un(der)developed [8].

Paper has not conductive, specific, or optical properties to apply in electronic, colorimetric, or (bio)sensing devices; however, several electronic and biologic composites, among other materials, can be used to satisfy the requirements of paper mentioned previously, which the paper contributes for a favorable further modification because of their cellulose structure. The common materials used as modifiers on paper surface can be classified into organic or inorganic materials and additionally may be conductors, semiconductors, and insulators. Although inorganic materials show greater performance when applied in electronic devices, e.g. their properties often lack flexibility, which makes the application of several devices difficult [10].

The paper can be patterned using several fabrication methods that commonly have been classified as batch procedure and roll-to-roll processing methods. In the first case, the device presents an improvement in the analytical resolution and performance compared to the second method. However, they require laborious procedures and different equipment, which makes the device costly [11]. One example of a high-resolution patterning technique is photolithography; however, one limiting factor in this technology is the use of a special mask and UV light radiation, which aggregates expensive cost. Therefore, roll-to-roll processing methods can be interesting processes for the functionalization of paper microfluidic devices, which include coating and printing approaches [10].

9.1.1 Fabrication Methods

Paper has numerous advantages over other substrates used in microfluidic devices, such as glass and polymeric matrices. Because of its great versatility associated with its low cost, the manufacturing techniques used in this material must be appropriate to the design and delicacy of the paper types. Because microfluidic paper-based analytical devices (μPADs) normally use small aliquots of aqueous samples, it is necessary to employ techniques that delimit the hydrophilic and hydrophobic regions, i.e. it is necessary to delimit and standardize hydrophilic channels through hydrophobic barriers or create hydrophilic zones on hydrophobic surfaces allowing the capillarity and reaction of the sample only in the regions of interest [12].

The choice of manufacturing methods to produce PADs depends on the objectives for which the device will be employed because the infrastructure required, cost, production scale, and the resolution of the patterned designs will affect directly the optimal patterning technique to be used. Currently, one of the techniques that have gained notorious prominence in the manufacture of these devices is wax printing and screen printing because it presents a low cost, high performance, speed, and facility. Another technique that has stood out in the development of these devices is the fused deposition modeling (FDM), using a 3D printer, where the deposition of the molten material (commonly a polymer) occurs, giving shape to the construction of the device. According to Gaal et al., this manufacturing approach brings with it several advantages, such as efficient and fast prototyping as well as its simplicity does not require handling in clean rooms, photoresists, or photomasks [13–15]. Detailed information and comparison between the major fabrication methods of PADs can be verified in some specialized reviews [16–19].

9.1.2 Main Detection Methods Coupled to PADs

μPADs have demonstrated great potentials such as portable and disposable devices since their popularization in 2007, whose characteristics are ideal for POC testing with important contributions in the field of health and clinical analyses [20]. Several sensing methods coupled to paper platforms are developed every day to detect different types of biological species, including biomarkers, viruses, bacteria, antibody, among others, whose specificity and selectivity depend basically on the detector and strategy used to determine the target molecules [4]. The first and more explored detection method is colorimetry, which presents a simpler and low-cost detection approach, allowing the rapid identification of qualitative results by direct visual inspection. Semiquantitative and quantitative results are commonly obtained by the register of the colors obtained in the paper assay using a smartphone camera, scanners, or other accessible equipment and analyzing the color patterns [Red, Green, and Blue (RGB); Cyan, Magenta, Yellow, and Black-Key (CMYK); grayscale; etc.] with appropriate image software.

The colorimetric method usually suffers from low selectivity and sensitivity; however, these parameters can be enhanced by the use of recognition elements such as enzymes and aptamers and preconcentration strategies directly on a lab-on-paper device. Besides, other common limitations and challenges to use the colorimetric approaches are the homogeneity and stability of color, color interpretation can be dependent on the analyst, the sample is already colored or present turbidity, and obtain reproducible acquisition of images of the test for quantitative purposes.

To circumvent the limitations of colorimetric approaches, other analytical techniques have been emerged as alternative detectors to be coupled on a paper platform enhancing the figure of merits and keeping the low cost and portability features. In 2009, Henry's group [21] introduced the electrochemical paper-based analytical devices (ePADs) as a viable, accessible, and cost-effective sensing method. Electrochemical methods provide higher selectivity, sensitivity, and accuracy compared to colorimetry. Besides, ePADs can be fabricated by the incorporation of conductive materials using a plethora of methods, their instrumentation can be miniaturized, and the instrumental analysis exempts the operator of the visual aspect identification [18, 22].

A recent trend is to combine different techniques on PADs to expand the range of analytes or even to obtain a confirmatory result in the same device [23, 24]. To improve the performance of the paper-based (bio)sensors, some nanomaterials such as quantum dots (QDs), metallic nanoparticles (e.g. gold, silver, and copper), graphene oxide (GO), carbon dots (CDs), among others have been used to provide a dual response, such as electrochemical and colorimetric when coupled to a target molecule.

Although colorimetry and electrochemistry are the two most used techniques in paper-based sensors, other techniques such as fluorescence, chemiluminescence, and surface-enhanced Raman scattering (SERS) have been widely explored [25, 26].

9.2 Current Outbreak and Coronavirus Biology

Since December 2019, worldwide attention is about coronavirus outbreak and COVID-19. How did the COVID-19 emerge? How does the disease spread? What is the difference between severe acute respiratory syndrome-related coronavirus (SARS-CoV) or Middle East respiratory syndrome-related coronavirus (MERS-CoV) and COVID-19? How to control the disease spread? These issues about the current outbreak remain unknown, but in the past, there were answers to questions like these.

In 1949, a coronavirus murine hepatitis virus (MHV) prototype was isolated [27]. This coronavirus has been used as a human model disease because it disturbs the gut as well as it causes respiratory disease, like the effect of coronavirus in human [28]. Although a disease caused by a virus may not be the only criterion to classify to which family each virus belongs to, it indicates what local the virus is hosted, showing which physicochemical and biological properties the virus has.

Many characteristics are used to classify viruses, and this arduous work involves several experts as ecologists, biologists, virologists, microbiologists, zoologists, and veterinarians. However, multidisciplinarity is not the only challenge: the criteria to classify a virus are continuously changing based on the novel studies of the evolutionary characteristics. Therefore, the existence of a large number of virus species is another challenge because it demands revision on the taxonomy criteria over the years [29–31]. Also, there is not a general method to classify viruses. However, there is a focus on factors as host tropism (determinant of the pathogenesis of viral infections) and receptor to which the virus binds to the host cell, and, the most recommended, the genetic mapping to compare with known viruses [32, 33].

In 2003 occurred the SARS-CoV outbreak in humans, and it affected more than 8000 people causing around 800 deaths worldwide [28]. This event led to a molecular study that has demonstrated the positive selection pressure on the spike (S) protein, which means that SARS-CoV may have originated from wild animals. The S protein is one of the main proteins of enveloped viruses responsible for its entrance into the host cell, the result of the attachment between the virus receptor-binding domain (RBD) and the host cell receptor. Moreover, the researchers observed genomic similarities between collected samples from the first SARS-CoV infection case on 31 January 2003 in China and SARS-CoV infection cases in Vietnam, Canada, Singapore, and the United States in the same year, evidencing a common transmission place, a Hotel M [34, 35]. So, it is possible to see that this molecular analysis is essential to understanding several pathogenesis issues of the COVID-19 outbreak. More severe than the already known human coronavirus OC43 and 229E (until that date) that causes common cold, the SARS-CoV is responsible for the acute respiratory syndrome as well as gastrointestinal symptoms [36, 37].

Another coronavirus outbreak in humans emerged between 2012 and 2013, named the MERS-CoV in Saudi Arabia. MERS-CoV causes the same symptoms that SARS-CoV, but the World Health Organization (WHO) reported more deaths than caused by SARS-CoV [38, 39]. Also, being of the same coronavirus family, both SARS-CoV and MERS-CoV have similar structural characteristics.

However, they have different affinities by host cell receptors: SARS-CoV binds to the angiotensin-converting enzyme 2 (ACE2) receptor and MERS-CoV to a cluster of differentiation 26 (CD26) (also known as dipeptidyl peptidase-4, DPP4) receptor [40].

In December 2019, SARS-CoV-2 emerged in Wuhan, China, leading to the COVID-19 disease outbreak. Besides acute respiratory syndrome, SARS-CoV-2 seems to cause RNAaemia, acute cardiac injury, and secondary infection [30, 41]. The case and death number released by WHO reveals that the SARS-CoV-2 spreading and lethality are the highest hitherto [29]. Because of these differences related to the symptoms and the disease severity, the current coronavirus is named SARS-CoV2. Another difference between SARS-CoV and SARS-CoV-2 refers to the S protein changes resulted from mutations in the RBD to ACE2: the critical amino acids for the binding to the host cells differ between them [42].

The knowledge of coronavirus biological characteristics is the key to a better understanding of the origin, spreading, and infection caused by viruses of this family, as well as is crucial to the managing of the vaccines and therapy development and the strategy establishment to control the outbreak successfully. The molecular and genetic analysis of coronavirus strain reveals that the origin of the SARS-CoV, MERS-CoV, and SARS-CoV-2 is a zoonotic virus isolated from bats that are the reservoir hosts, having different intermediary hosts, before infecting the humans [43]. This fact is related to the different affinity of coronavirus S protein by the cellular receptor to the entrance of the virus into the host cell. The S protein is one of four structural proteins of SARS-CoV2, being the other three: envelope (E), membrane (M), and nucleocapsid (N) proteins (Figure 9.1). N protein comprehends the viral genome constituted by ribonucleic acid (RNA) (+) [44].

Because of the first contact between SARS-CoV-2 and host cell involving S protein capacity to attach to the cell receptor, its role is crucial to the virus lifecycle. Structurally, S protein is composed of two subunits: S1 in the N-terminal domain that covers the S2 in the C-terminal fusion domain (Figure 9.2). The RBD is located in the S1, so when RBD binds to the ACE2 receptor of the host cell, S protein is submitted to the protease action and is cleaved, exposing the S2. This subunit allows the fusion of SARS-CoV2 to the host cell membranes [41, 45].

In SARS-CoV-2, the contact site between S1 and S2 subunits has several flexibility degrees that provide two conformations to S protein in which either the RBD is hidden or RBD is available, named "RBD down" or "RBD up," respectively [45, 46] (Figure 9.2).

The conformations heterogeneity of S1 and S2 domains that result from mutations are a challenge to vaccine and drug development. Thus, studying the molecular structures in detail allows advancement in this sense. Moreover, this knowledge is essential to sensing technique development that enables a rapid and accurate diagnosis to help with the decision about which therapy is suitable, improving the prognosis of patients.

Another biological mechanism deserves attention about SARS-CoV-2 infection: the immune response. The SARS-CoV-2 S protein binding to the host cell activates the immune system, starting the process of the cellular and humoral response. The

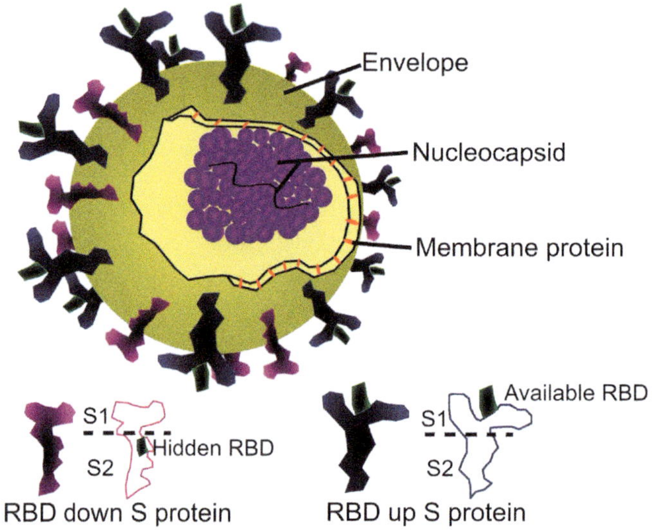

Figure 9.2 SARS-CoV-2 biological structure representation. SARS-CoV-2 is an enveloped virus comprehending two S protein conformations: RBD down (hidden RBD) and RBD up (available RBD). Inside of the envelope is the nucleocapsid formed by the capsid and single-stranded RNAs (+).

humoral response involves antibodies that attach to S protein and play the complement system to combat SARS-CoV-2 infection [47, 48].

9.3 Main Approach Used to COVID-19 Biosensing

The alert state of SARS-CoV-2 severity and its quick spreading caused worldwide researchers to search for solutions. Sensing SARS-CoV-2 is of great importance to proceed with other combat, therapy, and control strategies of the COVID-19.

To date, the gold standard assay to detect SARS-CoV-2 is the real-time reverse transcription-polymerase chain reaction (RT-PCR) method in collected samples of the upper airways with the nasopharyngeal swab. This COVID-19 biosensor is based on the SARS-CoV-2 genome, having single-stranded RNA loci as the target. So far, there is some commercial RT-PCR that uses different target genes, resulting in various sensitivity responses of the analysis [46, 49].

The biosensing sensitivity reflects directly on test accuracy, which means the correlation between true-positive and false-positive cases presents the potential to obtain a powerful diagnosis [50]. The aspects that can affect this accuracy concern some issues as the testing target, analysis, and response time, but the most important when it comes to the COVID-19 is the disease phase that the patient was tested [46].

Commonly, seroconversion in COVID-19 patients occurs after 15–20 days post-SARS-CoV-2 exposure but is known too that the seroconversion phase is highly variable in some individuals. The seroconversion phase increases antibodies, immunoglobulin M (IgM), or immunoglobulin G (IgG) levels in the body because of the response of the humoral and cellular immune system against SARS-CoV-2 [51]. In this case, it is necessary to apply serological tests in samples from venipuncture that perform the antibodies and immune cell detection to acquire the highest accuracy. The association between RT-PCR and the serological tests seems the better choice for successful diagnosis, even in asymptomatic cases [52–54].

To manage and suppress SARS-CoV-2 will likely require the rapid identification and isolation of infected individuals on an ongoing basis. RT-PCR tests are accurate but costly, making regular testing of every individual expensive. The costs are a challenge for all countries and particularly for developing countries [55]. Therefore, the development of inexpensive, rapid, selective, sensitive diagnostic systems for early-stage detection of COVID-19 is imperative to mass testing and facing the pandemic situation. Paper-based sensors can be a promising tool in this scenario.

9.4 Paper-Based Analytical Devices for COVID-19 Diagnostics

Because the SARS-CoV-2 outbreak pandemic started in the world, different strategies of COVID-19 diagnostics have been emerging with some similar goals, such as high sensitivity and selectivity, low cost, instrumental simplicity, and capability to use in point-of-care testing (POCT). In this context, as previously mentioned, PADs present as promising platforms to be used for COVID-19 diagnostics, taking account of all advantages offered by this substrate and the possibility to fabricate paper devices by using various techniques and also couple different detection methods. Taking into account that the COVID-19 pandemic is so recent, so far efforts are being made to develop and evaluate the applicability of LFIA tests based on NC membrane for coronavirus biosensing, once these tests allow the rapid and cost-effective detection of SARS-CoV-2-specific antibodies directly at point-of-need [41, 56–58].

In this section, we will present and discuss some recent advances in the development of LFIAs strip based on the use of the NC membrane to SARS-CoV-2 diagnosis. Li et al. [59] proposed a rapid IgM and IgG antibody test for SARS-CoV-2 infection diagnosis. The authors developed a rapid and simple POC LFIA that can detect IgM and IgG simultaneously against the SARS-CoV-2 virus in human blood samples within 15 minutes, which can detect patients at different infection stages. The test strip consists of five parts, including (i) plastic support, (ii) sample pad, (iii) conjugate pad, (iv) absorbent pad, and (v) NC membrane, as can be seen in Figure 9.3a. In summary, a total of three detection lines are on the strip. The control (C) line appears when the sample has flowed through the cartridge. The presence of anti-SARS-CoV-2 IgM and anti-SARS-CoV-2 IgG is indicated by a red/pink test line in the M and G region, respectively. However, if only the C line showed red, the sample is negative. It is important to highlight that if the C line does not appear red, the test is invalid and is necessary to repeat by using

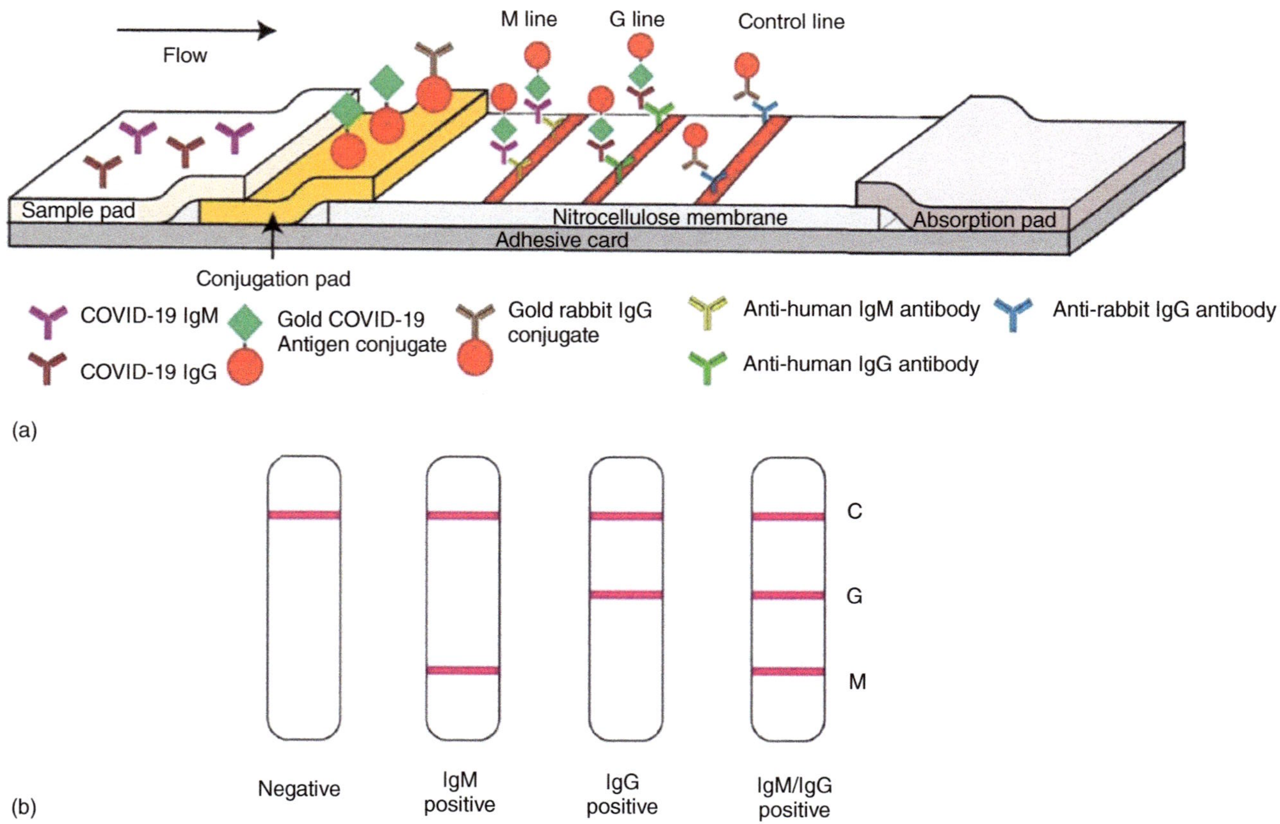

Figure 9.3 (a) Schematic illustration of rapid SARS-CoV-2 IgM–IgG combined antibody test developed. (b) Representation of different test results. Source: Li et al. [59]. Licensed under CC BY 4.0.

another cartridge. An illustration of different testing results can be observed in Figure 9.3b.

To demonstrate the feasibility of the proposed device, the authors tested 525 samples from multiple hospitals and Chinese CDC laboratories: 397 PCR confirmed COVID-19 patients and 128 negative patients. The overall testing sensitivity and specificity were 88.66% and 90.63%, respectively. It is important to emphasize that the authors evaluated clinical diagnosis results obtained from different types of venous and fingerstick blood samples. In summary, the authors demonstrated that the SARS-CoV-2 IgG–IgM combined antibody test through LFIA can be used as a rapid, simple, and reliable POCT for fast screening of SARS-CoV-2 infections and can also be performed by using fingerstick blood samples. Lastly, it is important to note that the mentioned test cannot confirm the virus presence, only provide evidence of recent infection, but it provides important immunological evidence for doctors to lead the correct diagnosis with other tests and thus start the patient treatment. The combination of the proposed test to nucleic acid RT-PCR can provide a more accurate SARS-CoV-2 infection diagnosis.

Chen et al. [60] developed an LFIA that uses lanthanide-doped polystyrene nanoparticles (LNPs) to detect anti-SARS-CoV-2 IgG in human samples. The goal of the mentioned study was to employ LNPs to overcome some problems related to fluorescence-based LFIAs, as an example of poor stability or quantum yield of the conventional fluorescent dyes. LFIA strips were prepared as previously reported by the same research group [61]. The authors dispensed a recombinant nucleocapsid phosphoprotein (NP) of SARS-CoV-2 onto an NC membrane to capture specific IgG. Mouse anti-human IgG (M-hIgG) antibody was labeled with self-assembled LNPs that served as a fluorescent reporter. The sample pad, conjugate pad, NC membrane, and absorbent pad were sequentially laminated together onto a backing card, as shown in Figure 9.4a. Briefly, serum samples were diluted with buffer and 100 μl of the diluted sample was added to the sample pad. As the sample flows, LNPs were captured at the test and control lines, as can be observed in Figure 9.4b. Finally, 10 minutes after the incubation period, the LFIA strip was put into the fluorescence reader. It is important to mention that the excitation of the functionalized LNPs produced a shiny fluorescent zone on the NC membrane and the results were expressed as the fluorescence peak area of the test line (At) and control line (Ac). Besides, the At/Ac ratio (R) was calculated to determine the concentration of anti-SARS-CoV-2 IgG in the serum sample.

The reproducibility of the LNP-based immunoassay was evaluated based on the coefficient of variation (CV) values measured of intra-assay and inter-assay. The CV values for intra-assay and inter-assay ranged from 7.71% to 9.69% and 11.51% to 14.63%, respectively. In this case, the authors considered that the proposed assay is reproducible, once all CV values were lower than 15%. To define the cutoff value of R, the authors tested 51 normal serum samples and 7 samples that were positively confirmed by RT-PCR. Based on the cutoff value estimated (0.0666), all the seven positive samples by RT-PCR had an R-value higher than 0.0666, thus indicating that the proposed LFIA can detect anti-SARS-CoV-2 IgG in positive samples. The feasibility of the assay was demonstrated in 7 positive samples and 12 negative

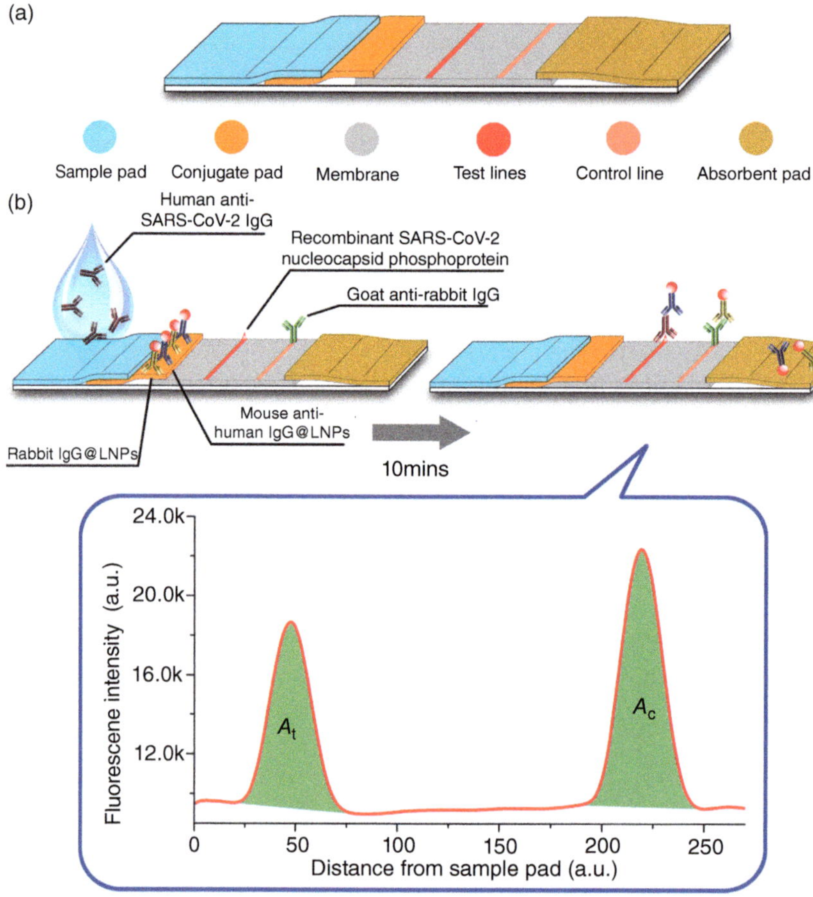

Figure 9.4 (a) Representation showing all consisted parts of the proposed LFIA strip containing LNPs. (b) Schematic illustration of the assay. Source: Chen et al. [60].

samples that were previously tested by RT-PCR. The data were statistically compared by using the McNemar test and no statistical difference was observed. In summary, the authors described a simple and rapid test using an LNP-based immunoassay for anti-SARS-CoV-2 IgG detection in human serum samples. It is important to highlight that the results of the validation tests meet the requirements for clinical diagnostics and the proposed LFIA can be used to monitor the COVID-19 progress and also the response to medical treatment.

Wen et al. [62] developed a serological POCT based on an LFIA strip for the rapid detection of the IgG antibody against the SARS-CoV-2 virus. The authors optimized some experimental parameters, such as coating antigen concentration, bovine serum albumin (BSA) blocking concentration, and pH for conjugation. The detection process took 15–20 minutes and required a low volume (10 μl) of the serum sample. Recombinant SARS-CoV-2 nucleocapsid protein was expressed and purified as previously published by the same research group [63] based on the recently reported SARS-CoV-2 sequence. The proposed strip consisted of four parts, including (i) a sample pad, (ii) a conjugated pad, (iii) an NC membrane, and (iv) an absorbent pad, and finally, all parts were placed over an adhesive polyvinyl chloride (PVC) substrate 4 mm width. The control line and test line on the NC membrane were marked with goat anti-mouse IgG polyclonal antibody and SARS-CoV-2 NP, respectively, with 4 mm of the distance between the two lines. Furthermore, the conjugated pad was pasted by overcrossing 2 mm with the NC membrane, the sample pad was also overcrossing 2 mm with the conjugated pad and the absorbent pad was placed on the opposite side of the strip. An illustration of the proposed LFIA strip and a mechanism of positive and negative response can be observed in Figure 9.5.

Aliquot of 80 μl diluted serum sample [10 μl serum and 70 μl Phosphate Buffer Saline (PBS)] was put on the sample pad and flows toward the absorbent pad. The presence of target proteins revealed a specific color tracer of AuNPs and the results were visible to the naked eye within 15 minutes. The authors evaluated the specificity of the LFIA by using human serum samples. It is important to emphasize that it used samples from patients previously diagnosed with COVID-19 and patients with severe fever with thrombocytopenia syndrome (SFTS) and avian influenza A(H7N9). The data demonstrated a good reproducibility and did not observe any cross-reactivity with other virus infections, such as SFTS and avian influenza A(H7N9), thus proving a great specificity of the proposed platform. The clinical feasibility was demonstrated using serum samples collected from 55 COVID-19 patients and the proposed LFIA presented a sensitivity of 69.1%. The data obtained from LFIA were compared by PCR, and the kappa statistic was calculated to measure the level of agreement between both methodologies. A value of 0.612 indicated a substantial agreement between the PCR and LFIA results. In summary, the authors reported the development, optimization, and validation of a POCT based on an LFIA strip IgG antibody detection for SARS-CoV-2 biosensing. Note that specific IgG detection is a complementary strategy to the molecular assays that can be helpful by doctors to make the correct diagnosis associated with other methods and thus provide suitable treatment to patients.

Gold nanoparticles (AuNPs) were also used as a labeling probe to develop an LFIA biosensor for rapid and simultaneous detection of IgM and IgG antibodies for SARS-CoV-2 diagnosis in blood samples [64]. The proposed biosensor contains five components, including a PVC backing card, an NC membrane, a sample pad, a conjugate pad, and an absorption pad. The assembly of the LFIA is illustrated in Figure 9.6. The running protocol is similar to other LFIA mentioned in this chapter. Briefly, the sample pad was pretreated with 0.01 M PBS (containing 0.5% BSA) and dried overnight at 37 °C. The AuNP-labeled rabbit IgG antibody and AuNP-labeled RBD were diluted with suspension buffer, sprinkled onto the conjugate pad, and dried at 37 °C for one hour. Furthermore, in the NC membrane were fixed one control line (C line) and two detection lines (G and M lines), being that anti-rabbit IgG antibody was immobilized on the C line, and anti-human IgM antibody and anti-human IgG antibody against SARS-CoV-2 were separately fixed on the M and G lines, respectively. Finally, the sample pad, conjugate pad, NC

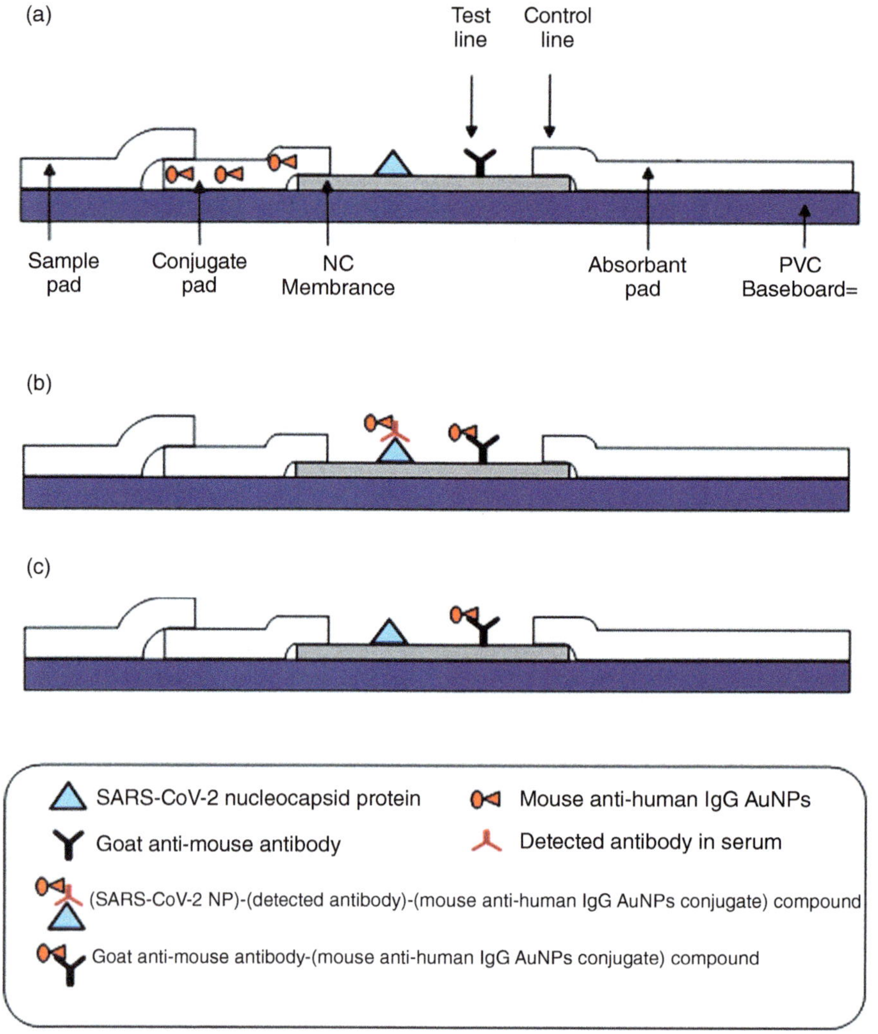

Figure 9.5 (a) Structure of SARS-CoV-2 IgG LFIA strip, (b) positive detection of SARS-CoV-2 IgG LFIA strip, and (c) negative detection of SARS-CoV-2 IgG LFIA strip. Source: Wen et al. [62].

membrane, and absorption pad were together laminated and pasted on the PVC backing card.

Forty blood serum samples were analyzed using a combined IgG–IgM AuNPs-LFIA strip and were possible to demonstrate the feasibility of the referred device for IgG, IgM, or both detection. To demonstrate the sensitivity and specificity of the referred strip, 80 serum samples were used, including 34 from SARS-CoV-2-positive patients, 6 from suspected cases, and 40 negative subjects (all tested by real-time RT-PCR). The data for combined IgM–IgG detection revealed 85.29% sensitivity and 100.00% specificity. In summary, a highly sensitive and

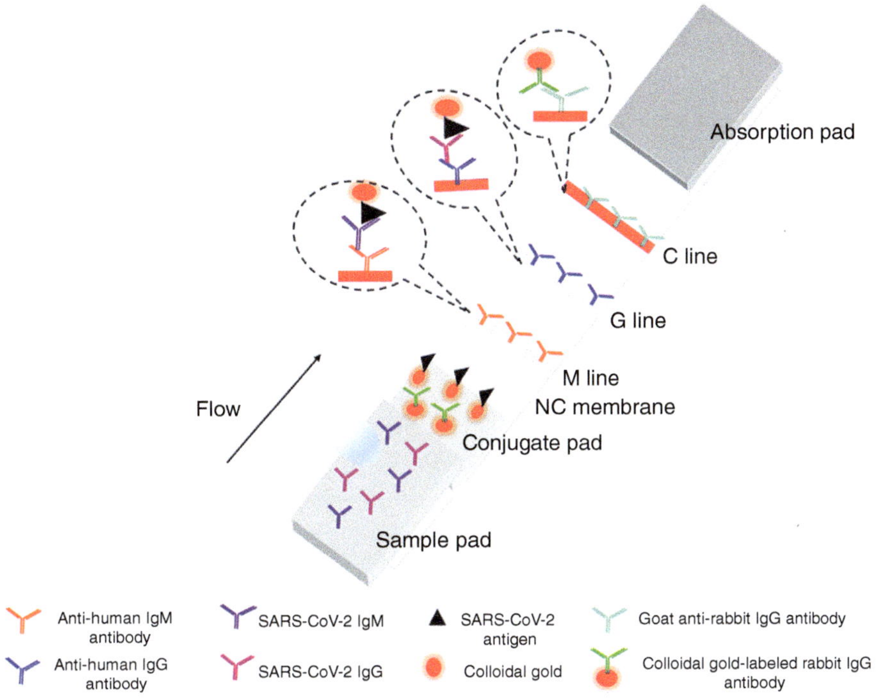

Figure 9.6 Schematic representation of the IgG–IgM combined AuNPs-LFIA strip for SARS-CoV-2 diagnosis. Source: Zeng et al. [64].

specific biosensor was proposed based on the AuNP-LFIA strip for combined detection of IgG–IgM in samples from confirmed COVID-19 patients as well as suspected subjects to perform a rapid screening of symptomatic and asymptomatic subjects. However, the authors highlighted that due to limited access to the patient's clinical information, it is necessary to perform a statistical analysis of the data and evaluate the correlation between the disease course and other diagnostic criteria.

In the same way, Wang et al. [65] employed selenium nanoparticles (SeNPs) as a labeling probe and couple with SARS-CoV-2 nucleoprotein to develop an LFIA as a rapid detection kit for anti-SARS-CoV-2 IgM and IgG. The goal of the mentioned study was to demonstrate the potential of the SeNPs as a labeling probe for this kind of application, once they present some advantages when compared with other probe types, such as higher levels of sensitivity, are not sensitive to electrolytes, and can be obtained at room temperature using a more economical procedure. The proposed LFIA consisted of a backing, a sample pad, a conjugate pad, a reaction pad, and an absorbent pad. An anti-human IgM antibody or IgG antibody was added on the test line and the anti-His antibody was added on the control line. The sample pad was previously immersed in a prepared solution (10 mM PBS, pH = 7.4, containing 0.05% Tween 20, and 5% serum sample) and the Se-NPs-SARS-CoV-2 nucleoprotein complex was immobilized in the conjugate pad.

Briefly, in the running protocol, plasma samples were added to the sample pad and then flowed to the conjugated pad to redissolve the Se-NP-labeled protein complex. Afterward, because of the capillary properties, the Se-NP-labeled nucleoprotein migrated with the sample toward the absorbent pad in the NC membrane. When the sample contains anti-SARS-CoV-2 IgM, the IgM conjugates to the Se-NP-labeled nucleoprotein and flows to the test line position, and then the IgM binds to the anti-human IgM antibody, revealing an orange color (visible to the naked eye) at the test line within five minutes. Therefore, when both lines (control and test) reveal color, the result is considered positive. On the other hand, when no color is revealed at the test line, the result is considered negative. This highlights that if the control line does not develop color, the result is considered invalid, and thus it is necessary to repeat the test using another LFIA platform. For the anti-SARS-CoV-2 IgG test, the anti-human IgG antibody was added at the position of the test strip, and the other conditions were the same used for IgM. In this case, the interpretation of IgG results is performed in a similar way for IgM. Clinical sample validation was demonstrated by using 60 plasma samples, being 41 negative and 19 positive (all previously clinically diagnosed by two specific local hospitals). The proposed kit presented a sensitivity of 94.74% and a specificity of 95.12%. In summary, the results are simply evaluated by the naked eye and the kit is suitable for use in POCT. However, note that the mentioned study corresponds to a preliminary development because of the low number of positive samples tested. Therefore, it is necessary to expand the sample size tested to validate the overall sensitivity and specificity of the proposed Se-NP-LFIA kit. Furthermore, it is also important to evaluate if the results can be affected by major patients factors, such as age and stage of COVID-19.

A nanozyme-based chemiluminescence paper test was developed for the rapid diagnosis of SARS-CoV-2 [66]. In this work, the authors proposed the use of Co–Fe@hemin nanozyme as a suitable alternative to horseradish peroxidase (HRP), thus combining traditional chemiluminescence immunoassay (CLIA) with lateral flow assay. Nanozymes are nanomaterials that mimic an enzymatic activity and thus can be used for enzyme-labeled probes. Co–Fe nanoparticles (Co–Fe NPs) with carboxyl modification were synthesized through the hydrothermal method, and in the last step, hemin (with a mass concentration ratio of 2.5 : 1) was dropwise added into the reaction system. Afterward, the nanocomposites were purified to obtain the Co–Fe@hemin. The nanozyme-based chemiluminescence consisted of a PVC backing card, an absorbing pad, a sample pad, an NC membrane, and a conjugate pad. Initially, the detection antibody of S-RBD (S-dAb)-conjugated nanozyme chemiluminescent probes was prepared by chemical coupling of carboxyl groups using 1-ethyl-3-(3-dimethylaminopropyl)carbodiimide/N-hydroxysuccinimide (EDC/NHS) agents. Then, the nanozyme probes were dispensed on the pretreated conjugate pad and the paired capture antibody of S-RBD (S-cAb) was immobilized on an NC membrane to form the test line (T-line), as well as the anti-human IgG antibody as the control line (C-line). Finally, the conjugate pad, NC membrane, sample pad, and absorbing pad were assembled on the PVC backing card, and paper boards were cut into 4 mm-wide strips.

The rapid testing using the proposed paper sensor was performed by the addition of 100 μl of a sample containing S-RBD protein on the sample pad. Then, the sample flows through the conjugate pad, and the target analytes were recognized specifically by nanozyme probes to form immunocomplexes. Afterward, nanozyme complexes were captured by S-dAb and aggregated at T-line, producing a specific brown color signal. On the other hand, nanozyme probes are uncombined with an antigen bound with the control IgG antibody at the C-line. Finally, a mixture consisted of luminol substrate, and the excitant was added onto the NC membrane after 15 minutes. The chemiluminescent signals of T-line and C-line were captured by using a smartphone camera or a CCD system, and then, the chemiluminescent signal intensity was analyzed using the Image-Pro Plus software. The entire detection process is completed within 16 minutes.

The nanozyme-based chemiluminescence paper test presented similar sensitivity when compared to ELISA; however, the proposed platform presents a linear range 32-fold wider compared to ELISA tests. The authors highlighted that the referred chemiluminescence paper test specifically recognized the SARS-CoV-2 S-RBD protein. Furthermore, to demonstrate the validity of the proposed paper test in actual viral infection, pseudo-SARS-CoV-2 expressing spike protein (S1 subunit) was used. In summary, the nanozyme-based chemiluminescence paper test combines the high sensitivity of chemiluminescence, high specificity of immunoassay, and short testing time. Additionally, because of the good analytical performance, remarkable portability, and low cost, the referred analytical device should provide a novel alternative for POCT aiming for early screening of SARS-CoV-2 infections.

A half-strip lateral flow assay for SARS-CoV-2 coronavirus nucleocapsid antigen detection using commercially available reagents was reported by Grant et al. [28]. It is important to highlight that half-strip LFIA is frequently used as the first step in the assay development for a "full" LFIA and has only an NC membrane and a wicking pad, without a conjugated pad or sample pad. In this type of LFIA, sample and conjugated are premixed in a recipient such as a 96-well plate before the use of the half strip. Briefly, the authors prepared all required chemical solutions using commercially available reagents, including antibodies for the detection of SARS-CoV-2. An NC membrane was striped with a test line and a control line, once the test line was striped at 1 mg/ml polystreptavidin and the control line was striped at 0.5 mg/ml goat anti-Chicken Immunoglobulin Y (IgY). Afterward, NC strips were blocked with a specific blocking solution and then were placed on a backing card. Finally, 20 mm Ahlstrom 320 was placed on top of the NC strip with a distance of 3 mm between both materials, and the excess of the backing card was cut off, resulting thus in an NC strip with 4 mm width, as can be seen in Figure 9.7.

For the LFIA running protocol, two commercial nucleocapsids (N) protein samples (from Genemedi and Genscript) were used. The control protein and nucleocapsid antibody conjugates were mixed as described in the referred manuscript. Thus, 75 μl of the mix was added to a 96-well plate and an LFIA was immersed. Twenty minutes after the immersion, LFIA was removed and read using an optical LFIA reader to construct the dose curve response for each commercially available SARS-CoV-2 nucleocapsid protein. The estimated limit of detection (LOD) from

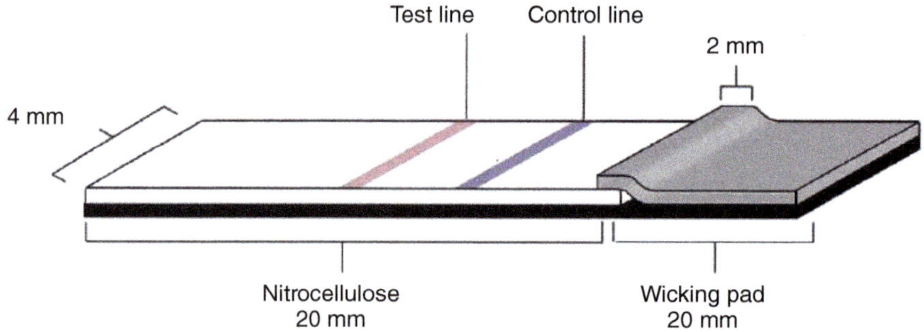

Figure 9.7 Half strip fabricated with a 20 mm NC membrane. Once is a half strip, no sample or conjugated pad is included on the strip. Source: Weiss and Navas-Martin [28].

Genemid N protein was 0.65 ng/ml and the Genscript N protein was 3.03 ng/ml. In summary, the authors reported a simple and interesting half-strip LFIA as a possible platform for SARS-CoV-2 detection using commercially available reagents, such as nucleocapsid protein samples. Note that none of the sample or conjugated pad was put on the half strip, once these materials were mixed in a specific recipient before the use of the proposed LFIA. Taking into account the use of the proposed half strip as a POC testing for SARS-CoV-2 detection, it is necessary to perform a study about the analytical sensitivity in either blood or nasal samples to meet an acceptable level of clinical sensitivity for SARS-CoV-2. Finally, the antibodies used in the mentioned assay are polyclonal, so great attention is required in further investigations to eliminate/prevent any cross-contamination with other active circulating coronaviruses and viruses with similar nucleocapsid proteins.

In this context, Liu et al. [67] developed a quantum dot lateral flow immunoassay (QD-LFIA) strip combined with a portable fluorescence smartphone to detect IgM/IgG for SARS-CoV-2 diagnosis. The proposed QD-LFIA strip was composed of a 25 mm sample pad, 20 mm NC membrane, and a 15 mm absorbent pad, and then all of those were pasted on an adhesive backing card. The recombinant RBD of SARS-CoV-2 spike protein (S-RBD) was used as an antigen to detect specific IgM and the mixture of recombinant nucleocapsid phosphoprotein (NP) and S-RBD were used to detect specific IgG, respectively. The authors used polystyrene-coated QD nanoparticles with 160 nm diameter and conjugated with M-hIgG or S-RBD. The M-hIgG@QD and S-RBD@QD conjugates were mixed and put onto the conjugate pad. The goat anti-mouse polyclonal immunoglobulin G (G-mIgG), a mixture of M-hIgG and NP, or mouse anti-human immunoglobulin M (M-hIgM) was spotted on the NC membrane to generate the C, T2, and T1 lines, respectively.

To test, firstly, 1 μl of human serum sample was mixed with 120 μl of PBS buffer (10 mM, pH = 7.4 containing 1% BSA and 0.05% Tween-20). Afterward, the solution was added to the sample pad and then the fluorescence signals were measured within 10 minutes by using a 3D-printed portable fluorescence strip reader. Finally, the signals were uploaded to the smartphone via WI-FI for data analysis. It is important to cite that the fluorescence lines were selected and the intensity of T1, T2, or C lines was converted to the peak area. The ratio value of IgG/control (T2/C) or IgM/control (T1/C) was calculated, and then, the response was reported as negativity or positivity after comparing it with the cutoff value by artificial intelligence (AI) processor.

The proof-of-concept of the developed QD-LFIA was demonstrated using 100 serum samples of COVID-19 patients (in different stages) and 450 plasma samples from healthy donors. Furthermore, a commercial colloidal gold lateral flow immunoassay strip (CG-LFIA) was used as a comparative assay. In comparison with CG-LFIA, the proposed QD-LFIA presented a better sensitivity (10–100-fold) for testing of anti-SARS-CoV-2 IgM and IgG, once the detection rate for IgM was 78% by QD-LFIA and 32% using CG-LFIA and for IgG was 99% using QD-LFIA and 71% by CG-LFIA. The specificity of QD-LFIA was equal to colloidal gold lateral flow immunoassay (CG-FIA), with values of 100% for IgM and 99.8% for IgG. In summary, the proposed QD-LFIA strip associated with a portable fluorescence smartphone system presented as a promising platform to be used for SARS-CoV-2 diagnostic with great analytical reliability.

Additionally, LFIA tests can be used to accompany and validate the development of SARS-CoV-2 assays. A CRISPR–Cas12-based assay was proposed for COVID-19 diagnosis and employed the use of LFIA strips to visualize the results from respiratory swab RNA extracts [36]. The developed assay is called SARS-CoV-2 DNA Endonuclease-Targeted CRISPR Trans Reporter (DETECTR). In summary, the mentioned assay performed simultaneous reverse transcription loop-mediated isothermal amplification (RT-LAMP) for RNA extracted from nasopharyngeal or oropharyngeal swabs, followed by Cas12 detection of predefined coronavirus sequences, after which cleavage of a reporter molecule confirms the virus detection. Optimization of experimental conditions was performed for the SARS-CoV-2 DETECTR assay on the envelope (E) and nucleoprotein (N) gene, and human RNase P gene as a control. Firstly, the RT-LAMP reaction was accomplished at 62 °C for 20–30 minutes and the Cas12 detection reaction at 37 °C for 10 minutes. Finally, the results were visualized by using an LFIA strip test. Therefore, all steps involved in the proposed DETECTR assay can be run in approximately 30–40 minutes.

The feasibility of the proposed technique was demonstrated using 11 extracted RNA samples collected from respiratory swab (including 6 from PCR-positive COVID-19 patients) and 12 nasopharyngeal swab samples from patients with influenza ($n = 5$) and common human seasonal coronavirus infections ($n = 7$). It is important to note that SARS-CoV-2 was detected in 9 of 11 patient swabs and the device did not present cross-reactivity with other respiratory viruses. Furthermore, two negative swabs from COVID-19 patients were confirmed to be below the established LOD. In this way, the proposed DETECTR assay presented comparable accuracy to quantitative reverse transcription polymerase chain reaction (qRT-PCR) and presents some advantages over qRT-PCR, such as single nucleotide target specificity, integration with accessible and easy-to-use LFIA strips, and not requiring sophisticated instrumentation.

9.5 Challenges and Perspectives

The current SARS-CoV-2 pandemic enforces the globe to immediately find opportunities and countermeasures in several fields [68]. From the diagnostic and monitoring techniques point of view and taking into account all discussion reported in this chapter, it is clear that to overcome the COVID-19 pandemic, it is necessary to develop rapid, reliable, and sensitive novel biosensors for the mentioned virus infection, which would be a single-step identification or sensing method that eliminates separation (extraction of nucleic acid), incubation, or use of any signal-reporting agents [69, 70]. As mentioned herein, immunoassays and antigen detection methods may play an outstanding role in the containment of the disease. Their power in early diagnosis is limited, as it has been shown that specific antibodies could be revealed only after a few days from the disease onset by the commercially available serological tests. In summary, the globe is in a constant and exhaustive run to develop strategies for SARS-CoV-2 diagnostics, and thus, a lot of challenges are expected in the research field. For this reason, in this special and critical moment of humanity, it is extremely necessary to join efforts to overcome this current pandemic [35, 71].

In this context, the global fight against novel coronavirus has encouraged many researchers to exploit every moment to develop and fabricate on-site, POC, easily portable, and affordable tests [72]. Moreover, the use of biosensors would represent one of the best choices against the actual emergency, allowing cheap and noninvasive monitoring systems, with fast response and, above all, a low background noise [73–76]. In this way, bio/nanomaterials can be used to modify the paper surface to obtain more specific, reliable, and robustness analytical devices. QDs [77], graphene [78], and molecularly imprinted polymers [79, 80] presented as promising materials that can be employed in the development of biosensors against COVID-19. Besides, wearable sensors are a class of analytical platforms that presents many advantages and can also be used focusing the same goal: obtain portable, reliable, effective, sensitive, and low-cost strategies to do remote patient monitoring and virtual assessments [81].

Given the recent outbreak of SARS-CoV-2, we believe that several other studies and devices using PADs will be demonstrated shortly. Therefore, we envision that the paper-based devices will fulfill the gap in portable and accessible POC diagnostics and help to face the COVID-19 pandemic. Additionally, the knowledge and tools developed in this critical situation will help prevent or minimize future challenges to global health.

Acknowledgments

We would like to acknowledge the Brazilian agencies CAPES (grant number: 88887.479793/2020-00), FAEPEX/UNICAMP (grant number: 3374/19), FAPESP (Grant Number: 2018/08782-1), and CNPq (grant number: 438828/2018-6) for supporting the research.

References

1 Cheng, C.M., Martinez, A.W., Gong, J. et al. (2010). Paper-based ELISA. *Angew. Chem. Int. Ed.* 49 (28): 4771–4774.

2 de Araujo, W.R., Cardoso, T.M.G., da Rocha, R.G. et al. (2018). Portable analytical platforms for forensic chemistry: a review. *Anal. Chim. Acta* 1034: 1–21.

3 Eltzov, E., Guttel, S., Low Yuen Kei, A. et al. (2015). Lateral flow immunoassays – from paper strip to smartphone technology. *Electroanalysis* 27 (9): 2116–2130.

4 Carrell, C., Kava, A., Nguyen, M. et al. (2019). Beyond the lateral flow assay: a review of paper-based microfluidics. *Microelectron. Eng.* 206 (January): 45–54.

5 Yagoda, H. (1937). Applications of confined spot tests in analytical chemistry: preliminary paper. *Ind. Eng. Chem. Anal. Ed.* 9 (2): 79–82.

6 Müller, R.H. and Clegg, D.L. (1949). Automatic paper chromatography. *Anal. Chem.* 21 (9): 1123–1125.

7 Martinez, A.W., Phillips, S.T., Butte, M.J., and Whitesides, G.M. (2007). Patterned paper as a platform for inexpensive, low-volume, portable bioassays. *Angew. Chem.* 119 (8): 1340–1342.

8 Salentijn, G.I.J., Grajewski, M., and Verpoorte, E. (2018). Reinventing (bio)chemical analysis with paper. *Anal. Chem.* 90 (23): 13815–13825.

9 Xu, Y., Liu, M., Kong, N., and Liu, J. (2016). Lab-on-paper micro- and nano-analytical devices: fabrication, modification, detection and emerging applications. *Microchim. Acta* 183 (5): 1521–1542.

10 Lin, Y., Gritsenko, D., Liu, Q. et al. (2016). Recent advancements in functionalized paper-based electronics. *ACS Appl. Mater. Interfaces* 8 (32): 20501–20515.

11 Wang, M., Wang, Y., Gao, B. et al. (2019). Fast strategy to functional paper surfaces. *ACS Appl. Mater. Interfaces* 11 (15): 14445–14456.

12 Mabey, D., Peeling, R.W., Ustianowski, A., and Perkins, M.D. (2004). Diagnostics for the developing world. *Nat. Rev. Microbiol.* 2 (3): 231–240.

13 Ho, C.M.B., Ng, S.H., Li, K.H.H., and Yoon, Y.J. (2015). 3D printed microfluidics for biological applications. *Lab Chip* 15 (18): 3627–3637.

14 Waldbaur, A., Rapp, H., Länge, K., and Rapp, B.E. (2011). Let there be chip – towards rapid prototyping of microfluidic devices: one-step manufacturing processes. *Anal. Methods* 3 (12): 2681–2716.

15 Gaal, G., Mendes, M., de Almeida, T.P. et al. (2017). Simplified fabrication of integrated microfluidic devices using fused deposition modeling 3D printing. *Sens. Actuators, B* 242: 35–40.

16 Jiang, X. and Fan, Z.H. (2016). Fabrication and operation of paper-based analytical devices. *Annu. Rev. Anal. Chem.* 9: 203–222.

17 Xia, Y., Si, J., and Li, Z. (2016). Fabrication techniques for microfluidic paper-based analytical devices and their applications for biological testing: a review. *Biosens. Bioelectron.* 77: 774–789.

18 Ataide, V.N., Mendes, L.F., Gama, L.I.L.M. et al. (2020). Electrochemical paper-based analytical devices: ten years of development. *Anal. Methods* 12 (8): 1030–1054.

19 Akyazi, T., Basabe-Desmonts, L., and Benito-Lopez, F. (2018). Review on microfluidic paper-based analytical devices towards commercialisation. *Anal. Chim. Acta* 1001: 1–17.

20 Peeling, R.W. and Mabey, D. (2010). Point-of-care tests for diagnosing infections in the developing world. *Clin. Microbiol. Infect.* 16 (8): 1062–1069.

21 Dungchai, W., Chailapakul, O., and Henry, C.S. (2009). Electrochemical detection for paper-based microfluidics. *Anal. Chem.* 81 (14): 5821–5826.

22 Noviana, E., McCord, C.P., Clark, K.M. et al. (2020). Electrochemical paper-based devices: sensing approaches and progress toward practical applications. *Lab Chip* 20 (1): 9–34.

23 Rattanarat, P., Dungchai, W., Cate, D. et al. (2014). Multilayer paper-based device for colorimetric and electrochemical quantification of metals. *Anal. Chem.* 86 (7): 3555–3562.

24 Ameku, W.A., De Araujo, W.R., Rangel, C.J. et al. (2019). Gold nanoparticle paper-based dual-detection device for forensics applications. *ACS Appl. Nano Mater.* 2 (9): 5460–5468.

25 Fu, L.M. and Wang, Y.N. (2018). Detection methods and applications of microfluidic paper-based analytical devices. *TrAC, Trends Anal. Chem.* 107: 196–211.

26 Lopez-Ruiz, N., Curto, V.F., Erenas, M.M. et al. (2014). Smartphone-based simultaneous pH and nitrite colorimetric determination for paper microfluidic devices. *Anal. Chem.* 86 (19): 9554–9562.

27 Cheever, F.S. and Daniels, J.B. (1949). A murine virus (JHM) causing disseminated encephalomyelitis with extensive destruction of myelin. *J. Exp. Med.* 90 (3): 181–210.

28 Grant, B.D., Anderson, C.E., Williford, J.R. et al. (2020). SARS-CoV-2 Coronavirus nucleocapsid antigen-detecting half-strip lateral flow assay toward the development of point of care tests using commercially available reagents. *Anal. Chem.* 92 (16): 11305–11309.

29 Borremans, B., Gamble, A., Prager, K. et al. (2020). Quantifying antibody kinetics and RNA shedding during early-phase SARS-CoV-2 infection. *Arch. Virol.* 162: 1441–1446.

30 Huang, C., Wang, Y., Li, X. et al. (2020). Clinical features of patients infected with 2019 novel coronavirus in Wuhan, China. *Lancet* 395 (10223): 497–506.

31 Cavanagh, D., Brien, D.A., Brinton, M. et al. (1994). Revision of the taxonomy of the coronavirus, torovirus and arterivirus genera. *Arch. Virol.* 135 (1–2): 227–237.

32 Kuhn, J.H. (2020). Virus taxonomy portal. *Ref. Modul. Life Sci.* https://linkinghub.elsevier.com/retrieve/pii/B9780128096338212314.

33 Kuhn, J.H., Wolf, Y.I., Krupovic, M. et al. (2019). Classify viruses – the gain is worth the pain. *Nature* 566 (7744): 318–320.

34 Chen, X., Zhao, B., Qu, Y. et al. (2020). Detectable serum SARS-CoV-2 viral load (RNAaemia) is closely correlated with drastically elevated interleukin 6 (IL-6) level in critically ill COVID-19 patients. *Clin. Infect. Dis.* 303 (5664): 1666–1669.

35 Guan, Y., Zheng, B.J., He, Y.Q. et al. (2003). Isolation and characterization of viruses related to the SARS coronavirus from animals in southern China. *Science*. 302 (5643): 276–278.

36 Peiris, J.S.M., Guan, Y., and Yuen, K.Y. (2004). Severe acute respiratory syndrome. *Nat. Med.* 10 (12S): S88–S97.

37 Leung, W.K., To, K.F., Chan, P.K.S. et al. (2003). Enteric involvement of severe acute respiratory syndrome – associated coronavirus infection. *Gastroenterology* 125 (4): 1011–1017.

38 Hui, D.S., Azhar, E.I., Kim, Y.J. et al. (2018). Middle East respiratory syndrome coronavirus: risk factors and determinants of primary, household, and nosocomial transmission. *Lancet Infect. Dis.* 18 (8): 217–227.

39 Drosten, C., Seilmaier, M., Corman, V.M. et al. (2013). Clinical features and virological analysis of a case of Middle East respiratory syndrome coronavirus infection. *Lancet Infect. Dis.* 13 (9): 745–751.

40 Song, Z., Xu, Y., Bao, L. et al. (2019). From SARS to MERS, thrusting coronaviruses into the spotlight. *Viruses* 11 (1): 59.

41 Chen, X., Zhao, B., Qu, Y. et al. (2020). Detectable serum SARS-CoV-2 viral load (RNAaemia) is closely correlated with drastically elevated interleukin 6 (IL-6) level in critically ill COVID-19 patients. *Clin. Infect. Dis.* 181 (2): 271–280.

42 Gorbalenya, A.E., Baker, S.C., Baric, R.S. et al. (2020). The species severe acute respiratory syndrome-related coronavirus: classifying 2019-nCoV and naming it SARS-CoV-2. *Nat. Microbiol.* 5 (4): 536–544.

43 Andersen, K.G., Rambaut, A., Lipkin, W.I. et al. (2020). The proximal origin of SARS-CoV-2. *Nat. Med.* 26 (4): 450–452.

44 Kim, D., Lee, J.Y., Yang, J.S. et al. (2020). The architecture of SARS-CoV-2 transcriptome. *Cell* 181 (4): 914–921. e10.

45 Henderson, R., Edwards, R., Mansouri, K. et al. (2020). Controlling the SARS-CoV-2 spike glycoprotein conformation. *bioRxiv* 18: 1–28.

46 Lieberman, J.A., Pepper, G., Naccache, S.N. et al. (2020). Comparison of commercially available and laboratory-developed assays for in vitro detection of SARS-CoV-2 in clinical laboratories. *J. Clin. Microbiol.* 58 (8): 1–6.

47 Li, G., Fan, Y., Lai, Y. et al. (2020). Coronavirus infections and immune responses. *J. Med. Virol.* 92 (4): 424–432.

48 Lan, J., Ge, J., Yu, J. et al. (2020). Structure of the SARS-CoV-2 spike receptor-binding domain bound to the ACE2 receptor. *Nature* 581 (7807): 215–220.

49 Lytras, T., Panagiotakopoulos, G., and Tsiodras, S. (2020). Estimating the ascertainment rate of SARS-CoV-2 infection in Wuhan, China: implications for management of the global outbreak. *Clin. Chem.* 66 (7): 977–979.

50 Bryan, A., Pepper, G., Wener, M.H. et al. (2020). Performance characteristics of the Abbott architect SARS-CoV-2 IgG assay and seroprevalence in Boise, Idaho. *J. Clin. Microbiol.* 58 (8): 1–19.

51 Lou, B., Li, T.D., Zheng, S.F. et al. (2020). Serology characteristics of SARS-CoV-2 infection since exposure and post symptom onset. *Eur. Respir. J.* 56 (2): 1–35.

52 Martin, C., Montesinos, I., Dauby, N. et al. (2020). Dynamics of SARS-CoV-2 RT-PCR positivity and seroprevalence among high-risk healthcare workers and hospital staff. *J. Hosp. Infect.* 106 (1): 102–106.

53 Wang, P. (2020). Combination of serological total antibody and RT-PCR test for detection of SARS-CoV-2 infections. *J. Virol. Methods* 283: 113919.

54 Long, Q.X., Liu, B.Z., Deng, H.J. et al. (2020). Antibody responses to SARS-CoV-2 in patients with COVID-19. *Nat. Med.* 26 (6): 845–848.

55 Mutesa, L., Ndishimye, P., Butera, Y. et al. (2020). A pooled testing strategy for identifying SARS-CoV-2 at low prevalence. *Nature* 52: 1–4.

56 Candel González, F.J., Viñuela-Prieto, J.M., Del Castillo, J.G. et al. (2020). Utility of lateral flow tests in SARS-CoV-2 infection monitorization. *Rev. Esp. Quimioter.* 33: 258–266.

57 Bernasconi, L., Oberle, M., Gisler, V. et al. (2020). Diagnostic performance of a SARS-CoV-2 IgG/IgM lateral flow immunochromatography assay in symptomatic patients presenting to the emergency department. *Clin. Chem. Lab. Med.* 58 (9): 159–161.

58 Rudolf, F., Kaltenbach, H.-M., Linnik, J. et al. (2020). Clinical characterisation of eleven lateral flow assays for detection of COVID-19 antibodies in a population. *medRxiv* 25: 1–19.

59 Li, Z., Yi, Y., Luo, X. et al. (2020). Development and clinical application of a rapid IgM–IgG combined antibody test for SARS-CoV-2 infection diagnosis. *J. Med. Virol.* 92 (9): 1518–1524.

60 Chen, Z., Zhang, Z., Zhai, X. et al. (2020). Rapid and sensitive detection of anti-SARS-CoV-2 IgG, using lanthanide-doped nanoparticles-based lateral flow immunoassay. *Anal. Chem.* 92 (10): 7226–7231.

61 Chen, Z., Liang, R., Guo, X. et al. (2017). Simultaneous quantitation of cytokeratin-19 fragment and carcinoembryonic antigen in human serum via quantum dot-doped nanoparticles. *Biosens. Bioelectron.* 91 (October 2016): 60–65.

62 Wen, T., Huang, C., Shi, F.J. et al. (2020). Development of a lateral flow immunoassay strip for rapid detection of IgG antibody against SARS-CoV-2 virus. *Analyst* 145 (15): 5345–5352.

63 Jiao, Y., Zeng, X., Guo, X. et al. (2012). Preparation and evaluation of recombinant severe fever with thrombocytopenia syndrome virus nucleocapsid protein for detection of total antibodies in human and animal sera by double-antigen sandwich enzyme-linked immunosorbent assay. *J. Clin. Microbiol.* 50 (2): 372–377.

64 Zeng, L., Li, Y., Liu, J. et al. (2020). Rapid, ultrasensitive and highly specific biosensor for the diagnosis of SARS-CoV-2 in clinical blood samples. *Mater. Chem. Front.* 4: 2000–2005.

65 Wang, Z.-Z., Zheng, Z., Wang, X.-C. et al. (2020). Rapid detection of anti-SARS-CoV-2 IgM and IgG using a selenium nanoparticle-based lateral flow immunoassay. *Res. Sq.*: 1–18.

66 Liu, D., Ju, C., Han, C. et al. (2020). Ultra-sensitive nanozyme-based chemiluminescence paper test for rapid diagnosis of SARS-CoV-2 infection. *bioRxiv* 28: 1–22.

67 Liu, B., Li, J., Tang, X. et al. (2020). Development of a quantum-dot lateral flow immunoassay strip based portable fluorescence smart-phone system for ultrasensitive detection of IgM/IgG to SARS-CoV-2. *medRxiv* 24: 1–18.

68 Ji, T., Liu, Z., Wang, G.Q. et al. (2020). Detection of COVID-19: a review of the current literature and future perspectives. *Biosens. Bioelectron.* 166 (March): 112455.

69 Feng, W., Newbigging, A.M., Le, C. et al. (2020). Molecular diagnosis of COVID-19: challenges and research needs. *Anal. Chem.* 92 (15): 10196–10209.

70 Santiago, I. (2020). Trends and innovations in biosensors for COVID-19 mass testing. *ChemBioChem* 21: 1–11.

71 Gupta, A., Kumar, S., Kumar, R. et al. (2020). COVID-19: emergence of infectious diseases, nanotechnology aspects, challenges, and future perspectives. *ChemistrySelect* 5 (25): 7521–7533.

72 Hussein, H.A., Hassan, R.Y.A., Chino, M., and Febbraio, F. (2020). Point-of-care diagnostics of COVID-19: from current work to future perspectives. *Sensors (Switzerland)* 20 (15): 1–28.

73 Bhalla, N., Pan, Y., Yang, Z., and Payam, A.F. (2020). Opportunities and challenges for biosensors and nanoscale analytical tools for pandemics: COVID-19. *ACS Nano* 14 (7): 7783–7807.

74 Moulahoum, H., Ghorbanizamani, F., Zihnioglu, F. et al. (2021). How should diagnostic kits development adapt quickly in COVID 19-like pandemic models? Pros and cons of sensory platforms used in COVID-19 sensing. *Talanta* 222: 1–11.

75 Samson, R., Navale, G.R., and Dharne, M.S. (2020). Biosensors: frontiers in rapid detection of COVID-19. *3 Biotech* 10: 1–9.

76 Cui, F. and Zhou, H.S. (2020). Diagnostic methods and potential portable biosensors for coronavirus disease 2019. *Biosens. Bioelectron.* 165 (May): 112349.

77 Manivannan, S. and Ponnuchamy, K. (2020). Quantum dots as a promising agent to combat COVID-19. *Appl. Organomet. Chem.* 34 (10): 17–22.

78 Palmieri, V. and Papi, M. (2020). Can graphene take part in the fight against COVID-19? *Nano Today* 33: 100883.

79 Nandy Chatterjee, T. and Bandyopadhyay, R. (2020). A molecularly imprinted polymer-based technology for rapid testing of COVID-19. *Trans. Indian Natl. Acad. Eng.* 5 (2): 225–228.

80 Jalandra, R., Yadav, A.K., Verma, D. et al. (2020). Strategies and perspectives to develop SARS-CoV-2 detection methods and diagnostics. *Biomed. Pharmacother.* 129 (May): 110446.

81 Seshadri, D.R., Davies, E.V., Harlow, E.R. et al. (2020). Wearable sensors for COVID-19: a call to action to harness our digital infrastructure for remote patient monitoring and virtual assessments. *Front. Digit. Health* 2 (June): 1–11.

Part IV

Commercialization and Standardization of Analytical Technologies

10

Nanobioengineering Approach for Early Detection of SARS-CoV-2

Sidra Rashid[1], Umay Amara[1,2], Khalid Mahmood[2], Mian H. Nawaz[1], and Akhtar Hayat[1]

[1] *COMSATS University Islamabad, Interdisciplinary Research Centre in Biomedical Materials (IRCBM), Lahore Campus 54000, Pakistan*
[2] *Bahauddin Zakariya University, Institute of Chemical Sciences, Multan 608000, Pakistan*

10.1 Introduction

The World Health Organization (WHO) has declared the outbreak of severe acute respiratory syndrome coronavirus-2 (SARS-CoV-2) "pandemic" on 11 March 2020. World has also familiarized with such corona pandemic about a decade ago when severe acute respiratory syndrome (SARS-2002 and SARS-2004) followed by the Middle East respiratory syndrome (MERS-2012) crossed the interspecies barriers and infected the humans. Unlike from the previous outbreaks of SARS-2002, SARS-2004, and MERS-2012, SARS-CoV-2 has spread to the whole world with many thousands of deaths and millions of cases, stimulating the international health organizations to announce the situation as a global health emergency [1]. This pandemic has induced severe difficulties to the total health system of the world such as higher percentage of infection rate, prolonged incubation interval, increased mortality rate, and saturation of hospital infrastructures along with serious financial crisis. The primary reason for the global rise of this pandemic situation was the instantaneous origin of SARS-CoV-2 along with the lack of awareness and technology to reduce spreading of SARS-CoV-2. These technological limitations made it essential to develop efficient methods for their detection and destruction [2].

Nanobioengineering, an interesting subject of biology, chemistry, physics, and medicinal disciplines, emerged as a feasible approach to improve the effectiveness of the health setup with minimal resources plus efforts. As the globe races to ramp up testing schemes for this pandemic situation, several bioengineered strategies are rolling out to make it quicker and easier for people to know whether they have been infected or not. For the exploration of inexpensive and rapid tests, various nanoengineered diagnostics based on biomaterials have been deployed by various organizations [3]. However, materials available and engineered for SARS-CoV-2 at present have been examined with conventional nanobioengineering. Although the outcomes with the conventional technology have not been found adequate,

Detection and Analysis of SARS Coronavirus: Advanced Biosensors for Pandemic Viruses and Related Pathogens,
First Edition. Edited by Chaudhery Mustansar Hussain and Sudheesh K. Shukla.

the integration of extensively researched modern nanoengineered materials such as quantum dots, low-dimensional semiconductors, and compounds with certain biomaterials such as DNA, RNA, antibodies (Abs), or antigens can provide possible alternatives to the pandemic [4, 5]. It has been reported that the detection of IgG and IgM antibodies can imply against SARS-CoV-2 by an immune detection approach, as both these antibody levels reported to have surged to a distinguished level after the prompt spread of the SARS-CoV-2 infection within the body [6]. The currently used rapid diagnostic tests (RDTs) worked on a principle where the target antigen from the respiratory tract will bind to specific antibodies generating a visual signal or through direct measurement of the antibodies in the blood sample of the SARS-CoV-2 infectant [7]. This method can provide the results within 30 minutes by using nasopharyngeal (NP), bronchoalveolar lavage (BAL), or oropharyngeal (OP) swab or even the blood and stool at the place of patient convenience. Such type of bioengineered diagnostic spectrum of SARS-CoV-2 from the respiratory tract samples is the prime key in the clinical diagnostics of SARS-CoV-2. Such bioengineered strategies can overcome the need of robust, fast, on-site point-of-care (POC) testing methods that do not necessarily require skilled personnel to perform them [8].

In the consideration of the above-mentioned information, various analytical strategies/tools have been recently established for the early detection of SARS-CoV-2 worldwide. Most of these diagnostic tests have been made practical directly in the clinics and hospitals for fast and effective screening and directly applied in various hospitals worldwide for fast screening of suspects [9, 10]. This chapter will enlighten the role of nanobioengineered techniques and the recently developed clinically practiced methods for the early detection of SARS-CoV-2.

10.2 Can Nanobioengineering Stand in the Battle Against SARS-CoV-2?

The highest transmissibility of SARS-CoV-2 is associated with its extraordinary viral load (VL) capacity from the upper parts of the respiratory system (RS) and also because of the fact that enough of the population stays asymptomatic, shedding and transferring this infection [11]. Serious interventions in the fight against this pandemic include the strategies to stop extensive spread of the infection and appropriate diagnostic coverage along with the immunization of the community against that particular pathogen involved. Thus, the early detection of SARS-CoV-2 is not only critical but also vital for people's safety. In this regard, scanning of patient's lungs with computed tomography (CT) has provided possible investigation as a handheld test for clinical diagnostics. However, these practices can only provide primary evaluation of the affected lungs. Thus, in-depth investigation is indispensable to authorize the intensity of the infection within the body [12, 13]. It has been stated in many studies that SARS-CoV-2 has close similarities with other known viruses including MERS (~50% similarity) and SARS-CoV-1 (~80% resemblance) [14, 15]. So, for the diagnostic purpose, molecular and structural data can play a significant role.

10.3 Sequential and Molecular Data Analysis

For the exploration of the actual nature of SARS-CoV-2, researchers targeted already reported information about β-coronaviruses (MERS and SARS-CoV-1) that are known to be highly pathogenic among humans and easily transmit in healthy individuals. Both these viruses were described as being zoonotic in origin [16]. The molecular sequential data (Figure 10.1) of this virus exposed various common reading frames to these viruses, for instance, ab1, that encodes for envelope (E) protein, nucleocapsid (N) protein, and the spike (S) protein [18]. Kim et al. carefully sequenced the viral genome using respiratory tract secretion from Korean infected patients during the December 19 outbreak [19]. This study showed >99.9% sequence homology to CoV-2, which was isolated from the infected individuals of other different countries. Also, Sah et al. observed a similar kind of homology in the samples taken from a 32-year-old Nepalian for SARS-CoV-2 [20].

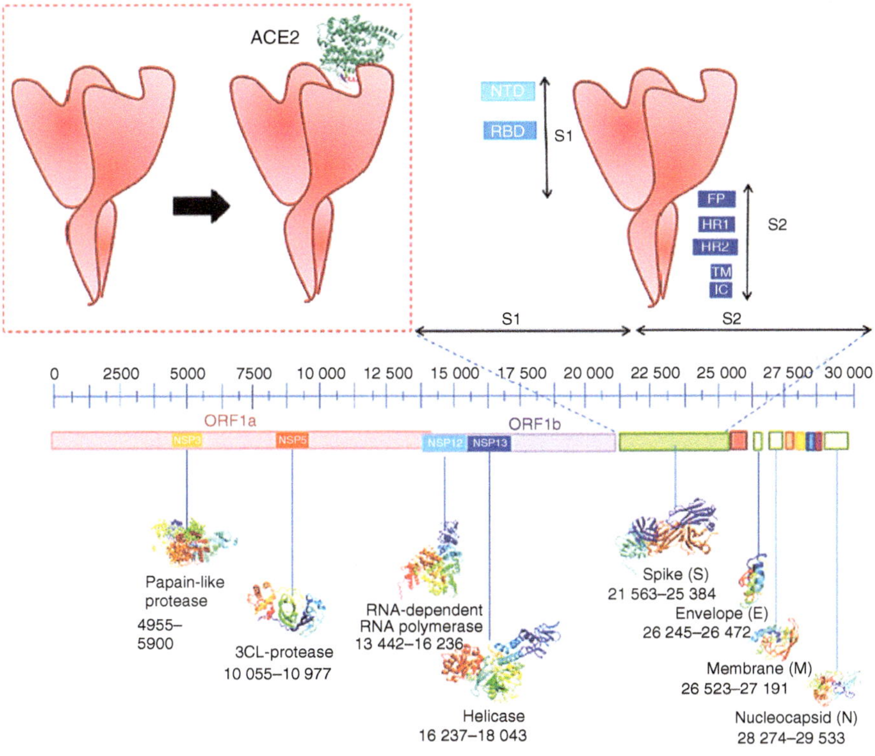

Figure 10.1 SARS-CoV-2 genome organization, codified proteins, and binding of spike protein to ACE2 receptor. Inset: illustration of ACE2 interaction with the RBD of SARS-CoV-2. Abbreviations: S1, receptor binding subunit; S2. membrane fusion subunit; NTD, N-terminal domain; RBD, receptor binding domain; FP, fusion peptide; HR1, heptad repeat 1; HR2, heptad repeat 2; TM, transmembrane anchor; IC, intracellular tail; NSP, nonstructural protein. Source: Chauhan et al. [17].

This sequence not only showed more than 99.99% homology to SARS-CoV-2 with two previously reported viral genomes presented at Gen-Bank from Wuhan, China, but also indicated seven additional genomic sequences. These additional sequences vary from that of SARS-CoV-1 and MERS. Based on the biochemical and structural experiments, a comparison study between α and β coronaviruses has revealed that SARS-CoV-2 may have the ability to bind effectively with a human receptor gene called angiotensin-converting enzyme-2 (ACE-2) [21]. The computational assessments do not confirm the interaction between ACE2 gene and spike proteins [22]. It led to the conclusion that higher affinity of CoV-2 to ACE-2 is mainly due to the natural selection or maybe due to the mutation to the human ACE2 gene [23]. It also suggests that any kind of mutation could alter the SARS-CoV-2 phenotype, which may influence the diagnosis of this virus by different sensing systems and biomolecular assays [24]. This finding confirms that the genomic sequencing of SARS-CoV-2 is absolutely justifiable for any succeeding outbreaks. Moreover, the identified location of the mutated region of the viral genome can predict the effects on the attachment between primer and probe's binding sites, which can greatly influence the diagnostic capabilities of the currently available commercialized kits [25].

10.3.1 Role of Nanobioengineering for SARS-CoV-2 Detection

In consideration of the previously mentioned vital genomic and sequential molecular data such as information about genetic biomarkers, primers, and molecular probes besides variable levels of antibodies (Abs) in patient's samples or in virus, the standard molecular technique that was used at the early stages of pandemic to detect SARS-CoV-2 was reverse transcription polymerase chain reaction (rRT-PCR). However, this technique requires the laboratory with biosafety level 2 or above along with sophisticated equipment [26, 27]. Further, three days are required to obtain the results. This time consumption is often extremely detrimental for community well-being during the crisis, for instance, the COVID-19 pandemic. In such situations, the development of low-cost, POC, and fast diagnostic tools with the ability to provide reliable outcomes is the immediate requirement. The tools needed to be developed for such a critical situation should be simple enough to be used in the field and on-site without the need of trained personnel. Various testing techniques have been developed, which target different fragments of the SARS-CoV-2 genetic profile. However, some serious glitches related to low accuracy and reliability of some of these kits were described, affecting the world health system. An additional major concern is the excessive use of reagents essential to execute the diagnostic tests, which has also become a bottleneck [28].

Nanobioengineering is the only discipline ready to provide the solution for the problems mentioned. For example, nanodiagnostics depends on the binding capacity among the nanoscale agents and the biomolecules under consideration to generate a quantifiable signal that allows the detection of responsible pathogens [29]. While implementing such nanodiagnostics, measures were carried out at the nanoscale, which lead to the development of handheld devices/tools that can be commercialized with additional benefits of stability, sensitivity, and high accuracy. This can help provide effective results with prompt and timely diagnosis of the disease [30]. The role of nanomaterials in bioengineering can be of significant importance for SARS-CoV-2 detection as this virus itself has a core–shell nanostructure with the size ranging from 60 to 140 nm [29, 31]. Thus, it allows the bioengineered nanomaterials to specifically bind with the virus [32], permitting to evaluate, engineer, and develop the procedures for the diagnosis, treatment, as well as the preventive measures of SARS-CoV-2 [31]. Further, variable bandgap, high luminescence, large surface-to-volume ratio, and easy integration of nanodimensional components with biomaterials offer extra benefits in comparison to the bulk for the fabrication of detection kits and vaccine for SARS-CoV-2 [33].

In Section 10.4, literature studies on the fabrication of sensing strategies and their utilization toward diagnostic applications for SARS-CoV-2 will be highlighted.

10.4 Nanobioengineering-Based Detection of SARS-CoV-2

Early detection of contagious diseases offers effective prevention of potential cases, whereas delayed reporting leads to minimal control of such problems [12]. Currently, various nucleic acid-based and protein-based tests are available for diagnostic purpose. However, there is still an urgent need for POC testing devices for prompt and self-diagnosis in the case of SARS-CoV-2 [34].

10.4.1 Nucleic Acid-Based Molecular Detection

Many analytical kits have been reported in the literature for the initial detection of SARS-CoV-2 based on nucleic acid data to detect SARS-CoV-2.

10.4.1.1 Reverse Transcription Polymerase Chain Reaction (RT-PCR)

RT-PCR is considered as a gold standard for SARS-CoV-2 detection based on the reverse transcription of RNA into DNA. It also amplifies the targeted DNA using polymerase chain reaction (PCR). For SARS-CoV-2 detection, sequences of envelope gene are screened, followed by confirmation of nucleocapsid (N) gene [35]. If negative results are obtained, further analysis is done using the RNA-dependent RNA polymerase (RdRp) gene. Positive results validate the existence of SARS-CoV-2 and the case is clinically termed as corona positive [36]. On the basis of RT-PCR principle, a number of commercialized testing kits available worldwide, which are highly sensitive, specific, and exploit this algorithm, are mentioned in Table 10.1. However, RT-PCR has some limitations including long turnaround time, lack of correlations with the viral load (VL), which is the numerical expression of the virus quantity in a given volume of a body fluid/blood plasma. Majority of commercialized kits mentioned are used in a specific system with specific operational manual; therefore, it can provide varying sensitivity in different PCR analyses [37].

Table 10.1 List of commercialized testing kits available worldwide for the detection of SARS-CoV-2.

(a) Nucleic acid-based diagnostic tests

No.	Brand	Developer	Method	Details	Samples	Time (min)	LOD
1	REALQUALITY RQ-2019-nCoV	AB ANALITICA	Single-step RT-PCR	Targets E and RdRp genes, amplification	Bronchoalveolar lavage, nasopharyngeal swabs	100	NR
2	COVID-19 Onestep RT-PCR Dual Target Gene	Pishtaz Te	One-step RT-PCR	Dual-target gene design targeting specific sequence encoding RdRp and N region	Nasal swab, sputum, throat, and bronchoalveolar lavage fluid	90	NR
3	A*STAR Fortitude Kit 2.0	A*ccelerate Technology Pte Ltd	One-step RT-PCR	Laboratory technique	Human clinical samples	90	25 copies/reaction
4	STANDARD M nCoV RT Detection Kit	SD BIOSENSOR Inc.	One-step RT-PCR	Detects CoV-19 associated ORF lab gene	Oropharyngeal swab, nasopharyngeal swab	90	NR
5	COVID-19/qPCR-Kit	1 copy™	RT-qPCR	Detect E, RdRp via RNA-extraction	Oropharyngeal swab, sputum	110	4 copies/reaction
6	Strong Step® Novel Coronavirus (SARS-CoV-2) Multiplex RT-PCR Kit	Liming BioProducts Co., Ltd.	Fluorescent probe-based Taqman® RT-qPCR	Detects ORF1ab gene, N-gene, E-gene, avoids nonspecific interference by SARS (2003) and Bat-SARS-viral strains	Clinical samples	NR	NR
7	SARS-CoV-2 nucleic acid test kit	Wuhan Easydiagnosis Biomedicine Co., Ltd	Fluorescent RT-PCR based	Qualitatively detect the new coronavirus 2019-nCoV infection in suspects by analyzing ORF1ab gene	Nasopharyngeal swabs, oropharyngeal swabs, and sputum samples from patients diagnosed	NR	NR
8	2019 nCoV RT-PCR Diagnostic Panel	US CDC	RT-PCR based	nCoVPC will yield a positive result with each assay in the systems diagnostic panel including RP	Human sera or pooled leftover negative respiratory specimen	80	$10^{2.5}$ copies/ml
9	Q-Sens® 2019-nCoV Detection Kit	Cancer Rop Co., Ltd	RT-PCR based	Detection and amplification of the genes specific to SARS-CoV-2	Sputum, nasopharyngeal smear	120	NR
10	Taq ManTM SARS-CoV-2 Assay Kit v2	Thermo Fisher Scientific	RT-PCR	2019-nCoV Assay (Orf1ab); 2019-nCoV Assay (spike [S] gene); 2019-nCoV Assay (nucleocapsid [N] gene); RNAse P	Human clinical samples	NR	NR
11	Genesig RT-PCR COVID-19	Novacyt/Primerde sign Ltd.	RT-PCR	Rapid detection and exclusive to the COVID-19 strain, high priming efficiency	Clinical samples	NR	NR
12	Abbott Real Time SARS-CoV-2	Abbott Molecular Inc.	RT-PCR	A dual-functional targeted assay for N and RdRp gene	Oropharyngeal swabs and nasopharyngeal	NR	$1–4 \times 10^2$ copies/ml
13	LyteStar 2019-nCoV RT-PCR Kit	ADT Biotech	RT-PCR	Dual-target kit for confirmation of SARS	Fresh or frozen human sample	NR	NR

(Continued)

(b) Protein-based diagnostic tests

No.	Product	Company	LOD diluted (% sensitivity)	Time (min)	Specificity (%)
1	Cellex/qSARS/CoV-2-IgG/IgM Rapid Test	Cellex	93.8	15–20	95.6
2	Roche's Elecsys IL-6	Roche Diagnostics	84	18	63
3	Elecsys Anti SARS-CoV-2	Roche	100	18	99.8
4	SARS-CoV-2 IgG	Abbott Laboratories	100	29	99.9
5	Anti-SARS-CoV-2 Rapid Test	Auto-bio-Diagnostics Co. Ltd. (in collaboration with Hardy Diagnostics)	99	15	99
6	SARS-CoV-2 Total-Assay	Siemens Healthcare Diagnostics Inc.	100	10	99.8
7	COVID-19 Antibody Rapid Detection Kit	Healgen Scientific LLC	96.7	10	97
8	Platelia-SARS-CoV-2 Total Ab assay	Bio-Rad	92.2	120	99.6
9	LIAISON-SARS-CoV-2 S1/S2-IgG	Dia-Sorin Inc.	97	35	98
10	VITROS-Immuno diagnostic Products Anti-SARS-CoV-2-IgG	Ortho Clinical Diagnostics, Inc.	87.5	48	100
11	NewYork-SARS-CoV-2 Micro-sphere Immunoassay	Wadsworth Center, Department of Health, New York State	90	<60	100
12	Right Sign COVID-19 IgG/IgM Rapid Test Cassette	Hangzhou Bio-test Biotech Co., Ltd.	92.5	—	99.5
13	SCoV-2 Detect IgG ELISA	In-Bios International Inc.	97.8	—	98.9
14	Vibrant COVID-19-Ab-assay	Vibrant America Clinical Laboratories	98.1	—	98.6

10.4.1.2 Loop-Mediated Isothermal Amplification (LAMP)

In order to avoid false-positive/negative results that can be obtained in standard detection methods such as RT-PCR, various other ultrasensitive methods have been developed. Loop-mediated isothermal amplification (LAMP) is a unique nucleic acid-based amplification technique that amplifies DNA with higher specificity, rapidity, and efficiency [25].

The emergence of this method is due to its specific primers, DNA polymerase enzyme, with chain displacement activity under isothermal conditions, without the need of thermocycler or electrophoresis equipment. In comparison to rRT-PCR, LAMP has more advantages of being sensitive (>95%) and more stable along with the ability of being used as a one-step virus detection technique and being applicable under isothermal conditions (Figure 10.2). Recently, the FDA approved an "automated assay" based on this technology. The combination of LAMP and RT technique (collectively known as RT-LAMP) has provided a one-step higher detection strategy for the SARS-CoV-2 genome. It is quite faster (30 minutes) and much simpler than RT-qPCR [38].

A current study validated the possibility to entirely automate LAMP methods called Simprova for the quick and effortless molecular diagnosis [39] of SARS-CoV-2. This consists of a central component that controls the complete system, a pretreatment unit that isolates and purifies nucleic acid from patient samples, and an LAMP component to identify and amplify the nucleic acid sequence (Figure 10.3(1)). Such systems can be economical and portable with effective generation of fast results using minimal reagents. Moreover, the paper-based LAMP strategy that has already been developed is called DETECTR by Yang et al. (Figure 10.3(2)) with few modifications, this strategy can be converted into a diagnostic tool for domestic use by any quarantined individual with inadequate access to the health care facilities [40]. Further, a strip-based test, such as a DETECTR platform in combination with a glucometer-like readout machine, can completely revolutionize the concept.

Recent investigations have been focusing on the utilisation of nanoparticles to simplify such multistep process into single step analysis. Such advances in LAMP technology suggest high sensitivity and specificity that will help in the swift diagnosis of a large number of suspected patients in pandemic situation such as SARS-CoV-2. However, some challenges are still there. The main disadvantage is the complexity of primer design in LAMP technique. Another important drawback is the production of false-positive results especially when upgraded in an automated network [41].

10.4.2 Protein-Based Detection

Protein-based SARS-CoV-2 diagnostics utilize either antibodies or antigens. However, in such cases, fluctuations in viral loading during the infectious course can impose complications in the recognition [17]. For example, higher viral load during the first week of the infection in salivary gland declines after the completion of the following week as shown in Figure 10.4.

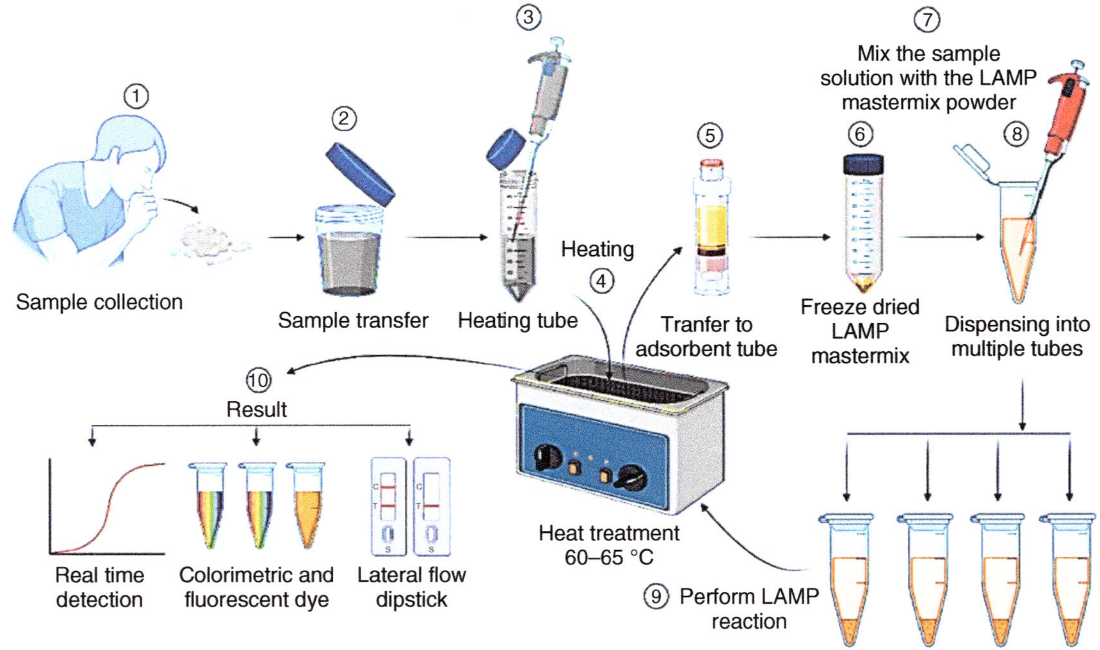

Figure 10.2 Steps in sample processing for the LAMP reaction, the LAMP reaction in a simple water bath, and the detection process.

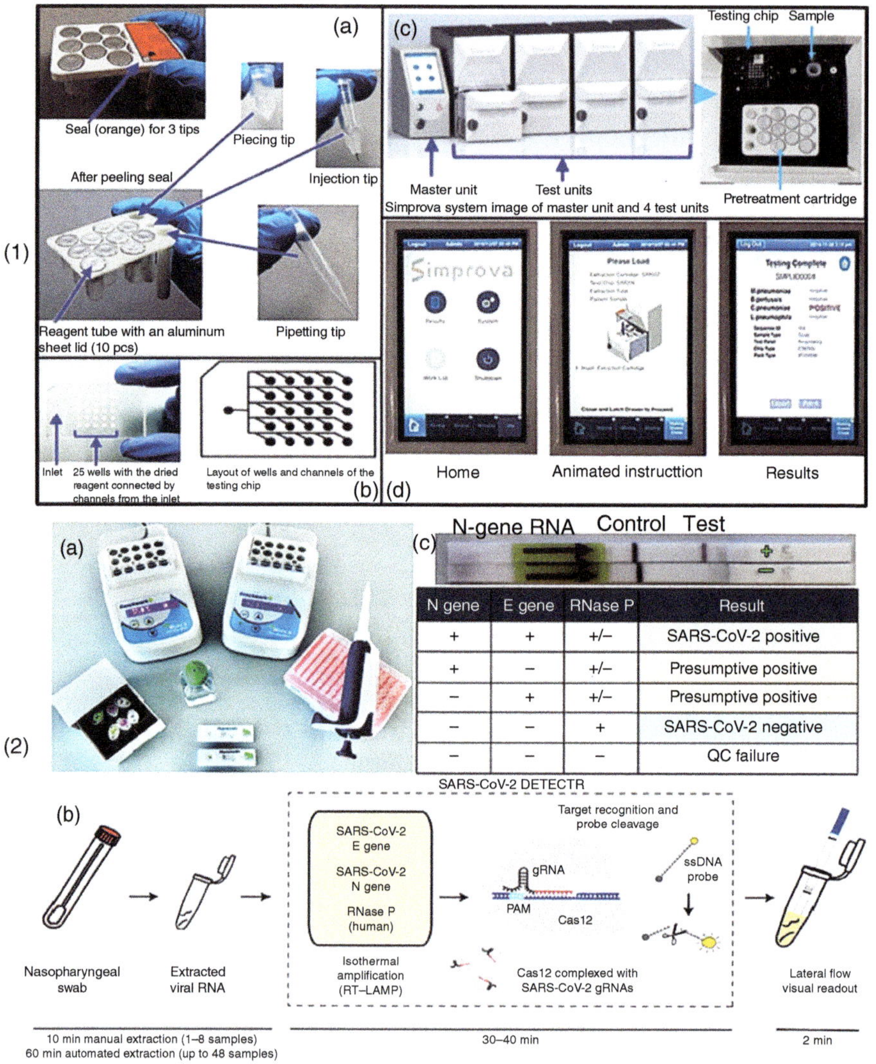

Figure 10.3 Simprova system. (1) (a) Pretreatment unit, (b) multiwell testing strip, (c) test unit with a complete master unit, and (d) example of readouts of the final master unit. (2) (a) DETECTR. Equipment required to perform DETECTR experiment: Eppendorf tubes, heat blocks/water bath at 37 and 62 °C, pipettes, lateral flow strips, and nuclease-free water. (b) Schematic representation of the SARS-CoV-2 DETECTR method. (c) Lateral flow strip assay readout. A positive result requires detection of at least one of the two SARS-CoV-2 viral gene targets (N gene or E gene, as indicated in the interpretation matrix). QC, quality control.

With these serological tests (ST), the level of antibodies or other proteins is measured in the blood as body responds to the SARS-CoV-2 infection. Actually, the immune response of the body is detected instead of the infection itself. These tests are executed by placing the recombinant proteins in the well plates, and after that, diluted sample serum is introduced to implement the ELISA test. Finally, to obtain

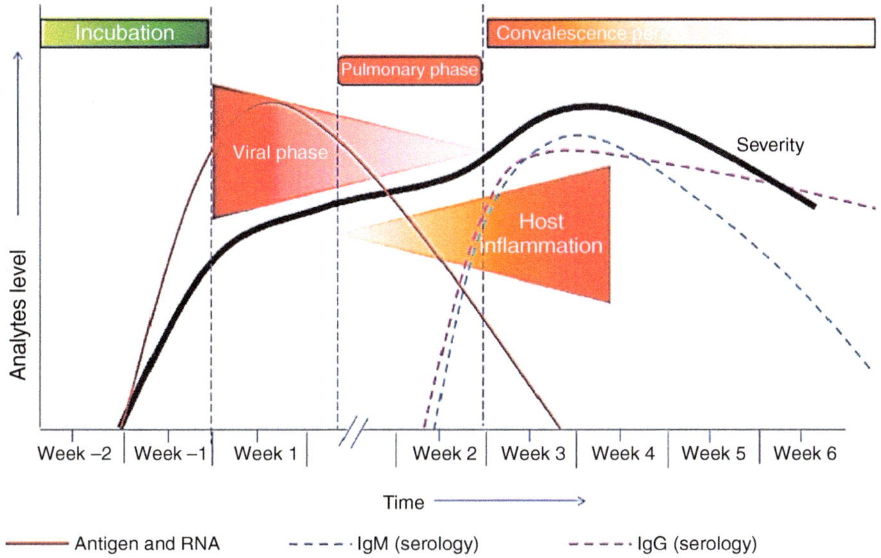

Figure 10.4 The standard relation between changes in analyte concentration with respect to the course of infection. Source: Chauhan et al. [17].

the signal, the IgG antihuman antibody functionalized by horseradish peroxidase is added. Anti-SARS-CoV-2-IgG will be sandwiched if present between antihuman IgG and the adsorbing protein. The same procedure can be performed with the IgM antibody [42]. Nucleocapsid proteins are highly immunogenic phosphor proteins abundantly found in the virus and commonly used as biomarkers in tests as they hardly mutate. Rp3 nucleocapsid protein of CoV-2 virus was used for the detection of IgM and IgG antibodies from CoV-2 patients because it has 90% similarity with SARS-CoV-2. Thus, a reliable ST will be of prime importance as it would provide significant information about the prevalence of SARS-CoV-2 for the community via recognizing the individuals with antibodies for CoV-19 pathogen [43]. Some of the serological tests approved by FDA have been listed in Table 10.2. Even though these serological tests can provide a wider frame to detect SARS-CoV-2, the cross-reaction between different antibodies can pose a serious problem.

10.4.3 Lymphopenia-Based Assessment

It has been observed that during the course of this disease, the percentage of lymphocytes was found to be the most significant and consistent parameter for disease diagnosis. For instance, the lymphocyte percentage was recorded about 5% from the sample of 12 death cases (due to CoV-19) aged above 75 after two weeks of disease incubation [17]. However, a downside and upside pattern of the lymphocytes percentage was noticed with an initial decline and then increased by 10% with the patients of an average age of 35. Further, in 11 patients aged 49 and a therapeutic time of 26 days, this percentage remained almost constant. These findings illustrate the reliability of lymphopenia as a detection strategy [44].

Table 10.2 List of different methods that can be used for the detection of SARS-CoV-2.

(a) Digital radiographical imaging

No.	Brand	Developer	Method	Details	Sample	Time	LOD
1	AiroStotleCV19	Canary Health Technology	PathSensors technology licensed by MIT-Lincolin Laboratory	Three instrumentation platforms based on PathSensors for rapid detection	Air monitoring from sensitive places, e.g. offices and hospitals	>5 min	NR
2	InferRead CT-Pneumonia	Beijing Infervision Technology Co. Ltd	AI-based detection using CT scan	Imaging-assisted diagnosis from different body areas	Physical presence of patient	>20 s	NR

(b) Laboratory tests for common clinical indications

Features		Observations
1	Expression	Bilateral, basal air spaces, normal chest radiography (15%)
2	Rare findings	Pneumothorax
3	Chronic phase	Unknown pleural effusion have not yet been reported
4	Poor prognosis	Consolidation
5	Follow-up imaging appearance	Persistence airspace opacities

Source: Based on Chauhan et al. [17].

10.4.4 Bioengineered Surfaces for SARS-CoV-2 Detection

Biointerfacial interactions among a surface and target protein play a significant role in diagnostic strategies. Surface characteristics such as chemistry, surface roughness, and hydrophobicity; energy with charge regulation; and the attachment or adsorption of the proteins are vital in this aspect [45]. Classically, as the surface hydrophobicity increases, the amount of adsorbed protein increases accordingly [46]. Surface hydrophobicity can be altered using the chemistry of surfaced functional groups of a self-assembled monolayer (SAM). Such type of protein adsorption is entropy driven, and the change in the conformations of the protein adsorbed greatly depends on the surface-layered chemistry, a prominent factor of any detection process [47]. In this regard, SAMs have been explored to fabricate nanostructured planes consisting wider hydrophobicity or wettability. These SAMs are designed by the covalent modification/attachment on the upper tops without affecting the overall possession of the materials. These surfaces can be created on a number of reagents including silica, different polymers, indium tin oxide (ITO), glass, gold, and other metallic components with appropriate surface modifiers [48]. In a recent report, biophysical aspects of SARS-2-S1 protein were analyzed using hydrophobic and positively charged amino acid residues. Different strategies to form mixed SAMs of hydrophobic (CH_3) and negatively charged (COOH) groups were discussed to be utilized for the specific and strong interactions with spike protein that may be helpful for early detection of SARS-CoV-2 [7, 49, 50]. Moreover, for the minimal biofouling of nanoengineered surface as well as random adsorption, poly-ethyleneglycol (PEG)-based modifiers such as PEG silanes and thiols can be investigated [51].

10.4.5 Nanobioengineered Prototypes

Besides the clinically practiced diagnostic methods, several nanobioengineered sensing prototypes have been advanced for SARS-COVID-2 detection. A highly sensitive and fast dually functional plasmonic sensor based on localized surface plasmon resonance (LSPR) and plasmonic photothermal (PPT) was developed by Qiu et al. on a 2D nano-Au chip [52]. With this configuration (Figure 10.5a), a label-free and fast diagnosis of SARS-2 was attainable with 0.22 pM limit of detection (LOD). Another biosensing device based on field effect transistor (FET) mechanism integrated with SARS-CoV-2 antibody was established by Seo et al. [53]. This strategy claims high specificity with an LOD of about 1 fg/ml with unique ability to distinguish between MERS-CoV and SARS-CoV-2. Thus, under the outbreak background of the CoV-19 pandemic, such bioengineered prototypes can establish a more reliable early diagnostic platforms (Figure 10.5b).

10.4.6 Digital Radiographical Biosensing Platforms

Analogous to pneumonia, the initial diagnosis of SARS-CoV-2 is based on the communal symptoms of fatigue, dry cough, fever, dyspnea, and myalgia. Thus, digital chest imaging (DCI) can be of particular importance for the evaluation of

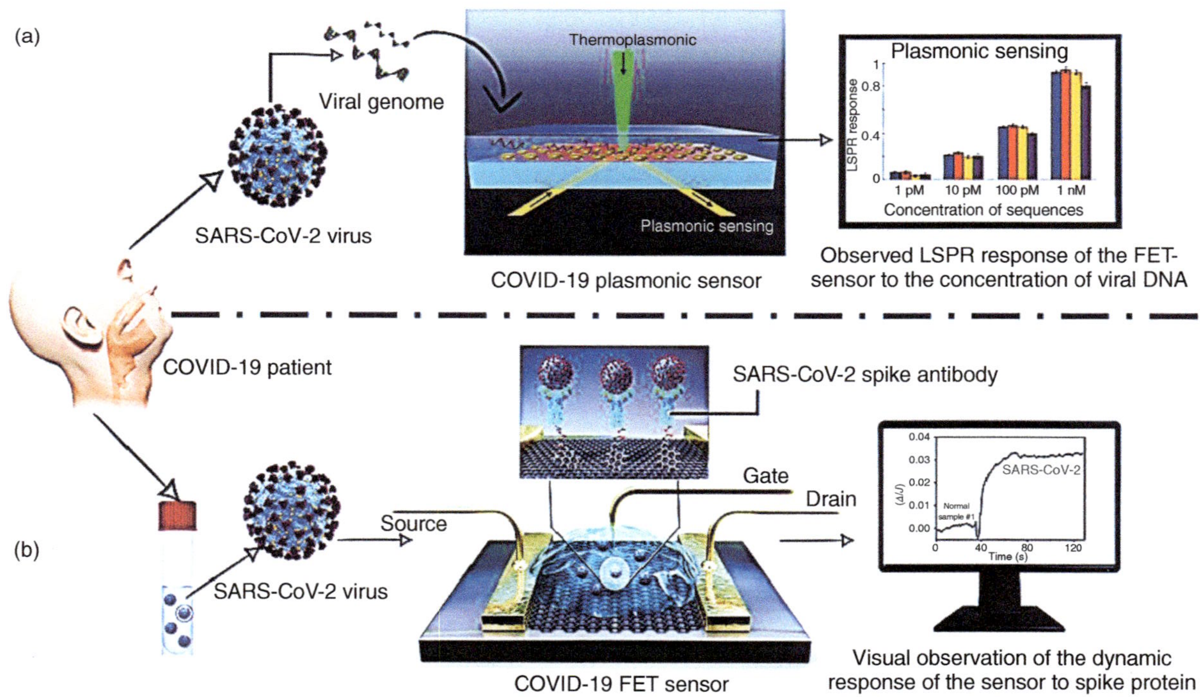

Figure 10.5 Schematic representation of (a) LSPR detection of nucleic acid sequences from SARS-CoV-2 virus and (b) COVID-19 FET sensor operation procedure.

the extent of the disease [54]. In a study, it was observed that about 40 out of 41 SARS-CoV-2-infected individuals showed involvement of bilateral lungs in CT scans. In another similar study, where a family of seven members was subjected to digitalized chest imaging, the same kind of bilateral patchy ground glass opacity (GGO) arrangement in the lungs was observed [55]. Furthermore, Chung et al. reported 57% of the patients with such a GGO design, 33% with opacities with rounded morphology, and another 33% with peripheral distribution of the disease. Cavitation or lymphadenopathy has also been reported in SARS-CoV-2 patients [56]. It was also confirmed by another finding that individuals with SARS-CoV-2 confirmed by swab test presented negative (−) fallouts after early CT findings for the CoV-19 detection [57]. Although the initial reports indicated the normal findings observed from digital imaging, however, only 15% of the patients seems to abide by this. However, recent reports have shown that the presence of typical CT image findings could be helpful for initial screening of the virus in suspected individuals [58]. Table 10.2(b) highlights the clinical observations after digital imaging of the lungs of the patients.

10.4.7 Other Methods for SARS-CoV-2 Detection

Other than direct diagnosis of SARS-CoV-2 in patients, some indirect means of biochemical monitoring has become essential to examine the presence of infection and its severity throughout the course of therapeutic intervention. Recently, few in vitro diagnosis (IVD) tests have been advocated during this pandemic, where different amounts of blood cells (BC) or different biomarkers had evaluated [59]. These tests can provide substantial evidence for SARS-CoV-2 load. In Table 10.2(b), some of the clinical tests based on advanced findings indicating major laboratory irregularities and few clinical interpretations have been listed. Apart from such communal lab experiments, few current reports correspondingly propose that people with intense SARS-CoV-2 condition can also be infected by cytokine storm syndrome. Thus, IL-6 cytokine test should be made compulsory for individuals affected by any infection [60]. Another research indicates that the viral infection severity can also be evaluated by counting the lymphocytes and measuring the level of procalcitonin and C-reactive protein [61].

10.5 Discussion

The agent responsible for "pneumonia of unknown etiology" on December 2019 in Wuhan, China, was called as "severe acute respiratory syndrome coronavirus 2" by the WHO [62]. Till date, no definite medicament modality has been entirely established for SARS-CoV-2. In such a pandemic situation, only early detection strategy can play a significant role to avoid life-threatening conditions. The two factors that are important to develop such an approach are sensitivity and specificity. RT-PCR is a time-tested technique wherein RNA from the swab specimen is amplified. LODs for such diagnostics display 95% of positive results with even an ultralow concentration of the coronavirus [63]. However, the disadvantage of this method is the lack of correlation of viral load and long processing steps. Alternative methodologies are hence required for prompt testing that can play a significant role. Immunosensing kits mainly work by detecting IgG and IgM antibodies present in the blood, plasma, and sputum specimens, and the outcomes obtained can be of visual interpretations. Two different detection lines with CoV-2 antibodies and antigen marker with fixed anti-IgM/IgG antibody can detect a higher level of coronavirus. Again, the 100% specificity of such approaches in comparison to the already existing methods is questionable as variable fallouts have been observed with these kits. A series of blood tests conducted in a Chinese 14-CDC hospital indicated that 352 positive RT-PCR tests out of 397 revealed 88.66% sensitivity. Therefore, an association should be made essential among clinical indications and serological data before applying this technique as a confirmatory test [27]. However, researchers are working for a more accurate and sensitive technique to detect COVID-19 such as RT-LAMP. An RT-LAMP study detected a synthetic strain of SARS-CoV-2 within the range of 20–200 aM [64]. This method of rapid investigation can be a better alternative to the procedures in the test center that lack complex possessions. An additional substantial part of early identification can also be accredited by digital imaging and radiography, as CoV-2 expresses distinctive radiographical outcomes. Such discoveries showed symptoms closely related to the other respiratory tract infections (RTIs) such as SARS-CoV-1 and MERS in infected lungs. Furthermore, these digital radiographs (DRs) double fold the probabilities of diagnosis of the symptoms, leading to fast and early detection [58]. The only disadvantage of this method is the overlapping of the results obtained from different RTIs. Moreover, the collaboration between bioengineering developments with nanotechnology can fulfill the demand of ultrasensitive, fast, and POC diagnostic tools. It can aid biosensors with extra specificity that can provide us faultless information in miniaturized and handheld devices as compared to other conventional clinical tests.

10.6 Conclusions

The above discussion concludes that nanobioengineering is the only discipline that can provide solution to SARS-CoV-2 pandemic situation by integrating engineered techniques with modern nanobioengineered materials. Thus, this integration has enough to offer to exploit the novel properties of nano-sized biosensors and nanobiocomposites not only for the early detection of coronavirus but also at multilevels in the battle against SARS-CoV-2. The following months will be critical to analyze such nanobioengineered techniques for early detection during the second wave of COVID-19 and then utilized them for other severe infectious diseases.

10.7 Expert Opinion

Without a doubt, nanobioengineering will play a crucial role against COVID-19 by advancing biosensors in such a way to identify the virus at early stages of the

COVID-19 infection and also by exploiting different nanosystems to disinfect the medical devices along with public places to immune the community. We strongly believe that the genomic segment or antibodies of SARS-CoV-2 can be integrated with innovative nanotransductors that cannot only produce an entirely new generation of biosensors but also amplify their function. Such bionanodevices can be coupled with former techniques, for instance, lab-on-chip methodologies, to simplify the molecular diagnosis of SARS-CoV-2, facilitating the expansion of POC diagnostics besides long-term assessments of infected individuals.

Even though various nanobiosystems have shown viral inhibition possessions for SARS-CoV-2, however, majority of the reports indicate the importance of early detection of the virus. The fate of the nanosystems has not been assessed irrespective of the final application (surface protection, water disinfection, therapeutics, etc.). Additionally, even after the development of the targeted detective nanobiosystem, it is critical to validate what could happen with this detective system if they are once administrated to the animal model. Finally, the administration for the establishment of the nanobiosystem that can detect and then simultaneously inhibit the growth of the infection should be the mandatory path to resist the spread of SARS-CoV-2. This will result in accessible and effective antiviral surface coatings for use in public infrastructure and medical devices.

10.8 Future Directions

Regarding the time-consuming techniques such as PCR, new innovative approaches are being reported in the urge of rapid diagnosis. Several desired molecular tests based on non-PCR methods have been proposed recently that include loop-mediated isothermal amplification technique and the nucleic acid sequence-based amplification technique (NAAT) for SARS-CoV-2 detection. Another possible rapid POC screening of SARS-CoV-2 involves the recombinant DNA technology. These POC or other near-POC-NAAT platforms can provide prospects in research laboratories for the execution of automated self-reliant trials with limited expertise, especially for endemic-prone zones. The expansion of such detection designs will also promisingly support within the country-way of dealing this infection. Further, more upgraded and subtle serological cost-effective assays with least cross-reactivity are highly desirable. In future, the integration of redox cyclic methods with signal amplification approaches utilizing nanobioengineered materials in SARS-CoV-2 immunosensing kits can further intensify the selectivity and sensitivity of the existing schemes. Finally, the support of trial biobanks equipped with well-characterized specimens' reference standards will simplify diagnostic advancements and quality assurance for worldwide SARS-CoV-2 detection.

References

1 Gorbalenya, A.E., Baker, S.C., Baric, R.S. et al. (2020). The species severe acute respiratory syndrome-related coronavirus: classifying 2019-nCoV and naming it SARS-CoV-2. *Nat. Microbiol.* 5 (4): 536–544.

2 Gupta, A., Kumar, S., Kumar, R. et al. (2020). COVID-19: emergence of infectious diseases, nanotechnology aspects, challenges, and future perspectives. *ChemistrySelect* 5 (25): 7521–7533.

3 Sheridan, C. (2020). Coronavirus and the race to distribute reliable diagnostics. *Nat. Biotechnol.* 38 (4): 382–384.

4 Westhaus, S., Weber, F.-A., Schiwy, S. et al. (2021). Detection of SARS-CoV-2 in raw and treated wastewater in Germany – Suitability for COVID-19 surveillance and potential transmission risks. *Sci. Total Environ.* 751: 141750.

5 Tang, B., Xia, F., Tang, S. et al. (2020). The effectiveness of quarantine and isolation determine the trend of the COVID-19 epidemic in the final phase of the current outbreak in China. *Int. J. Infect. Dis.* 96: 636–647.

6 Okba, N.M.A., Müller, M.A., Li, W. et al. (2020). Severe acute respiratory syndrome coronavirus 2-specific antibody responses in coronavirus disease patients. *Emerging Infect. Dis.* 26 (7): 1478–1488.

7 Chen, Y., Guo, Y., Pan, Y., and Zhao, Z.J. (2020). Structure analysis of the receptor binding of 2019-nCoV. *Biochem. Biophys. Res. Commun.* 525 (1): 135–140.

8 Feng, Z.-h., Cheng, Y.-r., Chen, J. et al. (2020). Chinese medical personnel against the 2019-nCoV. *J. Infect.* 80 (5): 578–606.

9 Mahapatra, S. and Chandra, P. (2020). Clinically practiced and commercially viable nanobio engineered analytical methods for COVID-19 diagnosis. *Biosens. Bioelectron.* 165: 112361.

10 Chan, J.F.-W., Yip, C.C.-Y., To, K.K.-W. et al. (2020). Improved molecular diagnosis of COVID-19 by the novel, highly sensitive and specific COVID-19-RdRp/Hel real-time reverse transcription-PCR assay validated *in vitro* and with clinical specimens. *J. Clin. Microbiol.* 58 (5): e00310-20.

11 Bai, Y., Yao, L., Wei, T. et al. (2020). Presumed asymptomatic carrier transmission of COVID-19. *JAMA* 323 (14): 1406–1407.

12 Nguyen, T., Duong Bang, D., and Wolff, A. (2020). 2019 novel coronavirus disease (COVID-19): paving the road for rapid detection and point-of-care diagnostics. *Micromachines* 11 (3): 306.

13 Corman, V.M., Landt, O., Kaiser, M. et al. (2020). Detection of 2019 novel coronavirus (2019-nCoV) by real-time RT-PCR. *Eurosurveillance* 25 (3): 2000045.

14 Zhang, T., Wu, Q., and Zhang, Z. (2020). Probable pangolin origin of SARS-CoV-2 associated with the COVID-19 outbreak. *Curr. Biol.* 30 (7), 1346–1351.e2.

15 Remuzzi, A. and Remuzzi, G. (2020). COVID-19 and Italy: what next? *Lancet* 395: 1225–1228.

16 Yip, C.C.-Y., Ho, C.-C., Chan, J.F.-W. et al. (2020). Development of a novel, genome subtraction-derived, SARS-CoV-2-specific COVID-19-nsp2 real-time RT-PCR assay and its evaluation using clinical specimens. *Int. J. Mol. Sci.* 21 (7): 2574.

17 Chauhan, D.S., Prasad, R., Srivastava, R. et al. (2020). Comprehensive review on current interventions, diagnostics, and nanotechnology perspectives against SARS-CoV-2. *Bioconjugate Chem.* 31 (9): 2021–2045.

18 Wang, C., Liu, Z., Chen, Z. et al. (2020). The establishment of reference sequence for SARS-CoV-2 and variation analysis. *J. Med. Virol.* 92 (6): 667–674.

19 Kim, J.-M., Chung, Y.-S., Jo, H.J. et al. (2020). Identification of coronavirus isolated from a patient in Korea with COVID-19. *Osong Public Health Res. Perspect.* 11 (1): 3.

20 Sah, R., Rodriguez-Morales, A.J., Jha, R. et al. (2020). Complete genome sequence of a 2019 novel coronavirus (SARS-CoV-2) strain isolated in Nepal. *Microbiol. Resour. Announce.* 9 (11): e00169-20.

21 Batlle, D., Wysocki, J., and Satchell, K. (2020). Soluble angiotensin-converting enzyme 2: a potential approach for coronavirus infection therapy? *Clin. Sci.* 134 (5): 543–545.

22 Zhou, Y., Hou, Y., Shen, J. et al. (2020). Network-based drug repurposing for novel coronavirus 2019-nCoV/SARS-CoV-2. *Cell Discovery* 6 (1): 1–18.

23 Tang, H. (2000). *Winners and Losers of EU Integration: Policy Issues for Central and Eastern Europe.* ACS-Publications.

24 Zou, J., Yin, J., Fang, L. et al. (2020). Computational prediction of mutational effects on SARS-CoV-2 binding by relative free energy calculations. *J. Chem. Inf. Model.* 60: 5794–5802.

25 Rabiee, N., Bagherzadeh, M., Ghasemi, A. et al. (2020). Point-of-use rapid detection of SARS-CoV-2: nanotechnology-enabled solutions for the COVID-19 pandemic. *Int. J. Mol. Sci.* 21 (14): 5126.

26 Chu, D.K., Pan, Y., Cheng, S.M. et al. (2020). Molecular diagnosis of a novel coronavirus (2019-nCoV) causing an outbreak of pneumonia. *Clin. Chem.* 66 (4): 549–555.

27 Li, Z., Yi, Y., Luo, X. et al. (2020). Development and clinical application of a rapid IgM-IgG combined antibody test for SARS-CoV-2 infection diagnosis. *J. Med. Virol.* 92: 1518–1524.

28 Wölfel, R., Corman, V.M., Guggemos, W. et al. (2020). Virological assessment of hospitalized patients with COVID-2019. *Nature* 581 (7809): 465–469.

29 Huang, C., Wang, Y., Li, X. et al. (2020). Clinical features of patients infected with 2019 novel coronavirus in Wuhan, China. *Lancet* 395 (10223): 497–506.

30 Ezhilan, M., Suresh, I., and Nesakumar, N. (2021). SARS-CoV, MERS-CoV and SARS-CoV-2: a diagnostic challenge. *Measurement* 168: 108335.

31 Chan, W.C. (2020). Nano research for COVID-19. *ACS Nano* 14: 3719–3720.

32 Lauster, D., Klenk, S., Ludwig, K. et al. (2020). Phage capsid nanoparticles with defined ligand arrangement block influenza virus entry. *Nat. Nanotechnol.* 15 (5): 373–379.

33 Lopez, A. (2020). Nano technology in the era of COVID-19. *Nanotechnol. Nanomater. Res.* 1: 1.

34 Park, G.-S., Ku, K., Baek, S.-H. et al. (2020). Development of reverse transcription loop-mediated isothermal amplification (RT-LAMP) assays targeting SARS-CoV-2. *J. Mol. Diagn.* 22: 729–735.

35 Sharma, S., Kumar, V., Chawla, A., and Logani, A. (2020). Rapid detection of SARS-CoV-2 in saliva: can an endodontist take the lead in point-of-care COVID-19 testing? *Int. Endodont. J.* 53: 1017–1019.

36 Udugama, B., Kadhiresan, P., Kozlowski, H.N. et al. (2020). Diagnosing COVID-19: the disease and tools for detection. *ACS Nano* 14 (4): 3822–3835.

37 Dong, Y., Wu, X., Li, S. et al. (2020). Comparative evaluation of 19 reverse transcription loop-mediated isothermal amplification assays for detection of SARS-CoV-2. *medRxiv* 11: 11.

38 Augustine, R., Hasan, A., Das, S. et al. (2020). Loop-mediated isothermal amplification (LAMP): a rapid, sensitive, specific, and cost-effective point-of-care test for coronaviruses in the context of COVID-19 pandemic. *Biology* 9 (8): 182.

39 Yonekawa, T., Watanabe, H., Hosaka, N. et al. (2020). Fully automated molecular diagnostic system "Simprova" for simultaneous testing of multiple items. *Sci. Rep.* 10 (1): 5409.

40 Yang, T., Wang, Y.-C., Shen, C.-F., and Cheng, C.-M. (2020). *Point-of-Care RNA-Based Diagnostic Device for COVID-19*. Multidisciplinary Digital Publishing Institute.

41 Wang, C., Horby, P.W., Hayden, F.G., and Gao, G.F. (2020). A novel coronavirus outbreak of global health concern. *Lancet* 395 (10223): 470–473.

42 Xiang, J., Yan, M., Li, H. et al. (2020). Evaluation of enzyme-linked immunoassay and colloidal gold-immunochromatographic assay kit for detection of novel coronavirus (SARS-Cov-2) causing an outbreak of pneumonia (COVID-19). *medRxiv* 395: 470–473.

43 Zhang, W., Du, R.-H., Li, B. et al. (2020). Molecular and serological investigation of 2019-nCoV infected patients: implication of multiple shedding routes. *Emerging Microbes Infect.* 9 (1): 386–389.

44 Tan, L., Wang, Q., Zhang, D. et al. (2020). Lymphopenia predicts disease severity of COVID-19: a descriptive and predictive study. *Signal Transduction Targeted Ther.* 5 (1): 1–3.

45 Pandey, L.M. and Pattanayek, S.K. (2013). Relation between the wetting effect and the adsorbed amount of water-soluble polymers or proteins at various interfaces. *J. Chem. Eng. Data* 58 (12): 3440–3446.

46 Disley, D.M., Cullen, D.C., You, H.-X., and Lowe, C.R. (1998). Covalent coupling of immunoglobulin G to self-assembled monolayers as a method for immobilizing the interfacial-recognition layer of a surface plasmon resonance immunosensor. *Biosens. Bioelectron.* 13 (11): 1213–1225.

47 Yang, S., Liu, J., Zheng, H. et al. (2020). Simulated revelation of the adsorption behaviours of acetylcholinesterase on charged self-assembled monolayers. *Nanoscale* 12 (6): 3701–3714.

48 Chen, W., Fu, M., Zhu, X., and Liu, Q. (2019). Protein recognition by polydopamine-based molecularly imprinted hollow spheres. *Biosens. Bioelectron.* 142: 111492.

49 Zuo, B., Li, S., Guo, Z. et al. (2004). Piezoelectric immunosensor for SARS-associated coronavirus in sputum. *Anal. Chem.* 76 (13): 3536–3540.

50 Pandey, L.M., Pattanayek, S.K., and Delabouglise, D. (2013). Properties of adsorbed bovine serum albumin and fibrinogen on self-assembled monolayers. *J. Phys. Chem. C* 117 (12): 6151–6160.

51 Maldonado, J., Estévez, M.C., Fernández-Gavela, A. et al. (2020). Label-free detection of nosocomial bacteria using a nanophotonic interferometric biosensor. *Analyst* 145 (2): 497–506.

52 Qiu, G., Gai, Z., Tao, Y. et al. (2020). Dual-functional plasmonic photothermal biosensors for highly accurate severe acute respiratory syndrome coronavirus 2 detection. *ACS Nano* 14 (5): 5268–5277.

53 Seo, G., Lee, G., Kim, M.J. et al. (2020). Rapid detection of COVID-19 causative virus (SARS-CoV-2) in human nasopharyngeal swab specimens using field-effect transistor-based biosensor. *ACS Nano* 14 (4): 5135–5142.

54 Kong, W. and Agarwal, P.P. (2020). Chest imaging appearance of COVID-19 infection. *Radiol. Cardiothorac. Imaging* 2 (1): e200028.

55 Qiu, H., Wu, J., Hong, L. et al. (2020). Clinical and epidemiological features of 36 children with coronavirus disease 2019 (COVID-19) in Zhejiang, China: an observational cohort study. *Lancet Infect. Dis.* 117: 6151–6160.

56 Chung, M., Bernheim, A., Mei, X. et al. (2020). CT imaging features of 2019 novel coronavirus (2019-nCoV). *Radiology* 295 (1): 202–207.

57 Sun, Q., Xu, X., Xie, J. et al. (2020). Evolution of computed tomography manifestations in five patients who recovered from coronavirus disease 2019 (COVID-19) pneumonia. *Korean J. Radiol.* 21 (5): 614–619.

58 Li, Y. and Xia, L. (2020). Coronavirus disease 2019 (COVID-19): role of chest CT in diagnosis and management. *AJR Am. J. Roentgenol.* 214 (6): 1280–1286.

59 Lippi, G., Lavie, C.J., and Sanchis-Gomar, F. (2020). Cardiac troponin I in patients with coronavirus disease 2019 (COVID-19): evidence from a meta-analysis. *Progr. Cardiovasc. Dis.* 63 (3): 390–391.

60 Rodriguez-Morales, A.J., Cardona-Ospina, J.A., Gutiérrez-Ocampo, E. et al. (2020). Clinical, laboratory and imaging features of COVID-19: a systematic review and meta-analysis. *Travel Med. Infect. Dis.* 34: 101623.

61 Henry, B.M., De Oliveira, M.H.S., Benoit, S. et al. (2020). Hematologic, biochemical and immune biomarker abnormalities associated with severe illness and mortality in coronavirus disease 2019 (COVID-19): a meta-analysis. *Clin. Chem. Lab. Med.* 58 (7): 1021–1028.

62 Cascella, M., Rajnik, M., Cuomo, A. et al. (2020). Features, evaluation, and treatment of coronavirus (COVID-19). In: *Statpearls* [Internet]. Stat Pearls Publishing.

63 Vukkadala, N., Qian, Z.J., Holsinger, F.C. et al. (2020). COVID-19 and the otolaryngologist: preliminary evidence-based review. *Laryngoscope* 130: 2537–2543.

64 Zhang, F, Abudayyeh, O.O., and Gootenberg, J.S. (2020). A protocol for detection of COVID-19 using CRISPR diagnostics, 8.

11

Development of Electrochemical Biosensors for Coronavirus Detection

Fulden Ulucan-Karnak[1], Cansu İ. Kuru[1,2], and Zeynep Yilmaz-Sercinoglu[3]

[1] *Ege University, Faculty of Science, Block E, Department of Biochemistry, Ankara Cd., Erzene Mah., Izmir, 35100, Turkey*
[2] *Buca Municipality Kızılçullu Science and Art Center, Doğuş Cd., Izmir, 35390, Turkey*
[3] *Marmara University, Faculty of Engineering, Department of Bioengineering, Fahrettin Kerim Gökay Cd., Istanbul, 34722, Turkey*

11.1 Introduction

Severe acute respiratory syndrome coronavirus 2 (SARS-CoV-2) or nCOVID-19 has become an enormous global health problem because of its rapid human to human transmission and lethality during 2020. Presently, there are few vaccine and treatment alternatives available for this disease; hence, early detection is still essential to overcome and control the pandemic. The existing detection methods have both advantages and disadvantages, such as low sensitivity, high cost, requirement of well-experienced personnel, and long time. Therefore, it is necessary to develop new detection methods. To make sure of the diagnosis, it is suggested to always use new methods together with the gold standard methods, such as reverse transcriptase polymerase chain reaction (RT-PCR), computer tomography (CT), etc.

11.2 Detection of Viral Infections

Diagnosis of viral infections is based on three main methods, which are detection of virus, viral DNA/RNA, and antibody production.

11.2.1 Detection of Virus

11.2.1.1 Electron Microscopy

Electron microscopy (EM) is a powerful tool and is accepted as a "gold standard" method to detect the morphology of the virus and the family the virus belongs to [1]. Although the molecular diagnosis is the leading technique today, EM is the mainstream technique for detection of viruses and unusual outbreaks. It might not detect the subfamily of the virus, but it can lead the way for more specific identification

Detection and Analysis of SARS Coronavirus: Advanced Biosensors for Pandemic Viruses and Related Pathogens,
First Edition. Edited by Chaudhery Mustansar Hussain and Sudheesh K. Shukla.

methods, e.g. biochemical assays. EM does not need organism-specific agents for recognition and specific solutions to keep the virus intact and alive.

Mutation of virus and effect of antiviral agents on virus can also be monitored via EM. For instance, elucidation of SARS spikes was achievable with EM. This kind of observation can help the scientists to understand the attachment of the virus to the host cell and how the virus departs from the host. Once these processes were enlightened in detail, these finding can pave the way for the discovery of new compounds blocking these processes and for vaccine development.

However, endogenous or extraneous viruses can confuse the diagnosis of the virus of interest. Mycoplasm contamination of viral cultures for instance, is another problem because mycoplasm infections may cause cytopathic effects similar to viral infections. Therefore, EM needs well-trained and experienced personnel, because there might be confusing interpretations. Effective concentration in order to obtain enriched virus is a necessity for easy detection and visualization of virus. Sample preparation, specimen source, and sampling time are important parameters [2]. Deformed viruses and cellular compartments may also lead to confusing interpretations [3].

11.2.1.2 Viral Culture

Viral culture is a good and effective way to diagnose COVID-19 infection, but unfortunately, it is not applicable and practical in large-extent diagnosis. Successful cultivation in human airway epithelial cells and in Vero E6 or Huh-7 cells takes almost three and six days, respectively. Special personal protective equipment (e.g. positive pressure suit), biosafety level 3 and 4 laboratories, and specialized/experienced personnel are needed for this purpose [4].

11.2.2 Detection of Viral DNA/RNA

11.2.2.1 Real-Time Reverse Transcriptase Polymerase Chain Reaction (RT-PCR)

The RT-PCR assay is recommended as a qualified and validated method by World Health Organization (WHO) to identify and confirm COVID-19 cases. This method detects unique RNA sequences for SARS-CoV-2 in the specimens of patients with COVID-19 symptoms. According to WHO recommendations, detection of three genes should be done by RT-PCR. These genes are E gene, N gene, and RNA-dependent RNA polymerase (RdRp) gene. E (envelope) gene is a conserved gene across betacoronaviruses [5]. N (nucleocapsid) gene is responsible for the expression of nucleocapsid protein, which is highly expressed during infection; thus, anti-N antibodies were found to be highly detected in the initial days of infection. Therefore, it is meaningful not only for the diagnosis of COVID-19, but also for the development of biological agents against the disease [6]. RdRp is an indispensible enzyme for the life cycle of RNA virus genome [7]. Detection of RdRp was stated to have highest analytical sensitivity and deployed in more than 30 laboratories in Europe. Combinative detection of these three genes guarantees the confirmed detection of infection while overcoming the risk of possible false-negative results, which can be derived from detection of a single gene for COVID-19 diagnosis [8, 9].

Despite being a qualified and sensitive detection method, RT-PCR is a tedious, expensive protocol and also needs special agents and experienced personnel. Mass screening of the population is not feasible with RT-PCR. Although the test completes in four to six hours, other aspects such as transportation of clinical samples increase this time period up to 24 hours [10]. Moreover, invasive sampling is a practical limitation and critical aspect for the outcome of the test. As stated in a number of studies [4, 8, 11, 12], the anatomic location of the sample is crucial for accurate diagnosis. Moreover, different viral loads, sampling time, and stage of the infection have non-negligible effect on the results of RT-PCR.

In order to avoid skipping possible COVID-19-positive patients, the following points should be kept in mind [13, 14].

(i) Good laboratory practice is needed,
(ii) High-quality PCR and extraction kits should be used to obtain a pure nucleic acid,
(iii) Proper sampling, sample handling, and transportation should be supplied,
(iv) Complementary action of CT, clinical symptoms, RT-PCR, and blood tests should be performed and the results should be interpreted together for a holistic evaluation about the disease, rather than depending on "one" test.

11.2.2.2 Microarrays

Microarrays developed for detection of viruses are based on the detection of products of multiplex PCR that uses specific primers. Selection of these primers affects the selectivity and specificity of microarrays. Therefore, investigation of genus- and/or species-specific sequences with bioinformatic approach is recommended. Moreover, sequences related to latent or lytic phase of the viral cycle should be taken into consideration for a more specific detection of the virus. For example, sequences expressed in the latent phase should be used to discriminate in the favor of the viruses on latent phase. This approach could also help to understand the disease course. It is reasonable to choose conserved/panviral sequences for microarray, but selection of sequences directly affects the discrimination property of the microarray.

Limitation of this method is its dependence of panviral sequences. Viruses without short regions of homology might be overlooked. Similar to EM, this technique is expensive because it needs specialized equipment. The demand of high-performance computing for bioinformatics is another disadvantage because most of the facilities do not have an access to such computational resources [15].

11.2.3 Detection of Post-infection Antibodies

Serological tests for detection of viral infections are pivotal for investigating the prevalence for SARS-CoV-2 infection among the population. It is an effective tool to detect the amount of infected (asymptomatic or oligosymptomatic) people who could be the carriers of infection for other people in the population. However, immunoglobulin G (IgG) antibodies against N-protein can be detected in the serum

4 days after the onset of COVID-19 at the earliest, and 14 days are needed to get a full seroconversion in most of the patients [16]. Host–pathogen interaction does not yield enough titers to be detected. Thus, serological tests cannot be used as a "primary" diagnostic tool for COVID-19 because of their time course-dependent limited sensitivity [10]. However, they can and must be used to evaluate whether the newly developed vaccine arises immune response or not.

N (nucleocapsid) proteins are synthesized intensively in course of viral infection, and S protein is essential for the attachment of virus to the host via binding to angiotensin-converting enzyme 2 (ACE2). Both N and S proteins are highly immunogenic, so this makes them target molecules in the serological methods and can be used for the development of diagnostic kits and vaccines [10].

Immunoglobulin M (IgM) antibodies are a "welcoming committee" for the viral infection. They reach their highest level in a week after the onset of the disease, but their titer in the blood decreases after three weeks. IgG, on the other hand, is cosynthesized with IgM, but their amount increases eight days after the start of the disease. IgG is in charge for ensuring the long-term memory and adaptive immunity against the infection [17].

So, most of the serological assays depend on the detection of IgM in the first place. However, it is better to examine the combination of IgG and IgM for trustworthy detection of the disease course, where IgM presence displays a recent exposure and IgG presence (in the absence of IgM) displays previous exposure to SARS-CoV-2 [18].

11.2.3.1 Lateral Flow Immunoassays (LFIAs)

Lateral flow immunoassays (LFIAs) are not much pronounced as RT-PCR and enzyme-linked immunosorbent assay (ELISA). Their principle of action is based on exploitation of selective and specific binding property of antibodies to certain antigens. Most of the LFIAs work as an on–off switch, which detects the presence or absence of a specific compound in the tested sample. So, the reading can be performed with "naked eye" and interpretation of the result is easy. No specific equipment is needed for detailed analysis of the results (Figure 11.1). There are also quantitative LFIAs being developed. Readers are kindly directed to a comprehensive review article for detailed information [19].

Their ease of application, quick outcome, and point-of-care usage make LFIAs to be widely used in many areas such as environment monitoring and investigation of food quality, besides their use in clinical application. They do not need refrigeration for prolonging the shelf life, which makes it very suitable to be used in the developing countries [21]. A small amount of sample is enough to use this test. LFIAs are being developed to detect COVID-19.

11.2.3.2 Enzyme-Linked Immunosorbent Assay (ELISA)

ELISA is a robust and mature technique capable of specific detection of numerous analytes in various samples. It is a routine protocol in clinics, but owing to its accurate, quantified, and reproducible results, it has been applied in various areas such as biotechnology, environment monitoring, and food quality control [22].

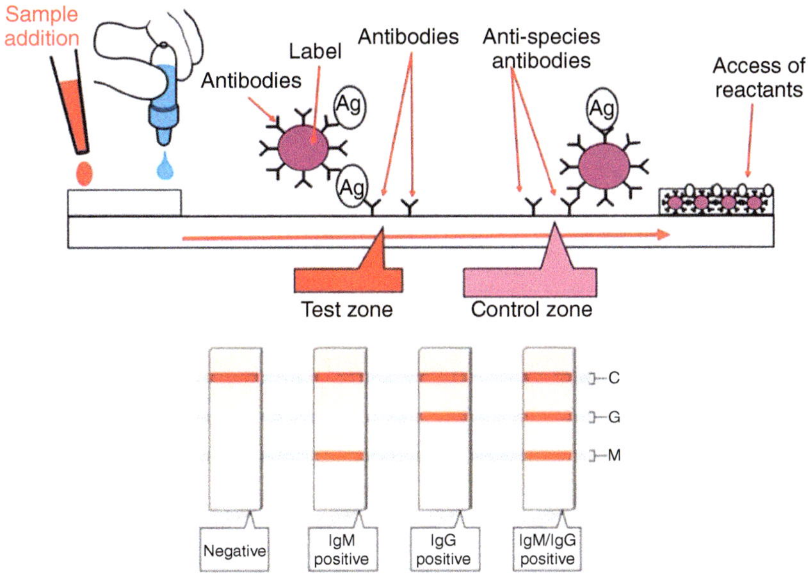

Figure 11.1 Universal configuration of lateral flow immunoassays (LFIA) and results of the LFIA strips. C is colored in all cases, which serves as a control line. Source: Adapted from Urusov et al. [19] and Ghaffari et al. [20].

Principle action of ELISA depends on the detection of specific antibody–antigen interaction via enzyme catalysis. Non-specific binding can be a disadvantage of the technique (Figure 11.2). Sufficient incubation time is needed to facilitate observable antibody–antigen interaction, so ELISA is a relatively slow process compared to LFIAs and chemiluminescent immunoassay (CLIA). Moreover, moderately sensitive detection of target proteins and non-heat-tolerant enzymes with short shelf life are affecting the performance and capability of ELISA. However, there is a recent study that covers the latest advances to improve the performance and sensitivity of ELISA. Readers are kindly directed to this article [23].

Practical drawbacks, such as requirement for experienced personnel to conduct the tests and special equipment (e.g. microplate readers) to determine the results, are common in ELISA protocol as in CLIA protocol. High sample volume is needed for ELISA compared to LFIA and CLIA.

11.2.3.3 Chemiluminescent Immunoassay (CLIA)

CLIA is based on detection of luminescence emitted as a result of chemical catalysis between the analyte and the luminescent molecule. Compared to absorbance, luminescence results in "absolute" measurement because the absorbance is relative to the "chosen" blank solution, which is mostly arbitrary and modified according to the study. Detection of analytes with low concentration with a high intensive, quick analytical signal (without interfering emission [e.g. high sensitivity]) is the key property of CLIA. Therefore, accurate and fast interpretation of the results is possible [24]. However, requirement of a specialized laboratory and educated personnel is the drawback of this method.

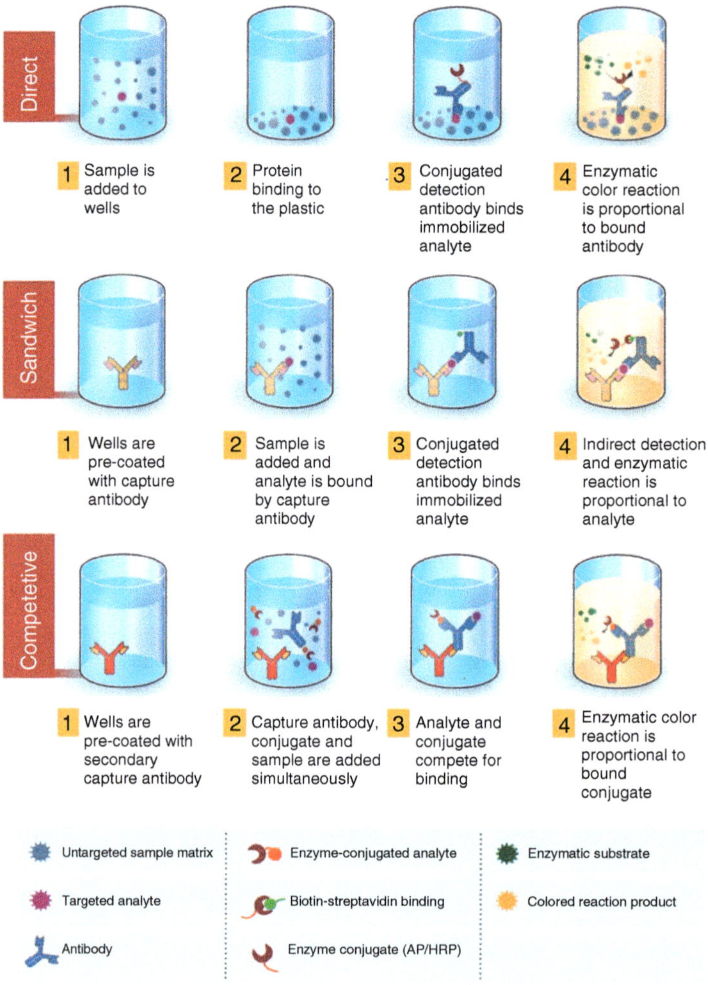

Figure 11.2 Schematic presentation of ELISA types. Source: Ghaffari et al. [20]. Licensed under CC BY 4.0.

Because it is good at detecting low amount of the target molecule, it can be used for detecting specific IgM in the early stages of COVID-19, but the sensitivity should be increased with further improvements in the assay [20]. In this respect, Lin et al. developed a CLIA for specific detection of both IgM and IgG in which the sensitivity and specificity of the assay were improved by using a recombinant nucleocapsid antigen and tosyl magnetic beads [25].

It is always needed to develop fast and reliable tests for COVID-19 diagnosis. Molecular methods can detect low concentration of viral particles, but they can be time-consuming and expensive and a well-equipped lab is needed, as mentioned already. Point-of-care-based biosensors are alternatives to specify viral infections with their portability, low response time, and high sensitivity properties. Early detection and diagnosis techniques are mandatory in order to enhance the control of infection, vaccine research, and treatment of diseases.

11.3 Current Biosensor Candidates for COVID-19 Detection

Biosensing technology focuses on detecting biological, chemical, and biochemical agents engaging biomimetic recognition or biologically derived materials, while either undergoing a biochemical reaction or highly specific binding of the target molecule. This binding can be transduced to a measurable signal either directly or employing signaling molecules such as fluorophores, enzymes, or optically/electrochemically active compounds [26].

Biosensor technologies are being used as diagnostic tools in health care facilities and they are emerging technologies that are more articulated nowadays. Biosensors can provide monitorization of biomarkers to diagnose COVID-19 patients in mild or critical stages and are helpful for the evaluation of anti-inflammation treatments. Early, sensitive, accurate, and rapid detection properties of biosensors make them available to be used in smartphones [27].

Viral biosensors detecting respiratory diseases are important alternatives to traditional diagnostic methods. They can offer low-cost, sensitive, rapid, and portable platforms than laboratory-based methods. Nucleic acid, CRISPR (Clustered Regularly Interspaced Short Palindromic Repeats)-Cas9-based paper strip, aptamer, and nanoparticle (NP)-based electrochemical, optical, and surface plasmon resonance biosensors are the established biosensors for the detection of RNA viruses. These systems provide more promising diagnosis in the COVID-19 pandemic that has already drastically affected all humanity and world economies [28].

11.3.1 Electrochemical Biosensors for SARS-CoV-2 Detection

Biosensors are analytical devices that have the capability of converting biological signals into electrical signals [29]. The biological element of a biosensor can be nucleic acids, antibodies, enzymes, cell receptors, microorganisms, organelles, and tissue. A biosensor can recognize an analyte via a transducer, and hence biosensors are named because of the type of the transducer. There are different biosensors available, such as electrochemical biosensors, physical biosensors, optical biosensors, and wearable biosensors [30]. The electrochemical biosensors to be developed have to meet the following requirements: (i) low limit of detection (LOD) for early stage detection, (ii) low detection time, (iii) selectivity, and (iv) sensitivity [31]. The schematic representation of an electrochemical biosensor is given in Figure 11.3.

Biosensor systems have accomplished selective and sensitive real-time detection of pathogens without the need of specific labeling and sample preparation stages. Moreover, the detection can be determined via several transducers or recognition

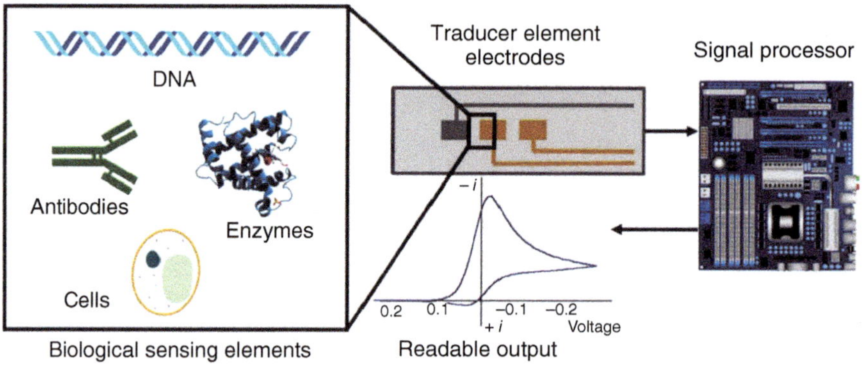

Figure 11.3 Schematic representation of an electrochemical biosensor. Source: Hernandez-Vargas et al. [32]. Licensed under CC BY 4.0.

mechanisms in differing matrices and environments (e.g. bodily fluids and surfaces of food and objects) [33].

Optical and electrical detection of viruses, especially influenza types, are realized to be fast and easy to use, and they provide inexpensive quantitative and qualitative analysis together with real-time monitoring [34]. In the literature, there are several electrochemical and optical biosensors with different measurement techniques and electrodes that detect respiratory viral infections. Although electrochemical transducers are mostly studied for viral determination, on the other hand optical transducers provide high sensitivity and efficiency.

Detection of SARS-CoV-2 is based on several detection methods, such as RT-PCR, LAMP (loop-mediated isothermal amplification), CRISPR, LFIA, ELISA, and so on. Every technique has some drawbacks; therefore, new detection methods are always imperative.

For instance, RT-PCR needs expensive thermocycler and professional staff. Moreover, false-negative results can be developed depending on the sampling time and the quality of the collected sample and inappropriate sample storage. LAMP methods showed lower sensitivity. The main disadvantage of LFIA is the elongated time of antibody production. LFIA and ELISA cannot be used for detection on early stage. As the existing tests are not adequate to determine infected people in public area, there is a great requirement of new point-of-care (POC) devices that can be able to detect infections without trained staff [35, 36].

Nanoparticles (NPs) can be synthesized with "molecular tailoring" for specific purposes, e.g. large surface area, small size, and unique surface properties. With the aid of these enhanced properties, nanoparticles can be used in detection methods, such as RT-PCR, ELISA, and RT-LAMP. Different kinds of NPs including metal, carbon nanotubes, silica, quantum dots (QDs), and polymeric nanoparticles have been used for detecting virus. Magnetic and metal and quantum dots and metal nanoislands (NIs) have been applied for detection of coronavirus because of their capability to plasmon resonance [37, 38].

Molecular imprinting technology can be used for creating synthetic receptors complementary to the shape and orientation of the target molecule. Because of its synthesis strategy, molecular imprinted materials have high sensitivity and selectivity. Therefore, they can be utilized in biosensors, especially to detect SARS-CoV-2. For this purpose, virus specific aptamers or antibodies can be used as a recognition element of the biosensor [39].

We have learned things of paramount importance during the COVID-19 outbreak. Even in countries that can be regarded as successful in controlling the pandemic such as Japan, South Korea, and Australia, expeditious development of rapid, portable, and highly sensitive testing and screening tools and massive production were prerequisites [40].

A new biosensor to be developed for detection of COVID-19 needs to be cost-effective, easy to use, disposable, and highly selective and sensitive. In addition it should respond in a short time. Paramount aspects of the development strategy are selectivity and sensitivity. Combinatory software with novel biosensors might be developed in order to determine viral pandemics [41]. Biomarkers or indicators of COVID-19 can be classified as RNA, whole virus, antigens, and antibodies. Artificial intelligence was used to discover other potential biomarkers of COVID-19 with clinical results. For example, white blood cell (WBC) count, eosinophil ratio and count, serum amyloid A (SAA), cytokines, and the level of interferon-γ (IFN-γ) and interleukin-6 (IL-6), IL-10, and IL-2 in the peripheral blood can be utilized as biomarkers [12].

As we still try to cope with COVID-19 pandemic, it is crucial to mention that the epidemiological characteristics of COVID-19 still have not been clearly understood. It is a big challenge for the detection of the disease. Developing artificial intelligence or smartphone controlling POC-based label-free sensing techniques can not only track the spreading of the disease around the world but also allow to form a data library. It is required that development of POC using electrochemical label-free technologies can be arsenal to combat COVID-19 and other infectious diseases [42]. Early diagnosis is the only way to manage the pandemic [27].

The electrochemical biosensors are based on electrochemically monitoring the reaction, which typically produces a measurable charge accumulation and current (amperometry) or potential (potentiometry) or the changes in the conductive properties (conductometry). The use of electrochemical impedance spectroscopy (EIS) (impedimetry)-based biosensors, which monitors both resistance and reactance, is in broad dissemination.

Electrochemical biosensors can be investigated as biocatalytic devices and affinity sensors, depending on their capability of biological recognition. Biocatalytic devices include whole cells and tissue slices or enzymes that recognize the target analyte and produce electroactive signals [43].

11.3.1.1 Impedimetry

EIS is an efficient method for detection of biological signals. It can detect any tiny changes on the solution–electrode interface. The interaction between the target and captured molecules on the electrode surfaces decreases the flux of the redox probe to the electrode surface, and these interaction increases the impedance [44]. EIS is commonly used to characterize materials, monitor binding, and detect samples without labeling. Impedimetric detection of COVID-19 is emerging. EIS method holds a potential to detect COVID-19, thanks to its label-free detection, rapid response, real-time monitoring, low LOD, and cost-effective properties [45]. EIS provides data with respect to diffusion, surface adsorption, charge transfer, and ion exchange [2].

In the study of Miodek et al. gold electrode was used with a specific anti-PB1-F2 antibody, which is related to "PB1-F2 Influenza pathogenesis" proteins such as PA-X, PB1-F2, NS1, and NS2. differential pulse voltammetry (DPV) and cyclic voltammetry (CV) measurements were resulted with two linear ranges as 50–300 nM to 0.5–1.5 mM of PB1-F2. LOD was on the 0.42 nM level [46].

Hushegyi et al. developed an impedimetric glycan biosensor for selective detection of H3N2 influenza viruses. The biosensor has a LOD of 5 aM. The biosensor can detect H3N2 viruses selectively, and it has a sensitivity ratio of 30 over H7N7 influenza viruses. Such glycan-based biosensors have great advantages over antibody-based detection of influenza viruses because glycans are natural viral receptors to detect subtypes specifically [47].

Earth is not inexperienced with viral pandemics. For example, Zika virus had raised great concern in the United States. For its detection, Ricotta et al. developed a POC diagnostic system that uses a gold chip and a 3D molecular imprinting technique. According to CV and EIS measurements, this developed system can detect 10^{-1} pfu/ml ZIKV in a buffered solution under 20 minutes without any pretreatment. This sensor was tested against dengue virus at clinical viral loads and showed no cross-reactivity. The high sensitivity and high selectivity were demonstrated in dengue virus detection and these results posed a great potential in order to use this biosensor for rapid and accurate screening of flaviviruses [48].

In the study of Rashed et al. a preprinted electrochemical impedance-based detector for rapid detection of SARS-CoV-2 antibodies was reported. This developed technique was successful when compared with standard ELISA. The R^2 value was 0.9. The applicability of using a quantitative EIS technique was demonstrated for accurate and rapid detection of SARS-CoV-2 antibodies [49].

11.3.1.2 Potentiometry

Potentiometric measurements are based on polymeric membrane ion-selective electrodes (ISEs). These type of sensors have advantages such as rapid response, low cost, small size, ease of use, and resistant to turbid and color interferences. ISEs have some unique features compared to other methods, such as providing information about the free ion concentration (activity) and independence of the sample volume [50].

In Hai et al.'s study, trisaccharide-grafted conducting polymers were developed for high sensitivity and specificity for detection of human influenza A virus (H1N1). Quartz crystal microbalance (QCM) and potentiometry-based analysis were used for detection of specific interaction of 2,6-sialyllactose with hemagglutinin of the human influenza A virus (H1N1), with a sensitivity of 2 orders of magnitude than commercial kits [51].

In a study of Mavrikou et al. human chimeric spike S1 antibody-based potentiometric electrochemical biosensor was developed in order to detect SARS-CoV-2. The device was designed with polydimethylsiloxane (PDMS) layer-covered gold screen-printed electrodes. This novel biosensor has response time as short as three minutes with a detection limit of 1 fg/ml and a semilinear range 10 fg and 1 μg/ml. Moreover, there was no cross-reactivity observed against the SARS-CoV-2 nucleocapsid protein [52].

11.3.1.3 Conductometry

The conductivity (κ) is a measurable property of the electrolyte solution. Conductivity can be measured with conductometric sensors. In most conductometric sensors, the electrodes that are electronic conductors are metallic [53]. Conductometric methods are non-selective, and they need modified surfaces for enhanced selectivity. These methods are appealing because of their simplicity and low cost, and in addition, there is no need for reference electrodes. Gas, humidity, and oxygen sensors can be detected with conductometry [54].

Luo et al. designed a biosensor which blends conductometric immunoassay and capillary separation for detection of bovine viral diarrhea virus (BVDV) and *Escherichia coli* O157:H7. According to the result of conductance vs. time, the biosensor detection time is eight minutes, and the detection limit is 61 CFU/ml for bacterial and 10^3 CCID/ml for viral samples [55].

Lee et al. designed a label-free, ultra-sensitive, low-cost electric immunoassay for swine influenza virus (SIV) H1N1 detection. The assay is grounded on single-walled carbon nanotubes (SWCNTs) and antibody–virus complexes. The sensor selectivity showed the normalized resistance shift of 12% as a background. The system gave a detection limit as 180 $TCID_{50}$/ml of SIV, and it can be interpreted as the developed system could be used as a lab-on-a chip system and point-of-care testing (POCT) [56].

Shen et al. developed a detection system using silicon nanowire (SiNW), which is selective for the 8 iso prostaglandin F2 alpha (PGF 2a) region of influenza A viruses. This system shows high selectivity and gives results in minutes [57].

11.3.1.4 Voltammetry

Voltammetry is generated from a voltamperometry term, and it expresses that the current is measured as a function of voltage. The system needs three electrodes: working electrode, auxiliary electrode, and reference electrode. In this system, the potential difference at the interface of the working electrode is controlled by a potentiostat. Most potentiostats apply a voltage between the working and auxiliary electrodes [58].

Voltammetry techniques include many types such as steady-state microelectrode, linear sweep, square wave, pulse, stripping, cyclic alternating current (AC), and hydrodynamic voltammetry depending on the mode of the potential control. Among them, cyclic voltammetry (CV) is used frequently [59].

In a study of Abad-Valle, a hybridization-based genosensor was developed with a gold film. This gold-based material is used for transduction and immobilization surface. As a target, a sequence that encodes a short lysine-rich region, unique to SARS virus, was chosen. Cyclic voltammetry, alternating current voltammetry (ACV), DPV, and square wave voltammetry (SWV) were used for analysis. An enhancement of the ACV and SWV signal was observed. LOD (6 pM) was obtained with amplified indirect electrochemical detection by interaction with alkaline phosphatase-labeled streptavidin [60].

Layqah and Eissa described an immunosensor for the determination of MERS-CoV based on a competitive assay of gold nanoparticle-modified carbon electrodes. Biomarker was selected as the recombinant spike protein S1. The electrode array provides multiplexed detection of different CoVs. Electrochemical measurements were recorded by using SWV. Linear responses of 0.001–100 ng/ml for MERS-CoV and 0.01–10 000 ng/ml for HCoV were observed. The detection time was 20 minutes and the detection limit was 0.4 for HCoV and 1.0 pg/ml for MERS-CoV. The method is highly selective, accurate, sensitive, and suitable for application to nasal samples and results in a single step [61].

Mahari's group reported a biosensor device (eCovSens) and compared it with a commercial potentiostat for the nCOVID-19 antigen (nCOVID-19Ag) detection in saliva samples. The eCovSens sensor was designed using gold nanoparticle (AuNPs) and fluorine-doped tin oxide (FTO) electrode and the sensor surface immobilized with nCOVID-19 monoclonal antibody (nCOVID-19Ab) on screen-printed carbon electrode (SPCE). The FTO-based immunosensor and eCovSens have high sensitivity for nCOVID-19Ag detection between the concentration rate of 1 fM to 1 μM within 10–30 seconds. As a result, a developed device can be able to detect nCOVID-19Ag at 10 fM concentration. LOD was 90 fM with eCovSens and 120 fM with potentiostat [62].

11.3.1.5 Amperometry

Amperometry is a technique that involves the oxidizing or reducing potential to a working electrode and the subsequent measurement of a steady-state current. Usually, the magnitude of the measured current is dependent on the concentration of the oxidized or reduced substance [63]. The applied potential is stepped to and then held at a constant value, and the resulting current is acquired as a function of time [64].

Liu et al. developed a sensitive, rapid, and real-time biosensor for rotavirus detection based on the micropatterned reduced graphene oxide based field-effect transistor (MRGO-FET). It resulted with current change as 31.45% with high sensitivity. An LOD of 10^2 pfu was determined for rotavirus and is superior to ELISA method [65].

Sayhi et al. reported a system for the detection and isolation of H9N2 influenza A virus. Firstly, anti-matrix protein 2 (M2) antibody was bound to magnetic nanoparticles for using virus isolation from allantoic fluid. Secondly, fetuin A was bound to gold nanoparticles and used to detect the virus by taking the advantage of fetuin–hemagglutinin interaction. Measurements were made with cyclic voltammetry and chronoamperometry and results showed that cathodic current decreases gradually in the range of 8–128 heamagglutination titer (HAU) with a LOD of more than 8 HAU [66].

Developing an early diagnosis system for COVID-19 poses crucial importance for containing the outbreak. Seo et al. developed a field-effect transistor (FET)-based biosensing device for SARS-CoV-2 detection in samples. The sensor was fabricated with SARS-CoV-2 spike protein-coated graphene sheets of the FET. Cultured virus, antigen protein, and nasopharyngeal swab specimens of COVID-19 patients were used to assess the performance of the sensor. The developed FET biosensor can be able to detect the SARS-CoV-2 spike protein at concentrations of 1 and 100 fg/ml of clinical transport medium without pretreatment or labeling. In addition, the FET sensor has an LOD of 1.6×10^1 pfu/ml in culture medium and an LOD of 2.42×10^2 copies/ml in clinical samples [67].

11.4 Conclusions

The COVID-19 pandemic continues and the necessity of diagnostic tests with high reliability, sensitivity, and specificity is pronounced more often by the scientific community. For braking the speed of the pandemic, detection of the people encountered SARS-CoV-2 is primary. In 80% of SARS-CoV-2 cases, a mild illness or asymptomatic carriage is observed. Therefore, diagnostic tests are imperative in identifying silent infections, evaluating the patient's immune response, predicting the progression of the disease, understanding the modes of transmission of the virus, and determining donors for the application of plasma therapy. Diagnosis of COVID-19 is possible via direct detection of the virus or by showing the host–pathogen interaction through specific antibodies produced. For detecting COVID-19 virus, different serological tests were developed. These tests were preferred by governments for large-scale seroprevalence studies. However, since there are still huge gaps in understanding the processes related to SARS-CoV-2 and immune system interactions, this may impose an unfavorable effect on utilization of these tests [20].

Biosensors play a crucial role in early diagnosis of COVID-19 infection by highly sensitive and rapid detection of virus antigens. Sensors are based on detecting surface genetic materials and proteins of virus. Multiplex biosensors combining detection of different biomarkers can be substituting tools to boost the accuracy of virus detection. The reproducibility and reliability characteristics of biosensors should be further enhanced by developing the machine learning-based signal processing platforms that have abilities to direct readout of results. For asymptomatic cases, POC biosensor devices should be developed to test the absence or presence of SARS-CoV-2 in each individual. Furthermore, use of nanomaterials combined with electrochemical diagnostic methods is encouraging in virus detection, in consequence of their excellent sensitivity and selectivity. Advanced smartphone-based biosensing devices, wearable biosensors, and calorimetric strips should be developed urgently not only to limit the quick spread of COVID-19 but also to monitor asymptomatic patients [27].

As of 2020, we have to learn and understand the symptoms, diagnosis, and treatment methods of a disease that we barely know. COVID-19 started as an epidemic and continued as a pan(dem)ic. The main strategy must be rapid testing, contact tracing, isolation of infected patients, and prudent self-isolation of contacts for reducing the number of new cases. This strategy can be called as "seek and destroy." All of these interventions have to be in coordination with normal life styles, because life should go on. During the pandemic, lockdowns and more strict prohibitions were discussed and applied according to the number of cases in each country.

Many diagnostic tests were developed as early as COVID-19 was declared as a pandemic. All developed tests are still in need for further optimization in order to reach the gold standard of specificity and sensitivity. Extensive clinical and epidemiological validation should be applied to catch the possible "bugs" of those tests (i.e. false-positive/negative results). Traditional techniques are time-consuming and may have false-positive outputs. Therefore, results of diagnostic methods should be interpreted together with other clinical data such as CT scan, complete blood count, and physical examination. In addition, they need to be combined with computational tools in order to expand data analysis [68, 69]. Moreover, a standardized clinical cutoff value should be implemented among all novel assays to determine the right choice of therapy and follow-up of successful therapy for survival.

Although genomic regions with low mutation frequency are used as targets in diagnostic tests, we are still oblivious about the mutation rates as the pandemic is going on. Regional and global virus typing is crucial in this respect. Similarity between viral strains and polymorphism should be kept in mind. Validation of the novel assays should be repeated by a third independent party in a worldwide population, not just with the clinically available samples from the most infected areas.

Almost all countries are working on several diagnostic methods for COVID-19 since December 2019. There are many literature studies, patents, and products in the industry. Scientific researchers and working groups focused on developing more rapid, sensitive, and selective methods and converting these works to the product such as POCT kits, lateral flow assays, biosensors, and so on. In the near future, emerging new technologies such as portable RNA extraction preps, CRISPR-Cas-based paper strip, aptamer-based systems, graphene-FET, nanoparticle-based electrochemical biosensors, optical biosensors, and surface plasmon-based innovative platforms can be developed. These developed systems could be efficient ways of rapid, sensitive, and promising biosensing diagnostic devices for COVID-19 and other pandemics [28]. More studies should be conducted on developing a low-cost single biosensor system, which has all reagents on chip and simplifies user steps. Specially, colorimetric detection-based systems represent the most common approach for rapid on-site diagnosis because of its simplicity and providing naked eye observation signal. Smartphone applications provide more accurate quantification readout in POC systems. Future studies should improve the function of smartphones and applications that enable data storage and on-site data analysis to track the health status of the patient. Emerging POC-based biosensors could identify and manage the spread of COVID-19 [70]. Along with these developments, the scientific standards should not be lowered and ethical issues should not be neglected. Standard quality controls and well-validated processes should always be the gold standard for the commercialization of diagnostic tests. COVID-19 pandemic can only be conquered by the complementary action of translational medicine. All parties in this field – academy, industry, and community – are responsible for suppressing the spread of the pandemic.

References

1 Yuan, X., Yang, C., He, Q. et al. (2020). Current and perspective diagnostic techniques for COVID-19. *ACS Infect. Dis.* 6: 1998–2016. https://doi.org/10.1021/acsinfecdis.0c00365.

2 Ekrami, E., Pouresmaieli, M., Barati, F. et al. (2020). Potential diagnostic systems for coronavirus detection: a critical review. *Biol. Proced. Online* 22 (1): 21. http://www.ncbi.nlm.nih.gov/pubmed/32884452.

3 Goldsmith, C.S. and Miller, S.E. (2009). Modern uses of electron microscopy for detection of viruses. *Clin. Microbiol. Rev.* 22: 552–563. http:/cmr.asm.org/.

4 Yan, Y., Chang, L., and Wang, L. (2020). Laboratory testing of SARS-CoV, MERS-CoV, and SARS-CoV-2 (2019-nCoV): current status, challenges, and countermeasures. *Rev. Med. Virol.* 30 (3). https://onlinelibrary.wiley.com/doi/abs/10.1002/rmv.2106.

5 Bianchi, M., Benvenuto, D., Giovanetti, M. et al. (2020). SARS-CoV-2 envelope and membrane proteins: structural differences linked to virus characteristics? *Biomed Res. Int.* 2020, 4389089. https://doi.org/10.1155/2020/4389089.

6 Zeng, W., Liu, G., Ma, H. et al. (2020). Biochemical characterization of SARS-CoV-2 nucleocapsid protein. *Biochem. Biophys. Res. Commun.* 527 (3): 618–623. /pmc/articles/PMC7190499/?report=abstract.

7 Kocabas, F., Turan, R.D., and Aslan, G.S. (2015). Fluorometric RdRp assay with self-priming RNA. *Virus Genes* 50 (3): 498–504. https://pubmed.ncbi.nlm.nih.gov/25749997/.

8 Chan, J.F.W., Yip, C.C.Y., To, K.K.W. et al. (2020). Improved molecular diagnosis of COVID-19 by the novel, highly sensitive and specific COVID-19-RdRp/Hel real-time reverse transcription-PCR assay validated in vitro and with clinical specimens. *J. Clin. Microbiol.* 58 (5) https://doi.org/10.1128/JCM.00310-20. /pmc/articles/PMC7180250/?report=abstract.

9 Peñarrubia L, Ruiz M, Porco R, Rao SN, Juanola-Falgarona M, Manissero D, et al. Multiple assays in a real-time RT-PCR SARS-CoV-2 panel can mitigate the risk of loss of sensitivity by new genomic variants during the COVID-19 outbreak. 2020. https://doi.org/10.1016/j.ijid.2020.06.027

10 Kubina, R. and Dziedzic, A. (2020). Molecular and serological tests for COVID-19. A comparative review of SARS-CoV-2 coronavirus laboratory and point-of-care diagnostics. *Diagnostics* 10. https://www.mdpi.com/2075-4418/10/6/434.

11 World Health Organisation (2020). Diagnostic testing for SARS-CoV-2. https://www.who.int/publications/i/item/diagnostic-testing-for-sars-cov-2 (accessed 10 March 2021).

12 Cui, F. and Zhou, H.S. (2020). Diagnostic methods and potential portable biosensors for coronavirus disease 2019. *Biosens. Bioelectron.* 165: 112349.

13 Tahamtan, A. and Ardebili, A. (2020). Real-time RT-PCR in COVID-19 detection: issues affecting the results. *Expert Rev. Mol. Diagn.* 20: 453–454. https://pubmed.ncbi.nlm.nih.gov/32297805/.

14 Feng, H., Liu, Y., Lv, M., and Zhong, J. (2020). A case report of COVID-19 with false negative RT-PCR test: necessity of chest CT. *Jpn. J. Radiol.* 38 (5): 409–410. https://pubmed.ncbi.nlm.nih.gov/32266524/.

15 Fischer, K., Urisman, A., and DeRisi, J. (2008). Diagnostic techniques: microarrays. In: *Encyclopedia of Virology*, 14–18. Elsevier. https://linkinghub.elsevier.com/retrieve/pii/B9780123744104007044.

16 Das, S.K. (2020). The pathophysiology, diagnosis and treatment of corona virus disease 2019 (COVID-19). *Indian J. Clin. Biochem.* 35 (4): 385–396. http://link.springer.com/10.1007/s12291-020-00919-0.

17 Caruana, G., Croxatto, A., Coste, A.T. et al. (2020). Diagnostic strategies for SARS-CoV-2 infection and interpretation of microbiological results. *Clin. Microbiol. Infect.* 26: 1178–1182. https://doi.org/10.1016/j.cmi.2020.06.019.

18 Hou, H., Wang, T., Zhang, B. et al. (2020). Detection of IgM and IgG antibodies in patients with coronavirus disease 2019. *Clin. Transl. Immunol.* 9 (5) https://doi.org/10.1002/cti2.1136. https://onlinelibrary.wiley.com/doi/10.1002/cti2.1136.

19 Urusov, A.E., Zherdev, A.V., and Dzantiev, B.B. (2019). Towards lateral flow quantitative assays: detection approaches. *Biosensors* 9 (3): 89. https://doi.org/10.3390/bios9030089. https://www.mdpi.com/2079-6374/9/3/89.

20 Ghaffari, A., Meurant, R., and Ardakani, A. (2020). COVID-19 serological tests: how well do they actually perform? *Diagnostics* 10 https://doi.org/10.3390/diagnostics10070453. https://www.ncbi.nlm.nih.gov/pmc/articles/PMC7400479/.

21 Koczula, K.M. and Gallotta, A. (2016). Lateral flow assays. *Essays Biochem.* 60 (1): 111–120. https://www.ncbi.nlm.nih.gov/pmc/articles/PMC4986465/.

22 Hosseini, S., Vázquez-Villegas, P., Rito-Palomares, M., and Martinez-Chapa, S.O. (2018). Advantages, disadvantages and modifications of conventional ELISA. In: *SpringerBriefs in Applied Sciences and Technology*, 67–115. Springer Verlag. https://doi.org/10.1007/978-981-10-6766-2_5.

23 Zhao, Q., Lu, D., Zhang, G. et al. (2021). Recent improvements in enzyme-linked immunosorbent assays based on nanomaterials. *Talanta* 223: 121722.

24 Cinquanta, L., Fontana, D.E., and Bizzaro, N. (2017). Chemiluminescent immunoassay technology: what does it change in autoantibody detection? *Autoimmun. Highlights* 8 (1). https://www.ncbi.nlm.nih.gov/pmc/articles/PMC5483212/.

25 Lin, D., Liu, L., Zhang, M. et al. (2020). Evaluations of the serological test in the diagnosis of 2019 novel coronavirus (SARS-CoV-2) infections during the COVID-19 outbreak. *Eur. J. Clin. Microbiol. Infect. Dis.*: 1–7. https://doi.org/10.1007/s10096-020-03978-6.

26 Morales-Narváez, E. and Dincer, C. (2020). The impact of biosensing in a pandemic outbreak: COVID-19. *Biosens. Bioelectron.* 163: 112274.

27 Asif, M., Ajmal, M., Ashraf, G. et al. (2020). The role of biosensors in COVID-19 outbreak. *Curr. Opin. Electrochem.* 23: 174–184. https://doi.org/10.1016/j.coelec.2020.08.011. https://www.ncbi.nlm.nih.gov/pmc/articles/PMC7500281/.

28 Samson, R., Navale, G.R., and Dharne, M.S. (2020). Biosensors: frontiers in rapid detection of COVID-19. *Biotech* 10 (3): 385. https://doi.org/10.1007/s13205-020-02369-0.

29 Mehrotra, P. (2016). Biosensors and their applications – a review. *J. Oral Biol. Craniofac. Res.* 6: 153–159. https://www.sciencedirect.com/science/article/abs/pii/S2212426815001323.

30 Bahl, S., Javaid, M., Bagha, A. et al. (2020). Biosensors applications in fighting COVID-19 pandemic. *Apollo Med.* 17: 221–223. https://doi.org/10.4103/am.am_56_20. https://www.apollomedicine.org/article.asp?issn=0976-0016;year=2020;volume=17;issue=3;spage=221;epage=223;aulast=Bahl.

31 Karbelkar, A.A. and Furst, A.L. (2020). Electrochemical diagnostics for bacterial infectious diseases. *ACS Infect. Dis.* 6: 1567–1571. https://doi.org/10.1021/acsinfecdis.0c00342.

32 Hernandez-Vargas, G., Sosa-Hernández, J.E., Saldarriaga-Hernandez, S. et al. (2018). Electrochemical biosensors: a solution to pollution detection with reference to environmental contaminants. *Biosensors* 8. https://pubmed.ncbi.nlm.nih.gov/29587374/.

33 Cesewski, E. and Johnson, B.N. (2020). Electrochemical biosensors for pathogen detection. *Biosens. Bioelectron.* 159: 112214. https://linkinghub.elsevier.com/retrieve/pii/S0956566320302116.

34 Dziabowska, K., Czaczyk, E., and Nidzworski, D. (2018). Detection methods of human and animal influenza virus – current trends. *Biosensors* 8 (4): 94. http://www.mdpi.com/2079-6374/8/4/94.

35 Sheikhzadeh, E., Eissa, S., Ismail, A., and Zourob, M. (2020). Diagnostic techniques for COVID-19 and new developments. *Talanta* 220: 121392. https://doi.org/10.1016/j.talanta.2020.121392.

36 Liang, K.H., Chang, T.J., Wang, M.L. et al. (2020). Novel biosensor platforms for the detection of coronavirus infection and severe acute respiratory syndrome coronavirus 2. *J. Chin. Med. Assoc.* 83: 701–703. https://covid19.elsevierpure.com/en/publications/novel-biosensor-platforms-for-the-detection-of-coronavirus-infect.

37 Medhi, R., Srinoi, P., Ngo, N. et al. (2020). Nanoparticle-based strategies to combat COVID-19. *ACS Appl. Nano Mater.* 3 (9): 8557–8580. https://pubs.acs.org/doi/abs/10.1021/acsanm.0c01978.

38 Adhikari, S., Adhikari, U., Mishra, A., and Guragain, B.S. (2020). Nanomaterials for diagnostic, treatment and prevention of COVID-19. *Appl. Sci. Technol. Ann.* 1 (1): 155–164. https://www.nepjol.info/index.php/asta/article/view/30295.

39 Nandy Chatterjee, T. and Bandyopadhyay, R.A. (2020). Molecularly imprinted polymer-based technology for rapid testing of COVID-19. *Trans. Indian Natl. Acad. Eng.* 5 (2): 225–228. https://link.springer.com/article/10.1007/s41403-020-00125-7.

40 Gooding, J.J. and Ligler, F.S. (2020). Virus detection: what were we doing before COVID-19 changed the world. *ACS Sens.* 5: 1503–1504. https://doi.org/10.1021/acssensors.0c01029.

41 Bhalla, N., Pan, Y., Yang, Z., and Payam, A.F. (2020). Opportunities and challenges for biosensors and nanoscale analytical tools for pandemics: COVID-19. *ACS Nano* 14 (4): 7783–7807. https://doi.org/10.1021/acsnano.0c04421.

42 Chandra, P. (2020). Miniaturized label-free smartphone assisted electrochemical sensing approach for personalized COVID-19 diagnosis. *Sens. Int.* 1: 100019.

43 Ronkainen, N.J., Halsall, H.B., and Heineman, W.R. (2010). Electrochemical biosensors. *Chem. Soc. Rev.* 39 (5): 1747–1763. https://pubs.rsc.org/en/content/articlehtml/2010/cs/b714449k.

44 Ji, T., Liu, Z., Wang, G.Q. et al. (2020). Detection of COVID-19: a review of the current literature and future perspectives. *Biosens. Bioelectron.* 166: 112455. https://www.ncbi.nlm.nih.gov/pmc/articles/PMC7371595/.

45 Khan, M.Z.H., Hasan, M.R., Hossain, S.I. et al. (2020). Ultrasensitive detection of pathogenic viruses with electrochemical biosensor: state of the art. *Biosens. Bioelectron.* 166: 112431. https://www.ncbi.nlm.nih.gov/pmc/articles/PMC7363606/.

46 Miodek, A., Sauriat-Dorizon, H., Chevalier, C. et al. (2014). Direct electrochemical detection of PB1-F2 protein of influenza A virus in infected cells. *Biosens. Bioelectron.* 59: 6–13. https://pubmed.ncbi.nlm.nih.gov/24686222/.

47 Hushegyi, A., Pihíková, D., Bertok, T. et al. (2016). Ultrasensitive detection of influenza viruses with a glycan-based impedimetric biosensor. *Biosens. Bioelectron.* 79: 644–649. https://www.ncbi.nlm.nih.gov/pmc/articles/PMC4883649/.

48 Ricotta, V., Yu, Y., Clayton, N. et al. (2019). A chip-based potentiometric sensor for a Zika virus diagnostic using 3D surface molecular imprinting. *Analyst* 144 (14): 4266–4280. https://pubs.rsc.org/en/content/articlehtml/2019/an/c9an00580c.

49 Rashed, M.Z., Kopechek, J.A., Priddy, M.C. et al. (2020). Rapid detection of SARS-CoV-2 antibodies using electrochemical impedance-based detector. *medRxiv*. https://doi.org/10.1101/2020.08.10.20171652.

50 Ding, J. and Qin, W. (2020). Recent advances in potentiometric biosensors. *TrAC, Trends Anal. Chem.* 124: 115803.

51 Hai, W., Goda, T., Takeuchi, H. et al. (2017). Specific recognition of human influenza virus with PEDOT bearing sialic acid-terminated trisaccharides. *ACS Appl. Mater. Interfaces* 9 (16): 14162–14170. https://pubmed.ncbi.nlm.nih.gov/28379685/.

52 Mavrikou, S., Moschopoulou, G., Tsekouras, V., and Kintzios, S. (2020). Development of a portable, ultra-rapid and ultra-sensitive cell-based biosensor for the direct detection of the SARS-CoV-2 S1 spike protein antigen. *Sensors* 20. https://www.mdpi.com/1424-8220/20/11/3121.

53 Janata, J. (2009). Conductometric Sensors. In: *Principles of Chemical Sensors. Springer*. Boston, MA: https://doi.org/10.1007/b136378_8.

54 Stradiotto, N.R., Yamanaka, H., and Zanoni, M.V.B. (2003). Electrochemical sensors: a powerful tool in analytical chemistry. *J. Braz. Chem. Soc.* 14:

159–173. http://www.scielo.br/scielo.php?script=sci_arttext&pid=S0103-50532003000200003&lng=en&nrm=iso&tlng=en.

55 Luo, Y., Nartker, S., Miller, H. et al. (2010). Surface functionalization of electrospun nanofibers for detecting *E. coli* O157:H7 and BVDV cells in a direct-charge transfer biosensor. *Biosens. Bioelectron.* 26 (4): 1612–1617. https://linkinghub.elsevier.com/retrieve/pii/S0956566310005324.

56 Lee, D., Chander, Y., Goyal, S.M., and Cui, T. (2011). Carbon nanotube electric immunoassay for the detection of swine influenza virus H1N1. *Biosens. Bioelectron.* 26 (8): 3482–3487. https://pubmed.ncbi.nlm.nih.gov/21354779/.

57 Shen, F., Wang, J., Xu, Z. et al. (2012). Rapid flu diagnosis using silicon nanowire sensor. *Nano Lett.* 12 (7): 3722–3730. https://pubs.acs.org/doi/10.1021/nl301516z.

58 Scholz, F. (2015). Voltammetric techniques of analysis: the essentials. *ChemTexts* 1 (4): 1–24. https://link.springer.com/article/10.1007/s40828-015-0016-y.

59 Jeremiah Aoki, K. and Chen, J. (2019). Tips of voltammetry. In: *Voltammetry* (eds. N.W. Maxakato, S.S. Gwebu and G.H. Mhlongo). IntechOpen https://doi.org/10.5772/intechopen.81341. http://www.intechopen.com.

60 Abad-Valle, P., Fernández-Abedul, M.T., and Costa-García, A. (2005). Genosensor on gold films with enzymatic electrochemical detection of a SARS virus sequence. *Biosens. Bioelectron.* 20: 2251–2260. https://europepmc.org/articles/PMC7126974.

61 Layqah, L.A. and Eissa, S. (2019). An electrochemical immunosensor for the corona virus associated with the Middle East respiratory syndrome using an array of gold nanoparticle-modified carbon electrodes. *Microchim. Acta* 186 (4): 1–10. https://doi.org/10.1007/s00604-019-3345-5.

62 Mahari, S., Roberts, A., Shahdeo, D., and Gandhi, S. (2020). eCovSens-ultrasensitive novel in-house built printed circuit board based electrochemical device for rapid detection of nCovid-19 antigen, a spike protein domain 1 of SARS-CoV-2. *bioRxiv*. https://doi.org/10.1101/2020.04.24.059204.

63 Adeloju, S.B. (2004). Amperometry. In: *Encyclopedia of Analytical Science*, 2e (eds. P. Worsfold, A. Townshend and C. Poole), 70–79. Elsevier Inc.

64 Roussel, T.J., Jackson, D.J., Baldwin, R.P., and Keynton, R.S. (2008). Amperometric techniques. In: *Encyclopedia of Microfluidics and Nanofluidics* (ed. D. Li), 39–47. Springer US. https://link.springer.com/referenceworkentry/10.1007/978-0-387-48998-8_26.

65 Liu, F., Kim, Y.H., Cheon, D.S., and Seo, T.S. (2013). Micropatterned reduced graphene oxide based field-effect transistor for real-time virus detection. *Sens. Actuators, B* 186: 252–257.

66 Sayhi, M., Ouerghi, O., Belgacem, K. et al. (2018). Electrochemical detection of influenza virus H9N2 based on both immunomagnetic extraction and gold catalysis using an immobilization-free screen printed carbon microelectrode. *Biosens. Bioelectron.* 107: 170–177. https://pubmed.ncbi.nlm.nih.gov/29455027/.

67 Seo, G., Lee, G., Kim, M.J. et al. (2020). Rapid detection of COVID-19 causative virus (SARS-CoV-2) in human nasopharyngeal swab specimens using field-effect transistor-based biosensor. *ACS Nano* 14 (4): 5135–5142. https://pubmed.ncbi.nlm.nih.gov/32293168/.

68 Vandenberg, O., Martiny, D., Rochas, O. et al. (2020). Considerations for diagnostic COVID-19 tests. *Nat. Rev. Microbiol.*: 1. https://www.nature.com/articles/s41579-020-00461-z.

69 Khanna, R.C., Cicinelli, M.V., Gilbert, S.S. et al. (2020). COVID-19 pandemic: lessons learned and future directions. *Indian J. Ophthalmol.* 68: 703–710. https://www.ijo.in/article.asp?issn=0301-4738;year=2020;volume=68;issue=5;spage=703;epage=710;aulast=Khanna.

70 Choi, J.R. (2020). Development of point-of-care biosensors for COVID-19. *Front. Chem.* 8: 517. https://www.frontiersin.org/article/10.3389/fchem.2020.00517/full.

12

Electrochemical Biosensor Fabrication for Coronavirus Testing

Monika Vats, Parvin, Mukul Taliyan, and Seema Rani Pathak

Amity University Haryana, Amity School of Applied Sciences, Department of Chemistry, Biochemistry & Forensic Science, Amity Education Valley Gurugram, Manesar, Panchgaon, Haryana 122413, India

12.1 Introduction

Coronavirus disease 2019 (COVID-19) is an infectious disease caused by an severe acute respiratory syndrome coronavirus 2 (SARS-CoV-2). The virus was first recorded in Wuhan City in December 2019 and has since spread to more than 200 countries worldwide [1]. The World Health Organization (WHO) declared the outbreak an international public health emergency on 20 January 2020. Overall, as of now, more than 25 million confirmed cases of COVID-19 have been reported to WHO, including more than 9 lakh deaths. In addition to the immediate effect on public health, the COVID-19 pandemic has had a substantial global social and economic effects, in part due to social distancing measures adopted and worldwide closures [2].

Coronavirus disease can spread primarily from an infected individual through direct or indirect contact. The virus has been shown to be able to spread by respiratory droplets, aerosols, and abiotic surface contact [3]. The clinical expression of COVID-19 varies from mild fever, cough, to other life-threatening syndromes such as pneumonia, respiratory failure, or death in several cases [4]. More specifically, it has shown asymptomatic infections and transmissions. It is posing major challenges to the health and thus survivability of human population across the world. Therefore, there is an urgent need to develop responsive, reliable, quick, and low-cost diagnostic tools for screening infected person to promote proper isolation and care. In such a scenario, which calls for a rapid response from the world community of scientists, biomarkers/specific indicators can be utilized as diagnostic tools to detect the disease. Diagnosis in such infected cases is generally done through molecular testing methods on respiratory secretions. The methods such as chest computed tomography (CT) scan combines with clinical symptoms, reverse transcription polymerase chain reaction (RT-PCR), surface-enhanced Raman scattering (SERS)-based biosensor, lateral flow immunochromatographic strip (LFICS), electrochemical (EC) biosensor, surface plasmon resonance (SPR)-based biosensor,

Detection and Analysis of SARS Coronavirus: Advanced Biosensors for Pandemic Viruses and Related Pathogens, First Edition. Edited by Chaudhery Mustansar Hussain and Sudheesh K. Shukla.

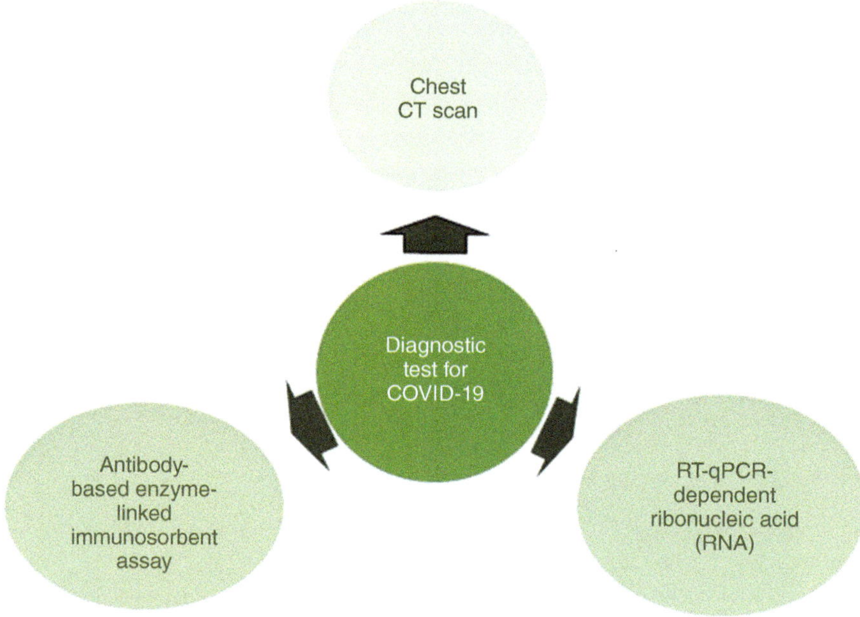

Figure 12.1 Different types of diagnostic tests for COVID-19.

etc., are few convenient techniques to detect low concentration of viruses. Different types of diagnostic tests for COVID-19 are shown in Figure 12.1. However, many of them have problems associated with them such as CT scan availability is limited and even it cannot differentiate between different viruses, nor can they recognize particular viruses, RT-PCR approach is time-consuming and sometimes may also produce false-negative results, and the antibody detection is also not appropriate for screening early and asymptomatic cases [5–7]. In such a condition where speed and accuracy are the prime requirements, electrochemical (EC) biosensors can be the best solution. EC is one of the techniques that uses an electrochemical transducer, which is ligand specific, and shows potential for detection of bioanalytes such as tissues, enzymes, coronavirus, and others. Electrochemical sensors work on the principle that (bio)analyte and sensor interactions result in chemical reactions, which in response generates electrical signals in proportional to (bio)analyte concentration.

The current chapter focuses on the fabrication as well as designing of electrochemical biosensors. There are various steps involved in the fabrication process and are discussed later in the chapter. The selection of an efficient material is quite essential for the fabrication of an electrochemical sensor. Various fabricating materials are also reviewed and described in Sections (12.3 and 12.4) of this chapter. The classification, working principle, and application of the electrochemical sensors in the detection of coronavirus testing are further elaborated in detail. Finally, the current and future prospects of the testing techniques in COVID-19 detection have also been included in the current chapter for comprehensive understanding of readers.

12.2 Application of Electrochemical Biosensors

Many biosensors are not just used in the laboratory but they also have extended their use as daily equipment. There are generally three types of biosensors: physical, optical, and electrochemical. Physical biosensors are further classified into two types: thermometric and piezoelectric. The electrochemical sensors are further classified into five types: potentiometric, amperometric, impedimetric, voltammetric, and immunosensor. A local equilibrium is achieved in potentiometric sensors at the interface of the sensor. The potential difference between the electrodes helps in the identification of samples in potentiometric sensors. For the amperometric sensors, the current produced due to oxidation or reduction reaction occurred when potential is applied between the reference and working electrodes. For the conductometric sensors, the conductivity is measured at a sequence of frequencies, whereas the electrochemical change observed due to antigen–antibody interaction (organism-specific immune response) is measured for an immunosensor. The types and applications of biosensors are presented in Table 12.1.

Biodevices such as electrochemical biosensors are fast and economic. These devises have been used for the detection of various biomolecules such as glucose,

Table 12.1 Different types of biosensors and their few applications.

S. no.	Biosensors	Type	Application	References
1.	Potentiometric biosensor	Electrochemical	Used to detect urea in blood serum	[8]
2.	Amperometric biosensor	Electrochemical	Detection of ethanol, glucose, and lactate	[9]
3.	Impedimetric biosensor	Electrochemical	Used in the field of microbiology to detect, quantify, and even identify bacteria	[10]
4.	Voltammetric biosensor	Electrochemical	Used for analyzing paracetamol	[11]
5.	Immunosensor	Electrochemical	Used to detect several pathogens such as viruses like COVID-19, influenza, etc.	[12]
6.	Piezoelectric biosensor	Physical	Used in rapid detection of HIV in biological fluids	[13]
7.	Thermometric biosensor	Physical	Used to measure enzyme activity, clinical monitoring, environmental monitoring, etc.	[14]
8.	Optical biosensor	Physical	Used in the study of transferrin binding proteins, lipo-oligosaccharide (LOS)–antibody interactions and serum responses to experimental vaccines	[15]

Source: Marchenkoa et al. [8], Goriushkina et al. [9], and Guana et al. [10].

pathogens, cholesterol, uric acid, lactate, DNA, hemoglobin, amino acids, blood ketones, etc. [16–18]. Recently, lots of research development took place in the past few years in the field of electrochemical sensors. The glutathione sensor has been developed using screen-printed carbon nanotubes (CNTs) as the sensing electrode [19]. The electrochemical acetylcholinesterase biosensor has been fabricated with the help of chitosan and impregnated with cerium oxide or carbon for detecting methamidophos as well as chlorpyrifos [20]. An electrochemical sensor has been developed for detection of glutathione thymine-Hg^{2+}-thymine using silica nanoparticles [21]. The graphite screen-printed electrode fabricated using ZnO/Al_2O_3 nanocomposite has been reported for ascorbic acid sensing [22]. An electrochemical sensor for detection of diazinon has also been reported [23]. A non-enzymatic sensor for the detection of ascorbic acid has been developed using $Cu(OH)_2$ nanorods [24]. The copper (II) detection heavy metal sensors have also been reported [25]. The nanocomposite of graphene CNTs coated with Bi nanoparticle sensor for detection of mercury (II) has also been reported in the literature [26].

Electrochemical sensors are low-cost methods and help in identifying human illness factors such as viruses, mutants, bacteria, influenza, fungi, and protozoa, etc., at the primary stage. The microorganisms have multiple transmitters, including humans, animals, and trees, so unchecked can cause pandemics. A quick, responsive, small size, cheap system is pivotal for these reasons. Early detection is one of the best means of avoidance, but because of high costs, stringent sample planning mechanisms, and long-term analysis, it is also complicated. Modern technology leads to the development of precise and quick biosensors [27].

In virus-related diseases such as influenza, HIV, and COVID-19, the key purpose is to identify them at very low levels, enabling the doctor to apply effective care at the onset of human or animal infections. Antibodies, peptides, aptamers, and nucleic acids are predominantly virus recognition receptors.

12.3 Fabrication of Electrochemical Biosensors

Electrochemical biosensors are produced by a variety of methods for various diagnostic purposes. An electrochemical sensor measures a response produced by analyte with the sensory layer. The transducer transforms the chemical response into a detectable electrical signal. A detector detects the signal and amplifies it and then displays the result. The electrochemical sensor is pictorially presented as a block diagram in Figure 12.2.

Nanomaterials such as CNT, inorganic and organic nanoparticles, nanosized clays, conductive and nanostructured polymers, and hybrids have been effectively used for increasing the electroanalytical efficacy of biosensors as well as for immobilizing the biorecognition. For constructing an electrochemical biosensor, one must consider these parameters: (i) measurement/detection parameter and matrix system, (ii) properties of a transducer, (iii) the model considered (chemical/biochemical), and (iv) area of application. Depending on biosensor applications,

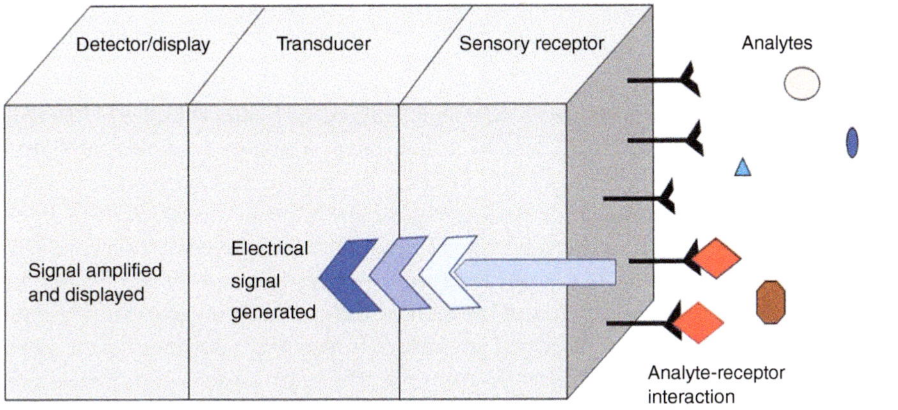

Figure 12.2 Pictorial representation of an electrochemical sensor.

the biorecognition elements or receptors can be biological or artificial biomimetic receptors. Biocatalytic recognition sensors are macromolecules or enzymes based, and are mainly used for biological elements. In biocomplexing/bioaffinity sensors, antibodies, chemoreceptors, and nucleic acids are used for providing selective interactions between analyte to form thermodynamically stable complex [28].

Popular processing methods use nanomaterials and polymeric materials. These materials are highly sensitive and stable, cost-effective, and have antibody seroprevalence properties. The biosensor unit should have high stability, good conductivity, and low operating potential. Polyaniline used in electrochemical sensors provides high stability, good conductivity, and low operating potential along with swift electrode–analyte electron transfer for sensing purposes. It is a challenge to increase the nanostructures and electrode surface adhesion, with no change reduction of electrochemical properties and biosensing capabilities of biosensors. The electrode polymerized in plasma by SnO_2@3D-rGO nanocomposite has been used to detect glucose electrochemically efficiently as compared to other sensors. Subsequently, several strategies mentioned by Muguruma et al. focus on the manufacture of plasma surface-modified polymer films, which efficiently provide electron-transfer-mediated biosensing such as acetonitrile mixed with vinyl ferrocene plasma-polymerized film. In this technique, vinyl ferrocene plasma-polymerized films were first deposited on a sputtered platinum electrode. The other sensors for glucose reported has plasma-modified dimethyl amino methyl ferrocene deposited on the electrode. The efficiency of the sensors was attributed to the bioelectrochemical reaction. The fabricated sensor has a plasma-polymerized acetonitrile film sandwiched between the immobilized enzyme and plasma-polymerized vinyl ferrocene [29]. The functionalization of the nanocomposite by nitrogen-containing groups provides good stability of the electrode materials in the aqueous solution. The porous 3D nanostructure formed from graphene oxide provides high loading capacity for glucose oxidase [30]. Al doped with ZnO nanostructures based electrochemical biosensor has also been fabricated using the sol–gel process [31].

12.4 Fabrication of Electrochemical Biosensors for COVID-19 (Immunosensors)

In electrochemical immunosensors, when the antigen–antibody complex is formed, a biological signal is produced, which is then converted into an electrical signal [32]. The electrochemical immunosensors may be used as labeled or non-labeled immunological biosensors on the basis of transduction mechanisms. For the identification of various viruses, both forms of immunosensors have been successfully used [33]. Non-labeled electrochemical immunosensors have a very simple, quick, and low-cost technique compared to the labeled technique. It involves simple synthetic tools and techniques; detection procedure is also very short, and even there is no need of secondary antibodies. During the immobilization of antibodies, the identification efficacy of label-free/non-labeled electrochemical sensors depends on antibody alignment [34]. There are different antibody immobilization methods that can be used, such as straight electrode surface adsorption, magnetic beads, the basic polymer matrices, etc. Self-assembled monolayers of alkanethiols are commonly applied techniques among these techniques because this technique offers a simple way to generate strong covalent bonds, regulated structure, ultra-thin, oriented, and arranged monolayers on the electrode surface [35].

An electrochemical immunosensor for the detection of the Middle East respiratory syndrome coronavirus (MERS-CoV) has recently been created. In this work, to improve the sensitivity of the sensor, carbon array electrodes were manufactured with electrodeposition of gold nanoparticles (AuNPs). As a biomarker for MERS-CoV, the authors used recombinant spike protein S1. The suggested sensor also demonstrated high selectivity against influenza A and B [36]. The high-performance COVID-19 biosensing miniature device has been reported using graphene oxide frame, which possesses high electrical properties. In addition, the extremely large basic surface area (two open sides), the plentiful surface functionalities containing oxygen, such as hydroxyl, epoxy, and carboxylic groups, and the high solubility of water make graphene oxide sheets a fantastic candidate for sensing applications [37]. A non-enzymatic electrochemical sensor for quick determination of chemical analytes of blood samples was created by a group of researchers in another study. To achieve high surface area, high conductivity, surface roughness, and high catalytic potential, a 3D gold-sputtered nanodendritic has been used. Another promising sensing surface known to have very high conductivity and electrocatalysis is dendrite-modified electrode combined with multi-walled carbon nanotube (MWCNT) [38]. The numerous identified biomarkers such as nucleocapsid protein, spike protein, envelope protein, and membrane protein or the open reading frame of viral RNA expressed in COVID-19 can be used for fabricating low-cost sensing probes [39]. These sensors benefit from their robustness and label-free operating mode, which is one of the most important metrics when used in hospital clinical diagnostics and/or customized disease tracking.

Immunosensors are biosensing devices that are used in body serum and many other media to detect specific antigens or antibodies through immunochemical reactions. A bioreceptor and a transducer are largely composed of immunosensors. A bioreceptor is used to identify the target antigen or antibody that transforms a transducer from the producing biological signal to the targeted signal. In immunological biosensors, various transducing mechanisms are used depending on signal production (such as an electrochemical or optical signal) or changes in properties (such as changes in mass) following the development of antigen–antibody complexes. It is an important pathogen detection strategy because antibodies are normally attached to antigens to form an antigen–antibody complex, which is the core immunosensor concept for antibody or antigen detection [40]. An electrochemical immunosensor for detection of MERS-CoV and human coronavirus (HCoV) has been developed using an array of nanostructured carbon electrodes with gold nanoparticles [36]. In a recent research, SARS-CoV-2 RNA-dependent RNA polymerase (RdRp)/helicase (H) genes were used for diagnostic purposes as a significant marker that does not display any cross-reactivity with other human coronaviruses or respiratory viruses. Therefore, because of its robustness and strong analytical efficiency, if anti-RdRp helicase is successfully immobilized on recently developed highly conductive surface, it can pave a new path to eventually detect the infection [41].

The label-free electrochemical sensors incorporated with metallic nanoparticles, nanodendroids, and graphene oxide nanocomposites based on spectroscopy have been reported for detection of biomarkers. The materials were deposited on the screen-printed carbon electrode in a sequence and then the specific biomarker antibodies were immobilized and functionalized using a bioconjugation procedure [42]. Graphene oxide has electrical property that can be utilized in the development of COVID-19 detection tool with high performance, which further shows potential of improvement by integrating new COVID-19 specific biomarkers [43]. Nanohierarchical 3D gold dendritic structure and MWCNT nanohybrid electrode have been fabricated, which has shown label-free biosensing of COVID-19 [38]. The above literature suggests that fabrication of any COVID-19 electrochemical sensor consists of majorly three components (represented in Figure 12.3): (i) COVID-19 specific sensors, (ii) a transducer which is produces electrical signal due to

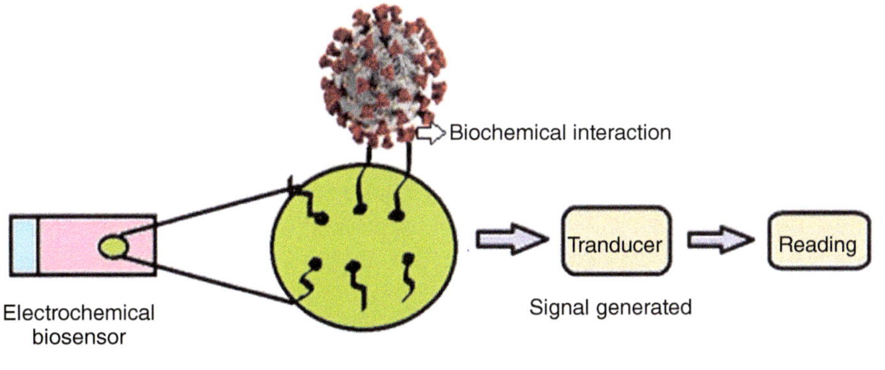

Figure 12.3 General fabrication and working of a COVID-19 electrochemical sensor.

chemical interaction, and (iii) display which give a reading in positive or negative value. The COVID-19 sensor is an immunosensor-type electrochemical sensing device. The function of body immune system is to identify and recognize harmful or non-harmful foreign substances (antigens) and produces immunoglobulins (antibodies) in response. The antigens bind specifically to the antigens and the phenomenon has been used in the fabrication of sensors known as immunosensors, where antibody act as a sensory layer or bioreceptor for antigens obtained from the sample of COVID-19-suspected person. The receptor (antibody) and analyte (antigens) interaction results in a coupling reaction by forming immunocomplex. This chemical signal is registered by a transducer converted into an electrical signal. The signal produced is recorded, amplified, and viewed on the display unit as reading [44].

12.5 Conclusion

COVID-19 is an infectious disease caused by an SARS-CoV-2. The virus was first discovered in Wuhan City in December 2019 and then spread to more than 200 countries worldwide. Coronavirus disease can spread primarily from an infected individual through direct or indirect contact. Biodevices such as electrochemical biosensors are one of the techniques to detect COVID-19. These electrochemical biosensors have shown their utility in the detection of COVID-19 efficiently. It is an economic, fast, precise, and early on-site identification method for detecting COVID-19. The electrochemical biosensors have recently been used for the detection of specific antigen–antibody immunochemical reactions. They are composed of a bioreceptor, a transducer, and a detector. A bioreceptor identifies the target antigen/antibody and then a transducer produces a biological signal. The transducing mechanisms are used depending on signal production (electrochemical, optical signal or change in mass) followed by the development of antigen–antibody complexes. It is an important pathogen detection strategy because antibodies are normally attached to antigens to form an antigen–antibody complex, which is the basis for immunosensor. The electrochemical biosensors have the potential to recognize biorecognition, they provide reliability results, and can be miniaturized for portable devices. However, their short life time, low stability electrode fouling, and non-specific interaction may limit their use. A more stable receptor for biorecognition can be prepared using multiple approaches such as genetic engineering and nanotechnology. The current research should majorly focus on the development of materials with desired properties such as effective immobilization and functionalization. Improvised electrochemical sensing-based smartphone-driven detection and monitoring devices may help in developing low-cost alternative for expensive COVID-19 for point-of-care diagnosis. The electrochemical sensors have shown its potential application not only for detection of COVID-19, but also it can be anticipated that the sensor has the capability to integrate with label-free technologies can be fabricated into personalized analytical devices to combat other infectious diseases.

References

1 Tymm, C., Zhou, J., Tadimety, A. et al. (2020). *Cellular and Molecular Bioengineering*. Switzerland: Springer Nature.

2 Zheng, Y.Y., Ma, Y.T., Zhang, J.Y. et al. (2020). COVID-19 and the cardiovascular system. *Nat. Rev. Cardiol.* 17: 259–260. https://doi.org/10.1038/s41569-020-0360-5.

3 Ong, S.W.X., Tan, Y.K., Chia, P.Y. et al. (2020). Air, surface environmental, and personal protective equipment contamination by severe acute respiratory syndrome coronavirus 2 (SARS-CoV-2) from a symptomatic patient. *JAMA* 323 (16): 1610–1612. https://doi.org/10.1001/jama.2020.3227.

4 Bedford, J., Enria, D., Giesecke, J. et al. (2020). COVID-19: towards controlling of a pandemic. *Lancet* 395: 1015–1018. https://doi.org/10.1016/S0140-6736(20)30673-5.

5 An, J., Liao, X., Xiao, T. et al. (2020). Clinical characteristics of the recovered COVID-19 patients with re-detectable positive RNA test. *Ann. Transl. Med.* 8 (17): 1–12. https://doi.org/10.21037/atm-20-5602.

6 Cui, F. and Zhou, H.S. (2019). Diagnostic methods and potential portable biosensors for coronavirus disease. *Biosens. Bioelectron.* 165: 112349. https://doi.org/10.1016/j.bios.2020.112349.

7 Dziabowska, K., Czaczyk, E., and Nidzworski, D. (2017). *Biosensing Technologies for the Detection of Pathogens – A Prospective Way for Rapid Analysis*. London, UK: IntechOpen Limited.

8 Marchenkoa, S.V., Kucherenkoa, I.S., Hereshkob, A.N. et al. (2015). Application of potentiometric biosensor based on recombinant urease for urea determination in blood serum and hemodialyzate. *Sens. Actuators, B* 207: 981–986. https://doi.org/10.1016/j.snb.2014.06.136.

9 Goriushkina, T.B., Soldatkin, A.P., and Dzyadevych, S.V. (2009). Application of amperometric biosensors for analysis of ethanol, glucose, and lactate in wine. *J. Agric. Food. Chem.* 57 (15): 6528–6535. https://doi.org/10.1021/jf9009087.

10 Guana, J.G., Miao, Y.Q., and Zhang, Q.J. (2004). Impedimetric biosensors. *J. Biosci. Bioeng.* 97 (4): 219–226. https://doi.org/10.1016/S1389-1723(04)70195-4.

11 Goyal, R.N., Gupta, V.K., and Chatterjee, S. (2010). Voltammetric biosensors for the determination of paracetamol at carbon nanotube modified pyrolytic graphite electrode. *Sens. Actuators, B* 149 (1): 252–258. https://doi.org/10.1016/j.snb.2010.05.019.

12 Aizawa, M. (1994). Immunosensors for clinical analysis. *Adv. Clin. Chem.* 31: 247–275. https://doi.org/10.1016/s0065-2423(08)60337-6.

13 Pohanka, M. (2016). The piezoelectric biosensors: principles and applications, a review. *Int. J. Electrochem. Sci.*: 496–506. https://doi.org/10.20964/2017.01.44.

14 Ramanathan, K. and Danielsson, B. (2001). Principles and applications of thermal biosensors. *Biosens. Bioelectron.*: 417–423. https://doi.org/10.1016/S0956-5663(01)00124-5.

15 Suker, J. and Charalambous, B.M. (2001). Application of optical biosensor techniques to the characterization of PorA-antibody binding kinetics. *Methods Mol. Med.* 66: 129–143. https://doi.org/10.1385/1-59259-148-5:129.

16 Beitollahi, H., Mohammadi, S.Z., Safaei, M. et al. (2020). Applications of electrochemical sensors and biosensors based on modified screen printed electrodes: a review. *Anal. Methods*. https://doi.org/10.1039/C9AY02598G.

17 Choi, J.W., Oh, B.K., Kim, Y.K., and Min, J. (2007). Nanotechnology in biodevices. *J. Microbiol. Biotechnol.* 17 (1): 5–14.

18 Hadi, B., Mohammadi, S.Z., Safaei, M., and Tajik, S. (2020). Applications of electrochemical sensors and biosensors based on modified screen-printed electrodes: a review. *Anal. Methods* 12: 1547–1560. https://doi.org/10.1039/C9AY02598G.

19 Lee, P.T., Goncalves, L.M., and Compton, R.G. (2015). Electrochemical determination of free and total glutathione in human saliva samples. *Sens. Actuators, B* 221: 962–968. https://doi.org/10.1016/j.snb.2015.07.050.

20 Fu, J., Zhang, Q., Shi, Z. et al. (2018). Sensitive acetylcholinesterase biosensor based on screen printed carbon electrode modified with cerium oxide chitosan/mesoporous carbon-chitosan for organophosphorus pesticide residue detection. *Int. J. Electrochem. Sci.* 13: 9231–9241. https://doi.org/10.20964/2018.09.146.

21 Wang, Y., Jiang, L., Chu, L. et al. (2017). Electrochemical detection of glutathione by using thymine-rich DNA-gated switch functionalized mesoporous silica nanoparticles. *Biosens. Bioelectron.* 87: 459–465. https://doi.org/10.1016/j.bios.2016.08.102.

22 Ganjali, M.R., Nejad, F.G., Beitollahi, H. et al. (2017). Highly sensitive voltammetric sensor for determination of ascorbic acid using graphite screen printed electrode modified with ZnO/Al_2O_3 nanocomposite. *Int. J. Electrochem. Sci.* 12 (4): 3231–3240. https://doi.org/10.20964/2017.04.07.

23 Hassani, S., Akmal, M.R., Salek-Maghsoudi, A. et al. (2018). Novel label-free electrochemical aptasensor for determination of diazinon using gold nanoparticles-modified screen-printed gold electrode. *Biosens. Bioelectron.* 120: 122–128. https://doi.org/10.1016/j.bios.2018.08.041.

24 Raveendran, J., Krishnan, R.G., Nair, B.G., and Babu, T.G.S. (2017). Voltammetric determination of ascorbic acid by using a disposable screen printed electrode modified with $Cu(OH)_2$ nanorods. *Microchim. Acta* 184 (9): 3573–3579. https://doi.org/10.1007/s00604-017-2391-0.

25 Bobrowski, A., Maczuga, M., Krolicka, A. et al. (2017). Determination of copper (II) through anodic stripping voltammetry in tartrate buffer using an antimony film screen-printed carbon electrode. *Anal. Lett.* 50 (18): 2920–2936. https://doi.org/10.1080/00032719.2017.1319850.

26 Jeromiyas, N., Elaiyappillai, E., Kumar, A.S. et al. (2018). Bismuth nanoparticles decorated graphenated carbon nanotubes modified screen-printed electrode for mercury detection. *J. Taiwan Inst. Chem. Eng.* 95: 466–474. https://doi.org/10.1016/j.jtice.2018.08.030.

27 Verma, M.S., Rogowski, J.L., and Jones, L. (2015). Colorimetric biosensing of pathogens using gold nanoparticles. *Biotechnol. Adv.* 33 (6.1): 666–680. https://doi.org/10.1016/j.biotechadv.2015.03.003.

28 Obert, S., Mihaela, T., Cecilia, C., and Ede, B. (2015). New materials for the construction of electrochemical biosensors. In: *Biosensors – Micro and Nanoscale Applications* (ed. T. Rinken). IntechOpen https://doi.org/10.5772/60510.

29 Muguruma, H., Hiratsuka, A., Kojima, K. et al. (2005). Electron transfer mediator micro-biosensor fabrication by organic plasma process. *Biosens. Bioelectron.* 21 (6): 957–964. https://doi.org/10.1016/j.bios.2005.03.002.

30 Wu, S., Su, F., Dong, X. et al. (2017). Development of glucose biosensors based on plasma polymerization-assisted nanocomposites of polyaniline, tin oxide, and three-dimensional reduced graphene oxide. *Appl. Surf. Sci.* 401: 262–270. https://doi.org/10.1016/j.apsusc.2017.01.024.

31 Gherab, K., Al-Dourib, Y., Hashim, U. et al. (2020). Fabrication and characterizations of Al nanoparticles doped ZnO nanostructures-based integrated electrochemical biosensor. *J. Mater. Sci. Technol.* 9 (1): 857–967. https://doi.org/10.1016/j.jmrt.2019.11.025.

32 Yadav, N., Mishra, A., and Narang, J. (2019). Electrochemical sensor method for food quality evaluation. In: *Evaluation Technologies for Food Quality*, 793–815. https://doi.org/10.1016/B978-0-12-814217-2.00031-7.

33 Oswald, B., Lehmann, F., Simon, L. et al. (2000). Red laser-induced fluorescence energy transfer in an immunosystem. *Anal. Biochem.* 280: 272–277. https://doi.org/10.1006/abio.2000.4553.

34 Shen, G., Zhang, X., Shen, Y., and Zhang, C. (2015). Immobilization of antibodies on aldehyde functionalized polymer/graphene films for the fabrication of a label-free electrochemical immunosensor. *J. Electroanal. Chem.* 759: 67–71. https://doi.org/10.1016/j.jelechem.2015.06.023.

35 Haji-Hashemi, H., Safarnejad, M.R., Norouzi, P. et al. (2019). Simple and effective label free electrochemical immunosensor for Fig mosaic virus detection. *Anal. Biochem.* 566: 102–106. https://doi.org/10.1016/j.ab.2018.11.017.

36 Layqah, L.A. and Eissa, S. (2019). An electrochemical immunosensor for the corona virus associated with the Middle East respiratory syndrome using an array of gold nanoparticle-modified carbon electrodes. *Microchim. Acta* 186 (4): 224. https://doi.org/10.1007/s00604-019-3345-5.

37 Lee, J., Kim, J., Kim, S. et al. (2016). Biosensors based on graphene oxide and its biomedical application. *Adv. Drug Delivery Rev.* 105: 275–287. https://doi.org/10.1016/j.addr.2016.06.001.

38 Purohit, B., Mahato, K., Kumar, A., and Chandra, P. (2020). Novel sensing assembly comprising engineered gold dendrites and MWCNT-AuNPs nanohybrid for acetaminophen detection in human urine. *Electroanalysis* 32: 561–570. https://doi.org/10.1002/elan.201900551.

39 Chu, D.K.W., Pan, Y., Cheng, S.M.S. et al. (2020). Molecular diagnosis of a novel coronavirus (2019-nCoV) causing an outbreak of pneumonia. *Clin. Chem.* https://doi.org/10.1093/clinchem/hvaa029.

40 Pollap, A. and Kochana, J. (2019). Electrochemical immunosensors for antibiotic detection. *Biosensors* 9: 61. https://doi.org/10.3390/bios9020061.

41 Chan, J.F.-W., Yip, C.C.-Y., To, K.K.-W. et al. (2020). Improved molecular diagnosis of COVID-19 by the novel, highly sensitive and specific COVID-19-RdRp/Hel real-time reverse transcription-polymerase chain reaction assay validated in vitro and with clinical specimens. *J. Clin. Microbiol.*: 1–33. https://doi.org/10.1128/JCM.00310-20.

42 Yadav, N., Mishra, A., and Narang, J. (2019). 31 - Electrochemical sensor method for food quality evaluation. In: *Woodhead Publishing Series in Food Science, Technology and Nutrition, Evaluation Technologies for Food Quality* (eds. J. Zhong and X. Wang), 793–815. Woodhead Publishing, ISBN 9780128142172.

43 Chandra, P. and Prakash, R. (2020). *Nanobiomaterial Engineering: Concepts and Their Applications in Biomedicine and Diagnostics*. https://doi.org/10.1007/978-981-32-9840-8.

44 Syazana, A.L. and Minhaz, U.A. (2019). Chapter 1: Introduction to immunosensors. In: *Immunosensors*, 1–20. RSC. https://doi.org/10.1039/9781788016162-00001.

Part V

Outlook

13

Effects of COVID-19: An Environmental Point of View

Kola Y. Kareem[1], Bashir Adelodun[1,2], AbdulGafar O. Tiamiyu[3], Fidelis O. Ajibade[4,5,6], Rahmat G. Ibrahim[7], Golden Odey[2], Madhumita Goala[8], Hashim O. Bakare[3], and Jamiu A. Adeniran[3,9]

[1] *University of Ilorin, Department of Agricultural and Biosystems Engineering, Ilorin, PMB 1515, Nigeria*
[2] *Kyungpook National University, Department of Agricultural Civil Engineering, Daegu 41566, Korea*
[3] *University of Ilorin, Department of Chemical Engineering, Ilorin, PMB 1515, Nigeria*
[4] *Federal University of Technology, Department of Civil and Environmental Engineering, Akure, PMB 704, Nigeria*
[5] *Chinese Academy of Sciences, Research Centre for Eco-Environmental Sciences, Key Laboratory of Environmental Biotechnology, Beijing 100085, PR China*
[6] *University of Chinese Academy of Sciences, Beijing, 100049, PR China*
[7] *Kwara State Ministry of Health, Ilorin, Kwara State, Nigeria*
[8] *Assam University, Nehru College, Labocpar Part II, Cachar, 788098, Assam, India*
[9] *Peking University, Atmospheric Chemistry and Modeling Group, Department of Atmospheric and Oceanic Sciences, Beijing, China*

13.1 Introduction

The emergence of Severe Acute Respiratory Syndrome Coronavirus 2 (SARS-CoV-2) (β-coronavirus) in Wuhan, China, in late December 2019 has created unprecedented devastating effects on social, economic, public health, and global environmental status with an alarming spike in mortality and morbidity. The novel coronavirus disease (COVID-19) is characterized as SARS-CoV-2 or simply put as CoV, which has generated such a global awareness in terms of the magnitude of infected patients, mortality rate, socioeconomic impact, public health exacerbation, and environmental degradation globally [1]. It has received much attention because of its impact on the environment, socioeconomic development, and health outcome [1, 2]. CoV belongs to the family "Coronaviridae" with four subfamilies, namely, alpha (α), beta (β), delta (δ), and gamma (γ). It has been reported that the α- and β-CoVs are pathogenic in human beings [2, 3]. The emergence of the virus was recorded in the Hubei province, Wuhan city, China, and it eventually spread across all the continents and was unanimously declared as a global pandemic by the World Health Organization [4]. The pathogenic β-coronavirus is novel in the sense that it is the first pandemic recorded in history to be caused by a coronavirus – the type that is rare and transmits rapidly

Detection and Analysis of SARS Coronavirus: Advanced Biosensors for Pandemic Viruses and Related Pathogens, First Edition. Edited by Chaudhery Mustansar Hussain and Sudheesh K. Shukla.

through contact with liquid, sputum, aerosols, and nasal fluid of infected individuals [4, 5]. Aerosols may penetrate the human respiratory system through inhalation through the nose or mouth [6, 7].

There is a long history of reported cases of similar global epidemics of the twenty-first century, and all share a common cause of pandemic events found in the close interaction between human populations with domestic and wildlife pathogens [8]. These pathogens are transmitted to the host community either by contact or aerosols or via consumption. Notable of the recorded epidemics were the SARS in 2003, which claimed 8000 infections with a mortality rate of 10% [9], Middle East respiratory syndrome (MERS) in Saudi Arabia and Jordan in 2012 – a zoonotic disease that recorded sporadic cases because of community crowding [10], the 2009 H1N1 influenza swine flu virus that stemmed from Mexico and the United States with a mortality rate of about 0.4% [11], and the Ebola virus that was first detected in 1976 in formerly known Zaire had boasted of about 28 000 confirmed cases with 11 000 deaths in Sub-Saharan African countries [12]. Other pandemic outbreaks include, but are not limited to the Zika fever from 2015 to 2016, the Avian flu [12–14].

The global spread engulfing almost 215 countries in all the continents of the world except Antarctica has generated much concern. It has spurred every country's National Disease Control unit to action on instituting COVID-19 prevention protocol and building capacity for research tailored toward understanding the novel virus, with the golden goal of developing an antiviral vaccine. According to the World Health Organization (24–30 August) weekly epidemiological update report that was assessed on Friday, 4 September 2020, about 1.8 million new COVID-19 cases and 38 000 new deaths were documented in the week ending 30 August [15]. The results indicated a 1% increase in the number of cases and a 3% decrease in the number of deaths than the previous week, i.e. 17–23 August. As at the end of August 2020, an alarming cumulative total of about 25 million cases and 800 000 deaths had been reported since the start of the pandemic. Figure 13.1 shows the number of cases and deaths reported weekly by the WHO from 30 December to 30 August 2020.

The preventive measures targeted at reducing and preventing the communal transmission of the virus varied from travel restrictions, contact tracing, social distancing etiquettes, good hygiene, ethanol sterilizer usage, face masks, and performance of massive COVID-19 screening tests. Informed measures on social distancing and isolation, when religiously adhered to by citizens, have offered a substantial decline in transportation/travel as a result of employees working remotely from the comfort of their homes. Groceries, basic needs, and essential items are ordered online. In highly developed countries such as China, Japan, and South Korea, drones and robots deliver packages with the highest hygiene level, strict adherence to COVID-19 protocol, while ensuring excellent service delivery. Travel restrictions enforced by many countries have been the reason for a significant reduction in the use and demand for oil and its products, consequently resulting in a lesser carbon footprint as a result of lesser smoke emission

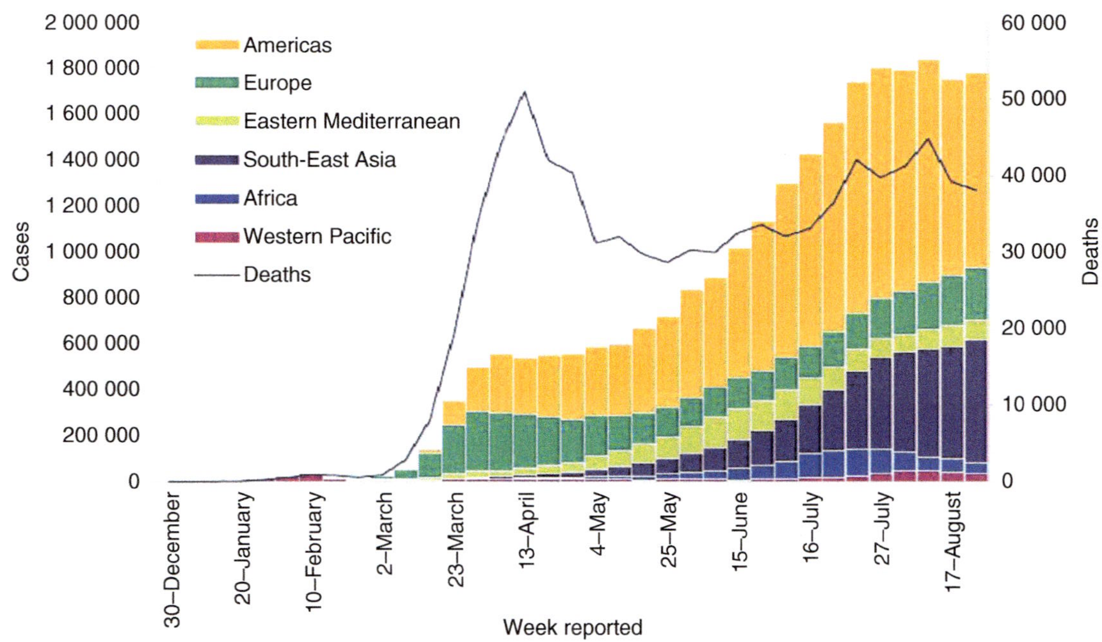

Figure 13.1 Weekly COVID-19 cases and deaths from 30 December to 30 August 2020. Source: WHO [15].

from manufacturing industries and other point sources and, of course, a cleaner atmosphere.

These practices indicate that priority has been solely hinged on public health safety, but little is being done on important environmental implications of the virus. It is highly likely that some countries may forgo sustainable environmental impact assessment of the virus and rather focus on pursuing the rapid economic growth of the nations. However, it is logical to state that an improvement of environmental quality and climate systems will generate a consequential improvement in the economy and hence the health and well-being of citizens. Therefore, the authors have taken a holistic approach in identifying numerous effects of COVID-19 on environmental systems, public health, and socioeconomic parameters of the world. The study aimed to analyze the myriad effects of COVID-19 on environmental parameters by evaluating the current global status of COVID-19 and proffering well-informed and sustainable solutions to assess the environmental effects of the COVID-19 pandemic.

13.2 Methodological Approach

This chapter has taken a holistic and thorough look at the effects of the COVID-19 pandemic from an environmental perspective. In order to address the objectives of this chapter, a three-stage approach was embraced. At first, the literature search was thoroughly conducted by the extraction of relevant peer-reviewed literature, papers, newspapers, book chapters, web scripts from Scopus (www.scopus.com), and Web of Science (www.webofknowledge.com) databases, and specifically, the WHO website. This was done by critically assessing a large volume of indexed documents from searches using relevant keywords such as "environment," "virus," "COVID-19," "air quality," "aquatic life," "wastewater quality," "environmental impact assessment," "environmental pollution," which were combined with the Boolean search words such as "AND," "OR," were used. Selected articles within the last 16 years (2004–2020) were filtered and streamlined based on relevance to subject matter and thoroughly studied to arrive at major themes that have been sectioned and elaborately discussed in this write-up. Other non-environmental impacts of COVID-19 were not considered.

Themes discussed in this chapter are the effects of COVID-19 on socioeconomic development, environmental management for COVID-19 transmission, environmental impact assessment of COVID-19 pandemic on global physical environment, air quality, environmental pollution, water resources and aquatic life, ecological parameters and soil systems, noise pollution, increased solid wastes and recycling, wastewater quality and sanitary systems, correlation of temperature with COVID-19, and other indirect effects of COVID-19 on the environment. A table of full library search of recently published papers in reputable journals with their concise findings within the scope of this chapter has been presented, and the conclusion was presented.

13.3 Effects of COVID-19 on Socioeconomic Development in the Environment

The assessment of the consequential effects of COVID-19 on socioeconomic development poses a very herculean task when adequate data are deficient and because each person responds differently to its effects. The socioeconomic parameters that are of great importance to the study include the household size, population density, rural and urban settings, level of education, people's lifestyle, tenants, and landowners. The size of a household greatly matters in the communal transmission of the virus. Large household members exhibit a higher tendency to contract and transmit the virus within their family setting than a small household. This is typical of large multi-generational homes that are prevalent in Italy and this explains one of the reasons for the outrageous morbidity and mortality rates experienced during the first quarter of the pandemic in Italy. It is also highly likely to witness infected people, who stay within the same neighborhood, to be adversely affected by the virus because the major means of transmission is via droplet contact from an infected person. Urban areas are noted for high population density, which increases contact and transmissibility. It has been recommended that special attention aimed at saving highly vulnerable populations, including children, healthcare workers, and older people be put in place to contain the transmission of COVID-19 in socioeconomically – disadvantaged settings [16].

Social distancing etiquettes are met with sheer defiance and lawlessness in most urban areas and large households, thereby promoting the transmission of COVID-19. Working remotely from home is negatively affected when smaller and more crowded homes are in use. Each occupant fights for limited space and ends up with no buffer zone for contamination protection since droplets fly around within such a small apartment. Human beings' disposition to smoking and other dangerous lifestyles increase their level of susceptibility to infection by the virus. Smokers are greatly at risk as a result of the apparent effects of smoking on the lungs, and this practice is predominantly in lower socioeconomic settings [17]. It has also been recorded that people with underlying medical conditions such as chronic respiratory diseases, cardiovascular diseases, diabetes, cancer, or high blood pressure are at a greater risk of COVID-19 [17]. Another perspective is the unfortunate reality that different socioeconomic groups have different access to the same healthcare services, with reasons bordering on illegal immigrant status, uninsured citizenship status, relative inaccessibility due to poor roads, and sheer disenfranchisement of right to good medical healthcare as a result of huge medical bills. This is a typical phenomenon prevalent in Sub-Saharan Africa and, of course, a worrisome field of study.

13.4 Environmental Management as an Important Factor for COVID-19 Transmission

The untoward modern lifestyle that encourages the construction of mega-cities, proliferation of slums, excessive deforestation, degradation of aquatic life, erratic exploitation of natural resources, desert encroachment, and heavy industrial emission have greatly affected the natural environment [18]. These activities created an artificial environment that offers a convenient platform for viruses, diseases, and poor sanitation to thrive. Through the principle of ecosystem services, environmental management offers a harmonious synergy in regulating water, air quality, pollutant purification, pest or disease control, and climatic condition [19]. Poor ecosystem services approach generates poor environmental management and inhibits biodiversity [19, 20]. Scientific findings have affirmed that the interaction between animals and man promotes pathogenic activities [21]. Enteric viruses have evolved such that they do not cause fatal health effects to the original host except secondary hosts to ensure continuity of life cycle and transmission dynamics [22]. With proper ecosystem services approach, efficient management of the natural environment and consequential reduction of negative anthropogenic effects are feasible to prevent the transmission of viruses by disallowing pathogens from leaving their natural environment and original hosts, with which they share symbolic co-evolution relationship. Table 13.1 shows a list of recent research findings on the effects of COVID-19 on the environment. These findings are fully explained in Section 13.5, while solutions and recommendations are presented afterward.

13.5 Environmental Impact Assessment of COVID-19

13.5.1 Environmental Variables Related to COVID-19

The spread and case fatality of COVID-19 have been reported to be greatly influence by environmental variables, including temperature, humidity, precipitation, particulate matter, and solar radiation [42–46]. Perone [42] found a correlation between COVID-19 mortality and some selected environmental variables including relative humidity and air pollutant concentrations such as NO_2, O_3, and particulate matters (PM_{10} and $PM_{2.5}$) in Italy. Temperature, solar radiation, and specific humidity were reported as the most relevant variables relating to COVID-19 in the Iberian Peninsula, with observed reduction in O_3 [43]. Ahmed and Ghanem [47] also noted a good link between precipitation and COVID-19 infection by relating it to a possible increase in the humidity of the ambient air and/or movement of the virus via environmental compartment by the running water. The particulate matter could serve as a safe haven for SARS-CoV-2 because of its complex mixture of fine particles, polycyclic aromatic hydrocarbon (PAH), and other organic matters by extending the virulence of the virus in the environment [44]. Rosario et al. [45] reported that the main environmental factor responsible for the suppression of the COVID-19 spread in the state of Rio de Janeiro in Brazil. There were also considerable impacts of COVID-19 measures on the environment and its variables. The major environmental variable that has received significant attention relating to the spread of COVID-19 pandemic is temperature, especially because of the wide gap of number of COVID-19 confirmed cases among the regional climatic conditions [46–48].

Table 13.1 Recent studies on COVID-19 impacts on the environment.

S/N	Authors	Findings (effects of COVID-19 on the environment)
1	Riou and Althaus [7] and Phan et al. [23]	• Aerosols affect human respiratory system via oral and nasal ingestion
2	Saadat et al. [16]	• Smoking and dangerous lifestyles increase the risk of contracting COVID-19
3	Zheng et al. [17]	• People with underlying medical conditions such as cardiovascular diseases, diabetes, cancer, etc., are at a greater risk of viral infection
4	Lal et al. [24]	• Air quality in New York greatly increased COVID-19 infection
5	Cheval et al. [25]	• Hypothesis stating pollutants as COVID-19 carriers is not found yet and that points of infections having similar demography, population density, and economic properties may experience similar viral infection and transmission curve – a typical case of Lombardy and Veneto regions in Italy
6	Adelodun et al. [26], Medema et al. [27], Gormey et al. [28], Mohan et al. [29], and Espejo et al. [30]	• Identification of SARS-CoV-2 in domestic sewage and wastewater has been documented and preventive measures such as ozonation, chlorination, and UV radiation technologies of water treatment for isolation holding facilities were suggested • Human enteric viruses have been spotted on surface water, public water supply, freshwater, groundwater, and sediments except sludge and wastewater samples
7	Atalan [10], Medema et al. [27], Saadat et al. [16], and Zambrano-monserrate et al. [31]	• Indiscriminate disposal of plastics and PPEs (made of polypropylene) in waterbodies in China, United States, and Hong Kong, thereby creating a surge in water pollution indices • Hospitals in Wuhan, China, produced an average 240 metric tons of medical waste per day during the pandemic as compared to their previous average of fewer than 50 tons pre-COVID
8	Núñez-Delgado [32]	• No scientific proof yet for the soil being a transmission medium for SARS-CoV-2
9	Chakraborty and Maity [33] and Zambrano-monserrate et al. [31]	• Drastic reduction in industrial wastes has been recorded in several countries as a result of lockdown • Sustainable waste management system is hampered by the use of single-use packaging since recycling programs are abolished
10	Wu et al. [34]	• A 1 $\mu g/m^3$ increase in PM2.5 generated an 8% increase in mortality rate in 3000 cities in the United States
11	Ogen [35]	• The long-term exposure of citizens of about 66 administrative regions in France, Germany, Spain, and Italy to nitrogen dioxide (NO_2) may contribute to mortality rate due to COVID-19
12	Xu et al. [36], Zhang et al. [37], Suhaimi et al. [38], and Pansini and Fornacca [39]	• Air pollution has improved transmission potential of COVID-19 • There is a direct correlation between air quality index (AQI) and confirmed cases of COVID-19
13	Zoran et al. [40] and Andree [41]	• Recorded new COVID-19 cases have a positive correlation with PM2.5 and PM10 • PM2.5 was found to be an appropriate predictor of the number of confirmed COVID-19 cases

Source: Saadat et al. [16], Riou and Althaus [7], and Phan et al. [23].

Several researchers have made concerted effort in analyzing the connection between temperature and COVID-19 infection. There have been diverse submissions from highly reputable scholars about how temperature affects the trend of confirmed cases of coronavirus and vice versa. Some researchers believe that there is a positive association between temperature and number of confirmed cases, while some believe that the two parameters are negatively correlated. There is the third category of scholars who believe that there is no interaction between the two at all [49]. Temperature and humidity affected daily COVID-19 cases in 30 Chinese provinces and it was discovered that a 1 °C increase in average daily temperature reading reduced daily rate of COVID-19 cases from 36% to 57% at a relative humidity range of 65–85.5% [50]. Similarly, a 1% rise in relative humidity reduced daily COVID-19 cases from 11% to 22%, while the average daily temperature was within the range of 5.04 and 8.2 °C [50]. Gupta et al. [51] confirmed that humidity and temperature can forecast COVID-19 transmissions in the United States. It was suggested that the hot and humid climates in India might have played an important role in the low transmission of COVID-19. However, this hypothesis has been contested that strict and early lockdown was accountable for the low transmission rate and a future rise is expected owing to the inability of the citizens to sustain the lockdown practice because of high population density, proliferation of slums, and inadequate housing facilities [52].

Other studies on inter-relationship of temperature and number of COVID-19 cases have been exclusively argued by Xie and Zhu [53], Chin et al. [54], and Shi et al. [55]. There was a direct linear relationship between temperature and number of COVID-19 cases when the temperature was lower than 3 °C [53]. This was refuted by Shi et al. [55] that the opposite effects were experienced in China, where temperatures about the range of 8–10 °C reduced daily transmission of confirmed cases. Chin et al. [54] added that the virus is highly stable at 4 °C but sensitive to heat. A study in Spain concluded that there is an insignificant effect of temperature on the transmission potential of SARS-CoV-2 by using spatiotemporal modeling techniques and also introducing socioeconomic parameters such as age, population density, number of firms, and number of travelers.

13.5.2 Effects of COVID-19 on Global Physical Environment: Air Quality and Environmental Pollution

The effects of COVID-19 on the physical systems in the global environment cannot be overemphasized. These effects offer beneficial possibilities as much as disadvantages to all the parameters that make up the physical system. Air quality significantly increased the spread of COVID-19 infection in New York City [24]. The institution of lockdown measures has, in a way, underlined the need for effective population check in terms of density and measure. Urban areas are notable for their highly populated architecture and sparingly ventilated settlements. Lockdown practice in these areas may be faced with an increased risk in terms of exposure to hazards and threats. Highly dense areas with more aeration also increase the dependency on clean air, thereby creating an unhealthy rivalry among inhabitants. In rural settings where there is proper ventilation and gardens, which ensure creation of clean air and scenic views, isolation measures are easily practiced with a net zero negative impact on the immediate physical system. Ordóñez et al. [43] carried out a study on the effect of strict lockdown measures (as a result of the first wave of COVID-19 in Europe) on air quality. They discovered that, within the study duration of 15 March 2020 to 30 April 2020, daily maximum NO_2 reduced consistently over the continents, with a reduction of 5–55% with respect to a base period of 2015–2019 for 80% of the study sites, with the highest reduction recorded in Spain, France, and Italy. Ozone concentration reduction was recorded in Iberia but increased elsewhere and was compared with expected NO_2 and O_3 concentrations in the absence of lockdown, using generalized additive models [43]. Their results pointed to the dominant role of meteorology on regional ozone anomalies in Europe during the lockdown and also determined ozone responses to emission changes or preferably, air quality. Also, there have been reported findings of a significant CO emission level of approximately 48.5–30.3% in Rio de Janeiro, Brazil [56]. Coal use in China caused a 25% reduction in CO_2, which was equivalent to about 6% of global emissions [57], and 10 cities in China recorded a 20% reduction of PM2.5 as a result of the pandemic [58]. Liu et al. [59] and Le Quéré et al. [60] reported some level of declining in country-level CO_2 emissions because of the energy consumption and human inactivity, especially ground transportation, as a result of lockdown measures imposed by each country to curtail the spread of COVID-19. About 9% global reduction of CO_2 emissions was reported for the first half year of 2020 compared to the year 2019 [59]. Table 13.2 presents the comparative assessment of CO_2 emission in some selected countries before and during the COVID-19 era.

It is a known fact that the level of air pollution of a typical area is an indication of the climatic condition of such a place and air pollution in the upward range may accelerate drastic changes in the ecosystem, thereby triggering pathogenic activity with increase in the outbreak of infectious diseases, vectors, hosts, and transmission dynamics [45, 46, 62, 63]. Hill [64] reported that pollutants such as organochlorines, organophosphates, heavy metals, carbamates, endocrine disruptors, dioxins, and methylmercury have greatly increased in concentrations and availability in the environment globally, as a result of anthropogenic activities and have been found in air, soil, water, vegetation, and animals. Furthermore, it has been reported that over 91% of the world population live in areas where poor air quality exceeds the allowable limits [65]. This finding has additionally offered a standpoint that inferred that air pollution contributes to about 8% of total global deaths, with the most badly-hit countries found in Asia, Europe, and Africa [65]. One of the response methodologies currently being practiced all over the world is partial/total lockdown, which has generated a reduction in carbon footprint in terms of greenhouse gas (GHG) emissions from manufacturing industries, aerosols, smoke emission from oil and its products' usage, and an obvious decline in nitrogen oxide (NO_2) concentration in European countries such as France, Germany, Spain, and Italy, as a result of anthropogenic-based activities. The pandemic has resulted into higher concentrations in atmospheric pollutants such as NO_2, which has been linked with diseases such as bronchoconstriction, lung infections, reduced

Table 13.2 Comparative assessment of carbon emissions between pre-COVID-19 and post-COVID-19.

Region	Country	Population (millions)[a)]		CO_2 emission (Mt per day)[b)]		
		2019	2020	2019 (1 January to 30 June)	2020 (1 January to 30 June)	% Reduction
North America	United States	329.06	331.00	2 543.61	2 205.31	13.3
Europe	United Kingdom	67.53	67.89	178.67	151.87	15.0
	Russia	145.87	145.93	764.15	723.65	5.3
	Germany	83.52	83.78	357.62	303.62	15.1
	France	65.13	65.27	151.41	129.91	14.2
	Italy	60.55	60.46	167.15	144.25	13.7
	Spain	46.74	46.75	122.87	99.77	18.8
Asia	Japan	126.86	126.48	574.67	531.57	7.5
	India	1 366.42	1 380.00	1 332.47	1 127.27	15.4
	China	1 433.78	1 439.32	5 059.46	4 872.26	3.7
South America	Brazil	211.05	212.56	215.83	189.93	12.0
Global	World	7 713.47	7 794.80	17 619.32	16 068.82	8.8

a) Worldometer [61].
b) Liu et al. [59].
Source: Liu et al. [59] and Worldometer [61].

immunity, and respiratory problems, thereby resulting in an increased susceptibility to colds and flu.

The effects of COVID-19 pandemic on air quality with respect to energy use, production and transportation companies, retail and personnel cannot be over flogged. A critical study of energy use for industrial and transportation purposes in European Economic Area (EEA) countries such as Norway, Iceland, and Liechtenstein showed that severe impacts on air quality were recorded in mid-April of the COVID-19 year, with about 40–50% reduction in economic activity [66]. This was as a result of strict COVID-19 protocols set in place to inhibit the transmission of the virus. Another notable effect of reduction of air pollution is the significant decline in global coal consumption, especially in China. Also, by the end of July 2020, there were reported reductions in air pollution in New York City, China, and Italy, and currently, there are predictions of declines in GHG emissions toward the end of the year [25]. However, it is imperative to know that the reduction in GHGs may offer an insignificant effect on total GHG concentrations, which have accumulated in the atmosphere for decades. The reduction may be temporal because once the pandemic ends, all countries may likely initiate extreme economic and industrial actions to resuscitate their economies, thereby increasing GHG emissions and reverting us to where we started or farther still.

The infection and mortality rates in Northern Italy became so worrisome that most researchers used the region as an epicenter case study for current and future research works. It is believed that the presence and subsequent inhalation of particulate matter such as PM_{10} and PM_{25} by citizens increased viral susceptibility of the respiratory systems of citizens even before COVID-19. So, when the coronavirus emerged, the elderly could not endure such viral load and yielded to the fatal tendency of the virus. Furthermore, it has been reported that there is a high level of correlation between COVID-19 viral spread and pollution levels [38], and even, further study has confirmed that stronger evidence abound on the idea that many COVID-19 patients in Italy were initially exposed to respiratory system threats and diseases. However, hypothesis stating pollutants such as COVID-19 carrier is not found yet [25].

Also, Cheval et al. [25] submitted that the two Italian regions (Lombardy and Veneto) that were mostly hit by COVID-19 shared a similar architecture in terms of demography, population density, and economic properties. Cheval et al. [25] feel that the two regions share these similar traits and that it is only logical that they experience similar COVID-19 adverse effect as a result of similar population densities within the range of 270–420 per km^2; gross domestic product of Veneto of £29 500/capita and Lombardy of £34 000/capita. Definitely, COVID-19 varies spatially.

13.5.3 COVID-19 Impacts on Water Resources and Aquatic Life

Lockdown activities have generated a great reduction in aquatic transportation and tourist activities; however, there is the worrisome report about increased disposal of sanitary items especially personal protective equipment (PPE) in open waterbodies.

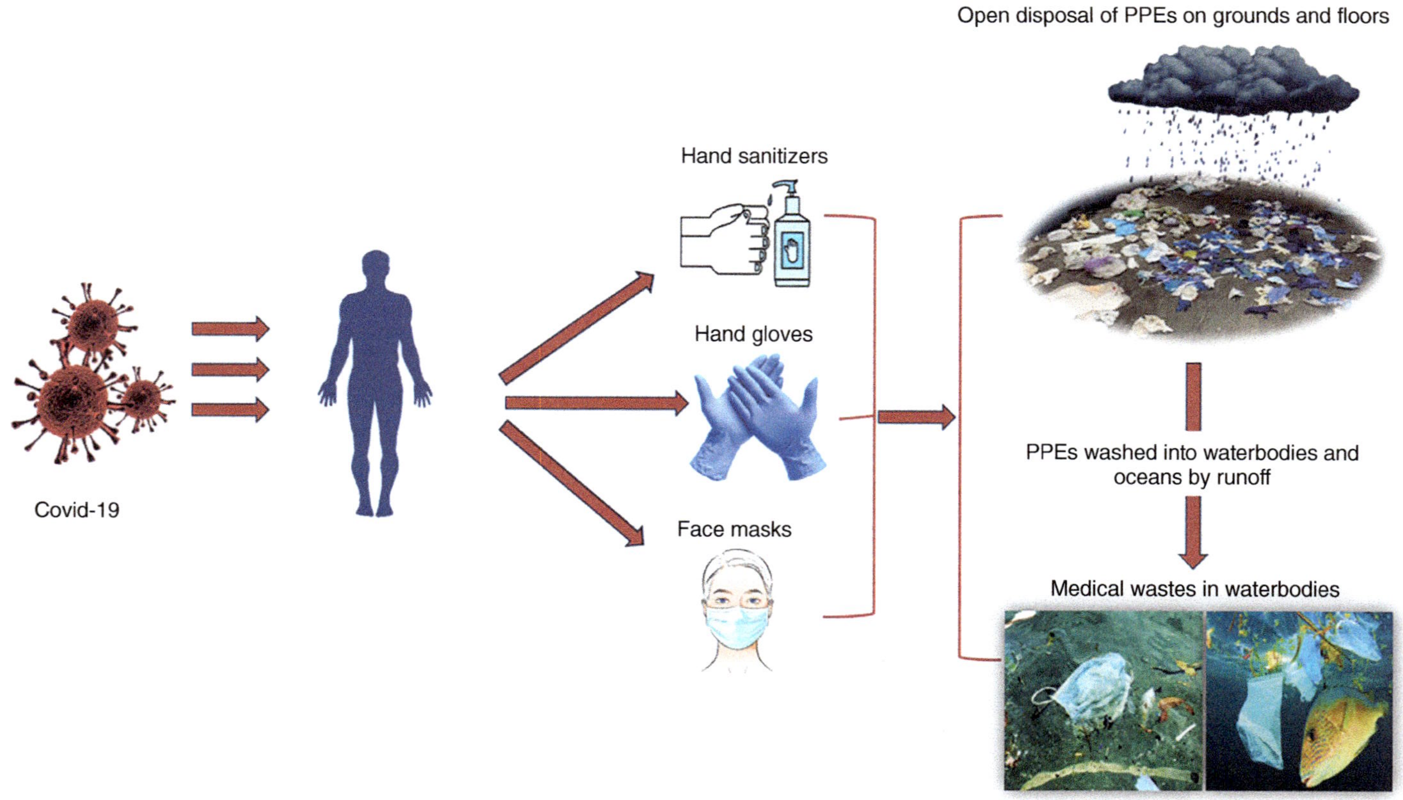

Figure 13.2 Medical wastes created during COVID-19 pandemic.

This may harm aquatic life and threaten the growth of riparian vegetation along water courses. Indiscriminate disposal of plastics and PPE have been reported in Canada, the United States, and Hong Kong [10, 27]. Plastic pollution of waterways was recorded in China where plastic-made medical masks (made of polypropylene) were collected from the sea by some environmentalists and concluded that COVID-19 measures created a surge in water pollution indices [27]. COVID-19 impacts on water resources may create a long-term impact with respect to depleting national reserves and nose-diving revenue generation status of every single country. Figure 13.2 shows the illustration of medical waste generation (PPE) during the pandemic.

In contrast, one of the beneficial effects of the pandemic on waterbodies as a result of lockdown was recorded in Venice, Italy. Water obtained from canals in Venice was cleaner and less turbid as a result of lesser or no tourist usage of waterbodies, lesser sediment agitation, and reduced water pollutants. The water was clean and aquatic life could be assessed visually.

Restriction of movement globally has generated more pressure on demand and supply of water for domestic use. In Austria, daily peak water consumption in the morning had skyrocketed from about 1.5 to 2 hours. The water authority had to compensate for this surge by increasing the supply and a normal curve of transactional econometrics was observed afterward. Industrial water consumption has drastically reduced because of the mandatory lockdown and work from home practices. Water management for agriculture must be centered on efficient and sustainable use of water resources during post-COVID-19 era. This will ensure the gradual recovery of water resources that have been lost during the pandemic.

13.5.4 COVID-19 Impacts on Ecological Parameters and Soil Systems

The soil offers a medium for the growth of food and food materials. The effects of COVID-19 on soil resources may not be too obvious in the short term for now; however, isolation and lockdowns have denied farmers access to soil and agriculture. National food supply grid has been greatly affected and every nation's economy is currently bearing the brunt. Available food and crops are becoming unbelievably limited and inflation of prices of commodities has become the order of the day in Sub-Saharan Africa. At the moment, there is no proof to confirm that the soil environment may host and transmit the virus as do solid surfaces and metal surfaces [32].

The effects of COVID-19 on the ecological systems cannot be overemphasized. The prevalence of the virus may pose greater threats to our society in years to come because the virus may eventually have an indirect effect on the climate change on the ecosphere. These effects may include changing species, alien invasion, and diminishing species habitat. There may also be an economic effect on conservation programs and ecology-oriented research globally, as funding for research deplete daily and people are only worried about their safety/survival than proffering solutions to the current situation.

13.5.5 COVID-19 Impacts on Noise Pollution, Increased Solid Wastes, and Recycling

During the lockdown, beaches and recreational centers were deserted, indicating a reduction in waste load generated by tourists and a significant noise reduction due to non-patronage of mass transit/public transportation systems, which may increase the transmission risk of COVID-19. Industrial waste emission has drastically reduced as a result of closure of industrial activities [33]. Gradual ease of lockdown protocols has required some firms to institute COVID-19 work ethics – one of which requires the repeal of disposable bags and containers while embracing single-use packaging and online ordering of necessities in order to reduce transmission [31]. As much as this offers a preventive approach to transmission and infection, however, it still poses a negative effect on people as recycling programs are being suspended and sustainable waste management is greatly inhibited.

The implementation of lockdown measures has resulted in the generation of increased demand for online orders for groceries and domestic supplies all across the world. This has created an enormous organic and inorganic waste overflow from households. Inorganic wastes are also generated from packaging packs of food supplies ordered on the internet and shipped to point of delivery by individuals or drones. Hospitals in Wuhan, China, have been reported to produce an average 240 metric tons of medical waste per day during the pandemic compared to their previous average of fewer than 50 tons post-COVID-19 era [31]. Water treatment plants in China have been enjoined to fortify the disinfection unit operation by increasing the dosage of chlorine with a view to preventing the spread of COVID-19 through wastewater, even though there has not been any verified reported case of COVID-19 transmission through wastewater [67].

13.5.6 COVID-19 Impacts on Wastewater Quality and Sanitary Systems

There have been reported cases of identification of SARS-CoV-2 in domestic sewage, and this has propelled the need to create life cycle assessment of the virus especially as it passes through waterbodies [26, 27, 68]. This potential transmission of the virus through wastewater is currently generating a growing concern in research areas of epidemiology, public health, wastewater engineering, and virology [69–71]. The possibility of transmission of SARS-CoV-2 through water infrastructure was reported to be a major concern, and its detection and subsequent inactivation will play a major role in curbing the spread of the virus to the community [29]. Furthermore, the detection of the virus in wastewater when the SARS-CoV-2 count was low indicated the functional role of wastewater monitoring, in order to assess the spread of the virus [29]. It has also been reported that human enteric viruses have been observed on surface water, groundwater, public water supply, freshwater, and sediments aside sludge and wastewater samples [1, 28, 29]. Recent studies have concluded that there are significant viral loads of SARS-CoV-2 in human feces from infected persons with severe symptoms [55, 56, 72, 73]. During conveyance of wastewater containing human excreta, there may be leakages in sewerage systems, thereby increasing the risk of sanitation-related diseases to man [27, 28, 74]. There is an urgent need to proffer proactive solutions to stem the transmission dynamics of the virus via contaminated wastewater. Adelodun et al. [26] highlighted several solutions to potential wastewater infection by SARS-CoV-2 especially in isolation homes, treatment facilities, slums, and low-income countries. Solutions proffered included decentralization of wastewater treatment facilities, practice of community – wide monitoring and testing of SARS-CoV-2 RNA in wastewater, improved sanitation and water quality by embracing ozonation, chlorination, and UV radiation operations in wastewater treatment plants, and enhanced policy intervention. These measures centered on instituting policies that will mitigate the communal outbreak of SARS-CoV-2 from waterbodies and streams, which serve as a primary source of water consumption and for bathing, to inhabitants of villages in low-income countries [26, 75, 76].

13.5.7 Socioeconomic Environmental Impacts of COVID-19

The event of COVID-19 has greatly altered the social, economic, and environmental indices, in which the impacts vary across the countries, sectors, and domains. For instance, although countries such as France, Germany, and the United Kingdom mandated the second lockdown because of the recent rising cases of COVID-19 [77], the majority of the countries in Africa and Asia are full of economic activities with only measures of face masking in public places. Similarly, the transportation sectors such as aviation and road transportation were found to experience significant decline because of the imposed lockdown measure in India, China, and the United States [59, 78]. Before the emergence of COVID-19, there was socioeconomic and environmental inequality among the citizenries in each of the countries [1]. The various measures to mitigate the spread of COVID-19, such as human and vehicular movement restrictions and the shutdown of industries, further aggravate the impacts of the socioeconomic activities and productivity of the people, the majority of which are from less-developed and developing nations [79]. Although there were some reported positive impacts gained from imposed human inactivity measures to address the COVID-19 spread, particularly from the environmental perspective such as pollution and CO_2 emission reductions [59, 60, 79], the global economy is, however, greatly impacted as indicated by the report from the International Monetary Fund of possibly 3% decline in the economic growth [80]. This indicates that the measures to combat the spread of COVID-19 have somewhat favored the reduction in environmental pollution at the expense of socioeconomic retrogression [2]. However, many countries have instituted economic assistance in monetary stimulus packages for their citizens [2]. The economic intervention also dubbed as social intervention programs in some countries [81].

13.5.8 Indirect Effects of COVID-19 on the Environment

The lockdown has reduced the number of tourists vising natural parks and other attraction centers, thereby reducing stress on wildlife [82]. Wild fauna have started encroaching and migrating back to the suburbs from which they had initially fled.

Forest reserves and wildlife policies should be implemented and focused on protection of wild animals and abolishment of indiscriminate hunting in less-developed countries after this COVID-19 era. In addition, there is a recorded global increase in solid wastes generated domestically and industrially as the lockdown relaxes and normalcy is being restored gradually. A few countries have resorted to a less sustainable approach of incinerating PPEs and contaminated medical wastes obtained from testing and isolation facilities in order to exterminate the virus, thereby affecting public health, soil life, water resources, and depleting the ozone layer [83–85]. Further harmful effects of this disastrous practice may result in unfavorable climate change, buildup of GHGs, acid rain, reduction in soil fertility, and poor agricultural produce. Since the wastes are biodegradable and the soil environment has not been proven to be a transmitting medium for SARS-CoV-2, it is recommended that such wastes should be decontaminated and used as landfills or compost for future agricultural use. The use of hand sanitizers is a common practice in this COVID-19 time and it is composed majorly of alcohol and a universal antimicrobial drug called Triclosan, which is tagged an emerging contaminant [86]. Too much use may create harmful effects on public health and environment.

Pérez-Espejo et al. [87] recommended the employment of Information and Communications Technology (ICT) knowledge in curbing the menace that comes with inaccessibility of data relevant to indoor air quality indices and human health as a result of lockdown and isolation. They concluded that the use of "smart home," "smart care," and "health monitoring gadgets" will go a long way in collating and analyzing meteorological data which affect the spread of the pandemic. Some developed countries such as the United States, South Korea, Japan, and Malaysia have health monitoring mobile applications that monitor COVID-19 symptoms of people. Travelers are subjected to COVID-19 testing and the mobile applications are installed on devices to monitor health progress of such people. This approach is safe and ensures less or no contact with potential COVID-19 victim, thereby reducing risk of infection, physical stress that comes with traveling to assess the travelers, and improving mental health of frontline workers in hospitals and holding facilities. Other indirect effects of COVID-19 pandemic on the environment include noise reduction, less turbid and cleaner waterbodies, cleaner air, etc. All these have been thoroughly explained in Section 13.5.8 of this book chapter.

13.6 Conclusion

This book chapter has taken a holistic approach in analyzing the effects of COVID-19 on the environment by taking a broad look at all environmental parameters as they affect the transmission potential of COVID-19 in the environment. The environmental effects of COVID-19 has cut across all phases of life and created a resonating effect on economy, sociocultural lifestyle, water resources, public health, many more. This unprecedented pandemic has unequivocally taught us a lesson of a lifetime and a need to further approach goal-driven research, regular climate change assessment, worthwhile economic surveys, forecasts, and sustainable ecosystem services, in a bid to creating a better world and an ever ready approach to survival. A proper evaluation of interrelationship and correlation of key meteorological parameters as they affect the transmissibility of COVID-19 and other respiratory viruses will go a long way in reducing the rapid infection by corona virus and providing a reliable viewpoint in formulating well-informed prediction policies of the effect of pollution and weather data on the transmission potential of viruses. All these can be set in motion while we anticipate the discovery and manufacturing of COVID-19 vaccines.

References

1 Adelodun, B., Ajibade, F.O., Ighalo, J.O. et al. (2021). Assessment of socioeconomic inequality based on virus-contaminated water usage in developing countries: a review. *Environ. Res.* 192 (September 2020): 110309.

2 Sarkodie, S.A. and Owusu, P.A. (2020). Global assessment of environment, health and economic impact of the novel coronavirus (COVID-19). *Environ. Dev. Sustainability*: 1–11, (0123456789).

3 Rothan, H.A. and Byrareddy, S.N. (2020). The epidemiology and pathogenesis of coronavirus disease (COVID-19) outbreak. *J. Autoimmun.* 109 (February): 102433.

4 WHO (2020). WHO Director-General's opening remarks at the media briefing on COVID-19-11 March 2020. *World Heal. Organ.*

5 de Wilde, A.H., Snijder, E.J., Kikkert, M., and van Hemert, M.J. (2017). Host factors in coronavirus replication. In: *Roles of Host Gene and Non-coding RNA Expression in Virus Infection. Current Topics in Microbiology and Immunology*, vol. 419 (eds. R. Tripp and S. Tompkins). Cham: Springer. https://doi.org/10.1007/82_2017_25.

6 Kitajima, M., Ahmed, W., Bibby, K. et al. (2020). SARS-CoV-2 in wastewater: state of the knowledge and research needs. *Sci. Total Environ.* 739: 139076.

7 Riou, J. and Althaus, C.L. (2020). Pattern of early human-to-human transmission of Wuhan 2019 novel coronavirus (2019-nCoV), December 2019 to January 2020. *Eurosurveillance* 25 (4): 1–5.

8 Woolhouse, M.E.J. and Gowtage-Sequeria, S. (2005). Host range and emerging and reemerging pathogens. *Emerg. Infect. Dis.* 11 (12): 1842–1847.

9 LeDuc, J.W. and Barry, M.A. (2004). SARS, the first pandemic of the 21st century. *Emerg. Infect. Dis.* 56: 38–42.

10 Atalan, A. (2020). Is the lockdown important to prevent the COVID-19 pandemic? Effects on psychology, environment and economy-perspective. *Ann. Med. Surg.* 56 (June): 38–42.

11 Al Hajjar, S. and McIntosh, K. (2010). The first influenza pandemic of the 21st century. *Ann. Saudi Med.* 30 (1): 1–10.

12 Malvy, D., McElroy, A.K., de Clerck, H. et al. (2019). Ebola virus disease. *Lancet* 393 (10174): 936–948.

13 Plourde, A.R. and Bloch, E.M. (2016). A literature review of zika virus. *Emerg. Infect. Dis.* 22 (7): 1185–1192.

14 Chatziprodromidou, I.P., Arvanitidou, M., Guitian, J. et al. (2018). Global avian influenza outbreaks 2010–2016: a systematic review of their distribution, avian species and virus subtype. *Syst. Rev.* 7 (1): 1–12.

15 WHO (2020). Weekly Epidemiological Update: Coronavirus Disease (COVID-19). *Glob. Epidemiol. Situat. reports.* Report Number 28 (28 September).

16 Saadat, S., Rawtani, D., and Hussain, C.M. (2020). Environmental perspective of COVID-19. *Sci. Total Environ.* 728: 138870.

17 Zheng, Y.Y., Ma, Y.T., Zhang, J.Y., and Xie, X. (2020). COVID-19 and the cardiovascular system. *Nat. Rev. Cardiol.* 17 (5): 259–260.

18 Boger, D.V. and Hart, B.C. (2008). Making an unsustainable industry more sustainable. In: *Proceedings of the Eleventh International Seminar on Paste and Thickened Tailings* (eds. R. Jewell, A.B. Fourie, P. Slatter and A. Paterson), 3–13. Perth: Australian Centre for Geomechanics, https://doi.org/https://doi.org/10.36487/ACG_repo/863_1 (https://papers.acg.uwa.edu.au/p/863_1_Boger/).

19 Ajibade, F.O., Adelodun, B., Lasisi, K.H. et al. (2020). Environmental pollution and their socioeconomic impacts. In: *Microbe Mediated Remediation of Environmental Contaminants*, 1e (eds. A. Kumar, V.K. Singh, P. Singh and V.K. Mishra), 321–354. Woodhead Publishing, Elsevier.

20 Fisher, B., Turner, R.K., and Morling, P. (2009). Defining and classifying ecosystem services for decision making. *Ecol. Econ.* 68 (3): 643–653.

21 Jones, B.A., Grace, D., Kock, R. et al. (2013). Zoonosis emergence linked to agricultural intensification and environmental change. *Proc. Natl. Acad. Sci. U.S.A.* 110 (21): 8399–8404.

22 Zohdy, S., Schwartz, T.S., and Oaks, J.R. (2019). The coevolution effect as a driver of spillover. *Trends Parasitol.* 35 (6): 399–408.

23 Phan, L.T., Nguyen, T.V., Luong, Q.C. et al. (2020). Importation and human-to-human transmission of a novel coronavirus in Vietnam. *N. Engl. J. Med.* 382 (9): 872–874.

24 Lal, P., Kumar, A., Kumar, S. et al. (2020). The dark cloud with a silver lining: assessing the impact of the SARS COVID-19 pandemic on the global environment. *Sci. Total Environ.* 732: 139297.

25 Cheval, S., Mihai Adamescu, C., Georgiadis, T. et al. (2020). Observed and potential impacts of the COVID-19 pandemic on the environment. *Int. J. Environ. Res. Public Health* 17 (11): 4140.

26 Adelodun, B., Ajibade, F.O., Ibrahim, R.G. et al. (2020). Snowballing transmission of COVID-19 (SARS-CoV-2) through wastewater: any sustainable preventive measures to curtail the scourge in low-income countries? *Sci. Total Environ.* 742: 140680.

27 Medema, G., Heijnen, L., Elsinga, G. et al. (2020). Presence of SARS-coronavirus-2 RNA in sewage and correlation with reported COVID-19 prevalence in the early stage of the epidemic in the Netherlands. *Environ. Sci. Technol. Lett.* 7 (7): 511–516.

28 Gormley, M., Aspray, T.J., and Kelly, D.A. (2020). COVID-19: mitigating transmission via wastewater plumbing systems. *Lancet Global Health* 8 (5): e643.

29 Mohan, S.V., Hemalatha, M., Kopperi, H. et al. (2021). SARS-CoV-2 in environmental perspective: occurrence, persistence, surveillance, inactivation and challenges. *Chem. Eng. J.* 405 (September 2020): 126893.

30 Espejo, W., Celis, J.E., Chiang, G., and Bahamonde, P. (2020). Environment and COVID-19: pollutants, impacts, dissemination, management and recommendations for facing future epidemic threats. *Sci. Total Environ.* 747: 141314.

31 Zambrano-Monserrate, M.A., Ruano, M.A., and Sanchez-Alcalde, L. (2020). Indirect effects of COVID-19 on the environment. *Sci. Total Environ.* 728: 138813.

32 Núñez-Delgado, A. (2020). What do we know about the SARS-CoV-2 coronavirus in the environment? *Sci. Total Environ.* 727: 138647.

33 Chakraborty, I. and Maity, P. (2020). COVID-19 outbreak: migration, effects on society, global environment and prevention. *Sci. Total Environ.* 728: 138882.

34 Wu, X., Nethery, R.C., Sabath, M.B. et al. (2020). Air pollution and COVID-19 mortality in the United States: strengths and limitations of an ecological regression analysis. *Sci. Adv.* 6 (45): eabd4049. https://doi.org/https://doi.org/10.1126/sciadv.abd4049.

35 Ogen, Y. (2020). Assessing nitrogen dioxide (NO_2) levels as a contributing factor to coronavirus (COVID-19) fatality. *Sci. Total Environ.* 726 (March): 138605.

36 Xu, H., Yan, C., Fu, Q. et al. (2020). Possible environmental effects on the spread of COVID-19 in China. *Sci. Total Environ.* 731: 139211.

37 Zhang, Z., Xue, T., and Jin, X. (2020). Effects of meteorological conditions and air pollution on COVID-19 transmission: evidence from 219 Chinese cities. *Sci. Total Environ.* 741: 140244.

38 Suhaimi, N.F., Jalaludin, J., and Latif, M.T. (2020). Demystifying a possible relationship between COVID-19, air quality and meteorological factors: evidence from Kuala Lumpur, Malaysia. *Aerosol Air Qual. Res.* 20 (7): 1520–1529.

39 Pansini, R. and Fornacca, D. (2020). COVID-19 higher morbidity and mortality in Chinese regions with lower air quality. *Front. Public Health* 8: 999. https://doi.org/https://doi.org/10.3389/fpubh.2020.597753.

40 Zoran, M.A., Savastru, R.S., Savastru, D.M., and Tautan, M.N. (2020). Assessing the relationship between surface levels of PM2.5 and PM10 particulate matter impact on COVID-19 in Milan, Italy. *Sci. Total Environ.* 738: 139825.

41 Andree, B.P.J. (2020). Incidence of COVID-19 and connections with air pollution exposure: evidence from the Netherlands. *medRxiv*. https://doi.org/10.1101/2020.04.27.20081562.

42 Perone, G. (2021). The determinants of COVID-19 case fatality rate (CFR) in the Italian regions and provinces: an analysis of environmental, demographic, and healthcare factors. *Sci. Total Environ.* 755: 142523.

43 Ordóñez, C., Garrido-Perez, J.M., and García-Herrera, R. (2020). Early spring near-surface ozone in Europe during the COVID-19 shutdown: meteorological effects outweigh emission changes. *Sci. Total Environ.* 747 (December 2019): 141322.

44 Wathore, R., Gupta, A., Bherwani, H., and Labhasetwar, N. (2020). Understanding air and water borne transmission and survival of coronavirus: insights and way forward for SARS-CoV-2. *Sci. Total Environ.* 749: 141486.

45 Rosario, D.K.A., Mutz, Y.S., Bernardes, P.C., and Conte-Junior, C.A. (2020). Relationship between COVID-19 and weather: case study in a tropical country. *Int. J. Hyg. Environ. Health* 229 (January): 113587.

46 Iqbal, M.M., Abid, I., Hussain, S. et al. (2020). The effects of regional climatic condition on the spread of COVID-19 at global scale. *Sci. Total Environ.* 739: 140101.

47 Ahmed, A.T. and Ghanem, A.S. (2020). A statistical study for impacts of environmental conditions on the rapid spread of new corona virus. *Int. J. Environ. Sci. Technol.* 17 (10): 4343–4352.

48 Cacho, P.M., Hernández, J.L., López-Hoyos, M., and Martínez-Taboada, V.M. (2020). Can climatic factors explain the differences in COVID-19 incidence and severity across the Spanish regions?: an ecological study. *Environ. Health* 19 (1): 1–8.

49 Martorell-Marugán, J., Villatoro-García, J.A., García-Moreno, A. et al. (2021). DatAC: a visual analytics platform to explore climate and air quality indicators associated with the COVID-19 pandemic in Spain. *Sci. Total Environ.* 750: 141424.

50 Qi, H., Xiao, S., Shi, R. et al. (2020). COVID-19 transmission in Mainland China is associated with temperature and humidity: a time-series analysis. *Sci. Total Environ.* 728: 138778.

51 Gupta, S., Raghuwanshi, G.S., and Chanda, A. (2020). Effect of weather on COVID-19 spread in the US: a prediction model for India in 2020. *Sci. Total Environ.* 728: 138860.

52 Paital, B. (2020). Nurture to nature via COVID-19, a self-regenerating environmental strategy of environment in global context. *Sci. Total Environ.* 729: 139088.

53 Xie, J. and Zhu, Y. (2020). Association between ambient temperature and COVID-19 infection in 122 cities from China. *Sci. Total Environ.* 724: 138201.

54 Chin, A.W.H., Chu, J.T.S., Perera, M.R.A. et al. (2020). Stability of SARS-CoV-2 in different environmental conditions. *Lancet Microbe* 1 (1): e10.

55 Shi, P., Dong, Y., Yan, H. et al. (2020). Impact of temperature on the dynamics of the COVID-19 outbreak in China. *Sci. Total Environ.* 728 (77): 138890.

56 Dantas, G., Siciliano, B., França, B.B. et al. (2020). The impact of COVID-19 partial lockdown on the air quality of the city of Rio de Janeiro, Brazil. *Sci. Total Environ.* 729: 139085.

57 Myllyvirta, L. (2020). Analysis: coronavirus temporarily reduced China's CO_2 emissions by a quarter. *CarbaonBrief*.

58 Wang, P., Chen, K., Zhu, S. et al. (2020). Severe air pollution events not avoided by reduced anthropogenic activities during COVID-19 outbreak. *Resour. Conserv. Recycl.* 158 (February): 104814.

59 Liu, Z., Ciais, P., Deng, Z. et al. (2020). Near-real-time monitoring of global CO_2 emissions reveals the effects of the COVID-19 pandemic. *Nat. Commun.* 11 (1): 1–12.

60 Le Quéré, C., Jackson, R.B., Jones, M.W. et al. (2020). Temporary reduction in daily global CO_2 emissions during the COVID-19 forced confinement. *Nat. Clim. Change* 10 (7): 647–653.

61 Worldometer (2020). Countries in the world by population in 2020. https://www.worldometers.info/world-population/population-by-country/ (accessed 10 January 2021).

62 Howard, C. and Huston, P. (2019). The health effects of climate change: know the risks and become part of the solutions. *Canada Commun. Dis. Rep.* 45 (5): 114–118.

63 Wu, X., Lu, Y., Zhou, S. et al. (2016). Impact of climate change on human infectious diseases: empirical evidence and human adaptation. *Environ. Int.* 86: 14–23.

64 Hill, M.K. (1997). *Understanding Environmental Pollution*, 4e. Cambridge: Cambridge University Press. https://doi.org/10.1017/9781108395021.

65 WHO (2016). *Air Pollution*. World Health Organization. https://www.who.int/health-topics/air-pollution#tab=tab_1 (accessed 16 October 2020).

66 Harris, K. (2020). *Tracking the Global Impact of the Coronavirus Outbreak: Bain's Situational Threat Report (SITREP) Index*. Bain Macro Trends Group analysis. https://www.bain.com/insights/tracking-the-global-impact-of-the-coronavirus-outbreak-snap-chart/ (accessed 16 October 2020).

67 WHO (2020). *Water, Sanitation, Hygiene and Waste Management for the COVID-19 Virus: Technical Brief*, 1–9. World Health Organization.

68 Gundy, P.M., Gerba, C.P., and Pepper, I.L. (2009). Survival of coronaviruses in water and wastewater. *Food Environ. Virol.* 1 (1): 10–14.

69 Venugopal, A., Ganesan, H., Sudalaimuthu Raja, S.S. et al. (2020). Novel wastewater surveillance strategy for early detection of coronavirus disease 2019 hotspots. *Curr. Opin. Environ. Sci. Health* 17: 8–13.

70 Nghiem, L.D., Morgan, B., Donner, E., and Short, M.D. (2020). The COVID-19 pandemic: considerations for the waste and wastewater services sector. *Case Stud. Chem. Environ. Eng.* 1 (April): 100006.

71 Ahmed, W., Angel, N., Edson, J. et al. (2020). First confirmed detection of SARS-CoV-2 in untreated wastewater in Australia: a proof of concept for the wastewater surveillance of COVID-19 in the community. *Sci. Total Environ.* 728: 138764.

72 Pan, X., Chen, D., and Xia, Y. (2020). Viral load of SARS-CoV-2 in clinical samples. *Lancet Infect. Dis.* 20 (4): 411–412.

73 Wang, J., Feng, H., Zhang, S. et al. (2020). SARS-CoV-2 RNA detection of hospital isolation wards hygiene monitoring during the coronavirus disease 2019 outbreak in a Chinese hospital. *Int. J. Infect. Dis.* 94 (January 2019): 103–106.

74 Behnke, N.L., Cronk, R., Shackelford, B.B. et al. (2020). Environmental health conditions in protracted displacement: a systematic scoping review. *Sci. Total Environ.* 726: 138234.

75 Adelodun, B., Ajibade, F.O., Ogunshina, M.S., and Choi, K.-S. (2019). Dosage and settling time course optimization of *Moringa oleifera* in municipal wastewater treatment using response surface methodology. *Desalin. Water Treat.* 167: 45–56.

76 Adelodun, B., Ogunshina, M.S., Ajibade, F.O. et al. (2020). Kinetic and prediction modeling studies of organic pollutants removal from municipal wastewater using *Moringa oleifera* biomass as a coagulant. *Water* 12 (7): 2052.

77 Bishop, K. and Feuer, W. (2020). *Coronavirus: France, Germany and England impose new lockdown orders as pandemic fatigue seeps in across Europe and cases soar*. Consumer News and Business Channel (CNBC) https://www.cnbc.com/2020/11/04/coronavirus-france-germany-and-the-uk-impose-new-lockdown-orders-as-pandemic-fatigue-seeps-in-across-europe-and-cases-soar.html (accessed 14 November 2020).

78 Bashir, M.F., Ma, B.J., Bilal et al. (2020). Correlation between environmental pollution indicators and COVID-19 pandemic: a brief study in Californian context. *Environ. Res.* 187 (April): 109652.

79 Mofijur, M., Fattah, I.M.R., Alam, M.A. et al. (2021). Impact of COVID-19 on the social, economic, environmental and energy domains: lessons learnt from a global pandemic. *Sustainable Prod. Consumption* 26 (September 2020): 343–359.

80 IMF (2020). *World Economic Outlook, April 2020: The Great Lockdown*. International Monetary Fund https://www.imf.org/en/Publications/WEO/Issues/2020/04/14/weo-april-2020.

81 Gentilini, U., Almenfi, M., Orton, I. et al. (2020). *Social Protection and Jobs Responses to COVID-19: A Real-Time Review of Country Measures*. Washington, DC: World Bank. © World Bank. https://openknowledge.worldbank.org/handle/10986/33635.

82 Corlett, R.T., Primack, R.B., Devictor, V. et al. (2020). Impacts of the coronavirus pandemic on biodiversity conservation. *Biol. Conserv.* 246 (January): 108571.

83 Tait, P.W., Brew, J., Che, A. et al. (2020). The health impacts of waste incineration: a systematic review. *Aust. N.Z. J. Public Health* 44 (1): 40–48.

84 Klemeš, J.J., Van Fan, Y., Tan, R.R., and Jiang, P. (2020). Minimising the present and future plastic waste, energy and environmental footprints related to COVID-19. *Renewable Sustainable Energy Rev.* 127 (April): 109883.

85 van Kasteren, P.B., van der Veer, B., van den Brink, S. et al. (2020). Comparison of seven commercial RT-PCR diagnostic kits for COVID-19. *J. Clin. Virol.* 128 (April): 104412.

86 Wang, F., Xu, R., Zheng, F., and Liu, H. (2018). Effects of triclosan on acute toxicity, genetic toxicity and oxidative stress in goldfish (*Carassius auratus*). *Exp. Anim.* 67 (2): 219–227.

87 Pérez-Espejo, R.H., Constantino-Toto, R.M., and Dávila-Ibáñez, H.R. (2016). Water, Food and Welfare. Water Footprint as a Complementary Approach to Water Management in Mexico. Volume 23, SpringerBriefs in Environment, Security, Development and Peace. https://doi.org/10.1007/978-3-319-28824-6.

14

COVID-19 Pandemic and CO_2 Emission in the United States: A Sectoral Analysis

Afees A. Salisu, Tirimisiyu F. Oloko, and Idris A. Adediran

University of Ibadan, Centre for Econometric & Allied Research, Ibadan, Oyo State, 0234, Nigeria

14.1 Introduction

Cleaner and safer environment is one of the key goals of the United Nation's Sustainable Development Goals (SDGs) to which 193 member states including the United States (US) are signatories. According to the United Nations, clean environment is paramount to the health of people and our planet with the claim that about 23% of all annual deaths worldwide are linked to environmental degradation (see [1]). Carbon dioxide (CO_2) emission is a major component of the greenhouse gases (GHGs), which abruptly increases the average temperature of earth's surface, thus increasing the global incidence of climate change.

The emergence of the COVID-19 pandemic and the contingency measures devised to curtail its spread (such as social distancing, restriction of movements, and lockdown of economic activities) have undoubtedly affected industrial and commercial activities globally and by extension the environment [2–4]. As noted by Zambrano-Monserrate et al. [4], COVID-19 pandemic has positive and negative indirect effects on the environment. This is apparent as the contingency measures of COVID-19 are found to improve air quality, stimulate clean beaches, and cause environmental noise reduction on the one hand and reduce recycling and increase waste on the other hand. In the same vein, a number of studies have also shown that the contingency measures have resulted in short-term improvement in the environment, particularly in terms of reduced CO_2 emission (see [1, 3, 5–16]).

Evidently, there is a considerably growing literature on the air quality effect of COVID-19 pandemic (see [6] for a review); however, very limited attention has been paid to the United States, which has become the new epicenter of the pandemic and a high CO_2 emitting country by virtue of its high level of industrialization. A cursory review of the literature reveals that majority of studies on the environmental effect of COVID-19 pandemic are predominantly on China (see [14–17]) and Italy (see [18–20]). With all the parameters of evaluating the severity or otherwise of COVID-19

Detection and Analysis of SARS Coronavirus: Advanced Biosensors for Pandemic Viruses and Related Pathogens,
First Edition. Edited by Chaudhery Mustansar Hussain and Sudheesh K. Shukla.

pandemic pointing toward the United States (US hereafter) as the new epicenter[1] of the pandemic, its consideration in this study is justifiable.

Thus, this study examines the effect of COVID-19 pandemic on CO_2 emission in the United States. There are two main reasons for choosing the United States: first, the country is the new epicenter of COVID-19 pandemic in the world, and second, it is a high CO_2 emitting country given that it is the largest economy and among the few top industrialized countries in the world. According to the recent country-based report on CO_2 emission by the World Population Review, the United States has the second highest level of CO_2 emissions of 499.75 million tons as well as the second highest CO_2 emissions per capita of 15.53.[2] Limited studies on the relationship between COVID-19 pandemic and CO_2 emission or quality air in the United States are Bashir et al. [22] and Wu et al. [23]. However, these studies dwell more on the aggregate CO_2 emission and limit their analyses to selected cities rather than the entire country using sectoral data that covers residential, industrial, commercial, and other sectors of the US economy. For example, Bashir et al. [22] cover only California while Wu et al. [23] cover 3000 cities. This implies that the relationship between COVID-19 pandemic and sectoral emission of carbon dioxide is yet to be analyzed, which necessitates the need for this study.

To control the spread of the coronavirus pandemic, the United States, like many other countries, introduced social distancing measures by closing public places such as schools, worship centers, restaurants, and international borders. As evident from the Oxford COVID-19 Government Response Tracker (Ox-CGRT) database [24], the stringency index[3] of the United States increased rapidly from about 5.56 index points in February 2020 to 45.77 points in March, 72.69 in May 2020 before reducing to 70.71 in June 2020. This suggests that the degree of the stringency measures against COVID-19 increased rapidly between February and May 2020. Paital [3] notes that the US carbon emissions dropped by about 40% during the lockdown because of lower traffic, indicating that the potential of the pandemic to influence environment and the CO_2 emission in the United States. Similarly, Gillingham et al. [26] report that CO_2 emission has declined in the United States by 15% due to COVID-19. There is however less restrictive stringency measures since June 2020, which will mean the return of CO_2 emission-induced activities. This may confirm the results of the earlier studies that the improvement in environment and quality air as a result of the pandemic would be temporary (see [4, 27]).

1 Since 26 March 2020, the United States has continued to record high reported cases surpassing China and Italy, which were hitherto considered high-risk countries for the pandemic (see [21]). As on 8 October 2020, the United States remained the most affected country by the COVID-19 pandemic, with more than 7.6 million confirmed cases, about 3.02 million recovered patients and 212 762 casualties.

2 See: https://worldpopulationreview.com/country-rankings/co2-emissions-by-country (accessed 10 October 2020).

3 Stringency index deals with the information on social distancing measures. It is coded from eight indicators including school closing, workplace closing, cancel public events, restrictions on gathering size, close public transport, stay at home requirements, restrictions on internal movement, and restrictions on international travel (see [25]).

Thus, this study contributes to the literature on the relationship between COVID-19 pandemic and CO_2 emission in the United States by considering sectoral data, and therefore, the inherent heterogeneity in using aggregate data is circumvented. Apparently, as social distancing and lockdown policies restrict movements and reduce transport, commercial and industrial activities by extension, CO_2 emissions from industrial, commercial, transportation, and electricity power sectors may be expected to reduce far more than that of the residential sector. This is apparent as the citizens' adherence to the "stay at home" policy by government would mean higher burning of fossil fuels at home. Hence, the CO_2 emissions by the residential sector may not likely reduce as much as other sectors if it does not increase. The potential distinction in the environmental effect of responsiveness of the residential sector to COVID-19 pandemic effect has been discussed in related studies such as Espejo et al. [6], Somani et al. [12], and Zambrano-Monserrate et al. [4], where low recycling and increase in waste have been found to constitute negative environmental effect of COVID-19 pandemic.

In addition, this study employs structural vector autoregressive (SVAR) approach to model the responses of the US sectoral CO_2 emission to the global pandemic. Virtually, all studies on the relationship between COVID-19 pandemic and air quality agreed that the pandemic only affects air quality indirectly through various stringency and contingency measures (see [4, 12]). SVAR accommodates this fact as it examines transmission channel from global pandemic and epidemic, through global oil price and US industrial production, to the US sectoral CO_2 emission. The methodologies applied in earlier studies are basically observation of weather satellite (see for example, [7, 8]). While this is a very good method, it does not capture the influence of the global economic dynamics. This is, however, well captured by the SVAR model as it accounts for the effect of the global pandemic on the global crude oil price and the US productivity, which might account for changes in the US sectoral CO_2 emission. Notably, the nature of this study requires that we use the SVAR model with monthly or quarterly time frequency, which might not be sufficient for time series analysis given the few months of the existence of the COVID-19 pandemic. We circumvent this problem by using the historical global pandemic and epidemic (infectious disease) tracker by Baker et al. [28], which exhibited dramatic increase in its value because of emergence of the COVID-19 pandemic.

Following this introduction, the remaining part of this study is organized as follows. Section 14.2 deals with the analysis of stylized facts about the US sectoral CO_2 emission, global pandemic, global oil price, and the US industrial production. Section 14.3 discusses data issues and methodology. Results are presented and discussed in Section 14.4, whereas Section 14.5 concludes the study.

14.2 Stylized Facts on the Effect of COVID-19 Pandemic on Sectoral CO_2 Emission

The novel coronavirus that turned out to be known as 2019-nCoV and popularly referred to as COVID-19 was first identified in Wuhan, China, in December 2019 and

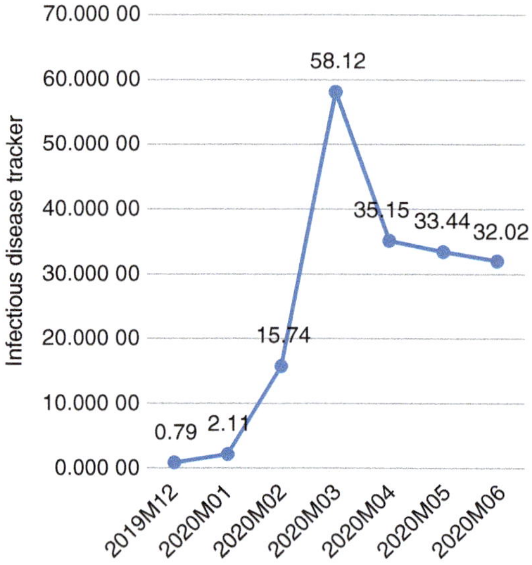

Figure 14.1 Trend in pandemic and epidemic tracker (December 2019 to June 2020). Source: Data from Federal Reserve Bank of St. Louis.

declared a Public Health Emergency of International Concern in January 2020 by the World Health Organization.[4] As the number of cases and deaths from the disease mounts, leading to institution of various response measures including lockdown, social distancing, travel restrictions, and economic activities began to be adversely affected as evidenced by increase in uncertainty in the global business space. This appears to be well captured in our choice of data for measuring the shock due to the pandemic, that is, the Equity Market Volatility (EMV)Infectious Disease Tracker. The EMV index, constructed by Baker et al. [28], documents and quantifies changes in economic uncertainty as a result of pandemic and epidemic.

We observe the movement of this variable in Figure 14.1. Apparently, Figure 14.1 shows drastic increase in uncertainty due to pandemic and epidemic between December 2019, when COVID-19 started off and March 2020 when it peaked. Specifically, as the incidence of COVID-19 pandemic deepens, the EMV increased rapidly from 0.79 points in December 2019, to 58.12 points in March 2020. Coincidentally, the stringency measures of the US government as measured by the stringency index (in Figure 14.2) also increased sharply during this period, rising from 5.56 points in 2 February 2020, to 72.69 points in 31 March 2020. This suggests that EMV truly explains pandemic generally and COVID-19 pandemic specifically. Even while COVID-19 pandemic is still on and the coronavirus cases in the United States is rising, the government has relaxed the associated stringency measures (see Figure 14.2). Accordingly, the EMV declined from its peak of 58.12 points in March 2020 to 32.02 points in June 2020 (see Figure 14.1). This suggests that as the economic uncertainty due to pandemic and epidemic is reducing, government's stringency measures are reducing. This further justifies our choice of EMV as a measure of economic effect of COVID-19 in this study.

---

4 See "Archived: WHO Timeline – COVID-19", online: https://www.who.int/news-room/detail/27-04-2020-who-timeline---covid-19.

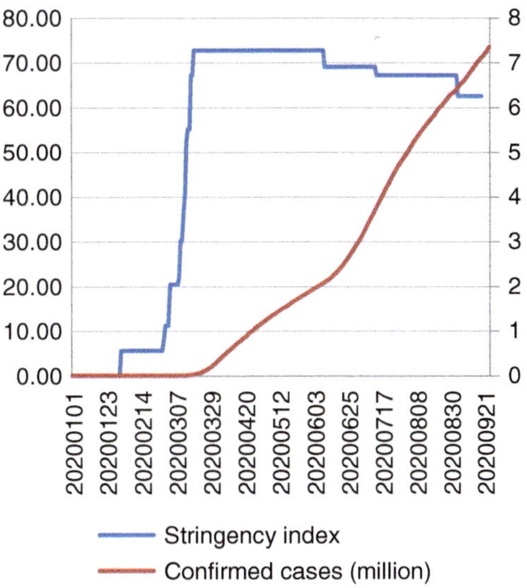

Figure 14.2 Trends in the US stringency measures and COVID-19 cases (January 2019 to October 2020). Source: Data from John Hopkins University and Ox-CGRT database.

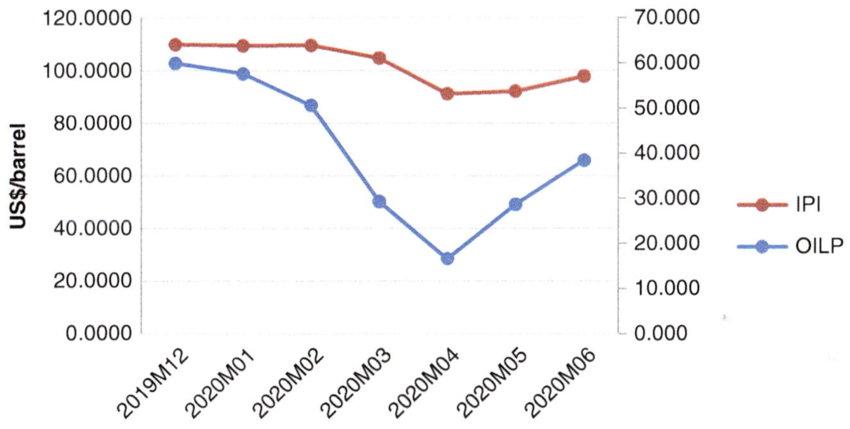

Figure 14.3 Trends in US IPI and International oil price, December 2019 to June 2020. Source: Data from Federal Reserve Bank of St. Louis.

More so, we observe closely identical movements in both the US industrial production and the global oil price (West Texas Intermediate [WTI]) data (see Figure 14.3). The decline in the two variables became steeper from February to March 2020 as also shown for the COVID-19 index when the disease moved from an epidemic to a pandemic as cases near 200 000 in 114 countries. As the uncertainty began to clear out and a number of countries relax stringent containment measures, production started to pick up as crude oil also rebounds in the international market.

During the period when the global economies were on hold as a result of the different containment measures that were strictly enforced, the expectation is that pollution levels would fall because economic activities, traveling and leisure, production, movements (hence, transportation), and many other related activities have

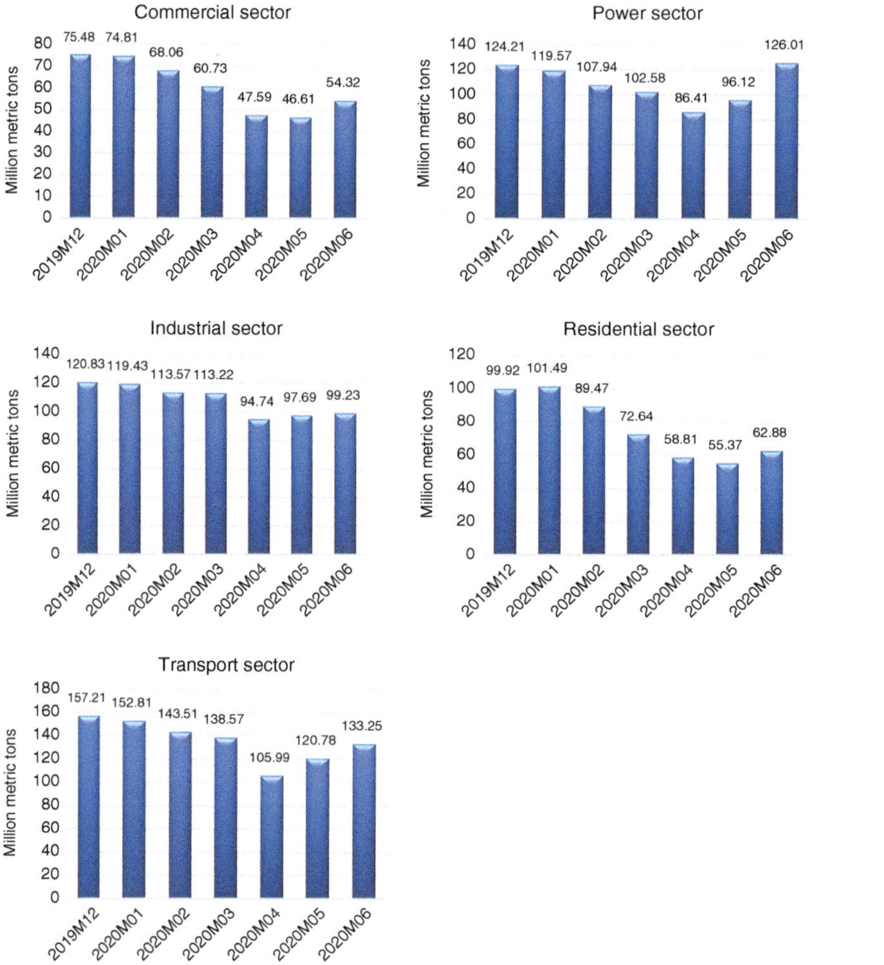

Figure 14.4 Sectoral CO_2 emission of the United States (December 2019 to June 2020). Source: Data from Federal Reserve Bank of St. Louis.

been contained. This position appears to be validated by several observation-based studies (for example, [1, 11, 27]) that observe drop in pollution levels as indirect implications of the COVID-19 pandemic. Thus, we observe trends in the sectoral CO_2 emission of the United States, given the contribution of the present study. This is presented in Figure 14.4.

Specifically, we document the dynamics of sectoral CO_2 emission across the five energy sectors in the United States: commercial, industrial, residential, transportation, and electric power sectors. Figure 14.4 indicate reductions in the CO_2 emission from December 2019 up to May 2020 for the commercial and residential sectors and up to April 2020 for the transport, power, and industrial sectors. In terms of magnitude, residential emissions reduced the most by 44.6% (December–May), followed by commercial, 38.2% (December–May), transport, 32.6% (December–April), electric power, 30.4% (December–April), and industrial, 21.6% (December–April), in that order.

In addition, the graphs revealed that activities picked up earlier in some sectors than the others, but by June 2020, all sectors have resumed activities as emissions have started increasing in all the sectors. This may not be unconnected with reduction economic uncertainty and the associated reduction in stringency measures by the US government (see Figures 14.1 and 14.2). This suggests that the conviction that the environmental gains of the pandemic are not sustainable may also be true for the United States.

14.3 Data Issues and Methodology

We apply the SVAR modeland define a vector of four endogenous variables as follows:

$$z_t' = [\text{EMV}_t \;\; \Delta\text{OILP}_t \;\; \Delta\text{IPI}_t \;\; \Delta\text{CO}_{2t}]' \tag{14.1}$$

where the observables are described as equity market volatility: infectious disease tracker (EMV_t) as the COVID-19 index, the WTI (OILP_t) as a measure of oil price, US industrial production index (IPI_t) as a measure of production, and CO_2 emissions (CO_{2t}), representing the sectoral emissions across the commercial, residential, industrial, transportation, and electric power sectors of the US economy. The variables are thus captured in a different form to reflect the unit root properties of the series in Table 14.1, showing that three out of four are integrated of order 1.

Based on the foregoing, we specify a four-endogenous variable VAR model where each of the five sectoral CO_2 emission series features distinctly in the model as follows.

$$z_t = \alpha_0 + \sum_{i=1}^{k} \phi_i z_{t-i} + \varepsilon_t \tag{14.2}$$

where α_0 represents a 4×1 vector of constants, ϕ_i are square matrices of coefficients associated with the i-th lagged endogenous variable, and k is the lag length of the optimal VAR model determined with the Schwartz Information Criterion.

We build on Eq. (14.2) to specify a structural VAR (hereinafter, SVAR) using a recursive structure to define the 4×4 matrix of contemporaneous parameters A_0 as follows.

$$A_0 = \begin{bmatrix} 1 & 0 & 0 & 0 \\ a_{21} & 1 & 0 & 0 \\ a_{31} & a_{32} & 1 & 0 \\ a_{41} & a_{42} & a_{43} & 1 \end{bmatrix}; \quad \text{and} \quad \Sigma = \begin{bmatrix} \sigma_1 & 0 & 0 & 0 \\ 0 & \sigma_2 & 0 & 0 \\ 0 & 0 & \sigma_3 & 0 \\ 0 & 0 & 0 & \sigma_4 \end{bmatrix} \tag{14.3}$$

The recursive structure is designed in line with the ordering of the variables as defined in vectorz_t; Π is the variance–covariance matrix of the SVAR model that conforms with the recursive construct ($A_0 \Pi A_0' = \Sigma\Sigma'$); Σ is a diagonal matrix useful

Table 14.1 Preliminary results.

Series	Mean	Standard deviation	Skewness	Kurtosis	ADF_Level	ADF_Diff	$I(d)$
			Energy sectors CO_2 emissions				
Commercial	76.0487	11.7128	0.0251	2.4876	0.5028	−6.4542***	$I(1)$
Industrial	134.912	12.2246	−0.3559	2.4198	−2.0281	−4.5948***	$I(1)$
Power	168.298	29.0122	0.2050	2.7387	0.3204	−6.6168***	$I(1)$
Residential	89.9344	21.9241	0.7647	2.9747	−0.6930	−11.835***	$I(1)$
Transport	147.358	14.4888	−0.3468	2.4314	−0.6056	−4.9252***	$I(1)$
			COVID-19 index				
EMV	0.9851	4.0035	10.783	129.508	−4.6719 ***	—	$I(0)$
			Macroeconomic variables				
Oil price (OILP)	43.6267	28.7341	0.9179	2.7304	−3.3417 *	−13.583***	$I(1)$
IPI	86.9826	16.9700	−0.4997	1.7641	−0.9594	−5.7571***	$I(1)$

The unit root test equations include trend components across the series except the EMV series, which does not exhibit trends.
*, **, and *** indicate 1, 5, and 10% levels of statistical significance.

to impose restrictions on the heteroscedastic error term (ε_t) to obtain the vector of structural shocks, $\Sigma A_0^{-1}\varepsilon_t$.

In addition to the contemporaneous coefficients defined in matrix A_0, we are concerned with the impulse responses of the sectoral co_{2t} among the endogenous variables included in z_t to the shocks to emv_t due to COVID-19 pandemic. The impulse response functions are captured in the relation defined as follows: $\Sigma A_0^{-1}\varepsilon_t$.

The data for the analysis are sourced from various sources including the energy information administration (EIA) and the US Federal reserve. The sectoral carbon dioxide emission (co_{2t}) data are obtained from the US EIA (see https://www.eia.gov/totalenergy/data/monthly/index.php#environment), the index for measuring COVID-19 pandemics is the EMV: Infectious Disease Tracker from the Federal Reserve (FRED) Database (see https://fred.stlouisfed.org/series/INFECTDISEMVTRACKD), the US industrial production index is also sourced from FRED (see https://fred.stlouisfed.org/series/INDPRO), and the WTI oil price data also from FRED (see https://fred.stlouisfed.org/series/WTISPLC). The series are obtained in monthly frequency from 1985M1 to 2020M6.

14.4 Empirical Results

14.4.1 Preliminary Results

We present preliminary discussion of observations in preparation for the analysis of the main results in Section 14.4.2. In Table 14.1, we present the summary statistics of the variables under investigation. The nature of these variables has been discussed in Section 14.3. Using measure of location, we observe that the power sector followed by the transport and then the industrial sector are the largest emitters of carbon dioxide in the United States. The commercial sector is the least behind the residential sector. Wang et al. [29], using data from 1997 to 2016, arrived at a similar relationship by showing transportation sector as the major contributor to the US CO_2 emissions. Also, on the average, the US industrial production index is about 86.98 points and the WTI oil price sells for US$43.63/barrel in the review period, 1985M1–2020M6. During the same period, the average value of the pandemic index, EMV, is 0.985 points. This value increased astronomically during the COVID-19 pandemic period, suggesting its appropriateness as a measure of pandemic.

Some statistical effects that cannot be ignored also emanate from the pre-tests. One of such is the skewness statistics, which show that transportation sector CO_2 emission is negatively skewed as well as the industrial production index (IPI), while CO_2 emission by other sectors, is positively skewed. The second is that sectoral CO_2 emission, IPI, and crude oil price are flat in relation to normal distribution, that is, they are platykurtic (kurtosis < 3). The EMV is, however, leptokurtic and peaked in relation to the normal distribution (kurtosis > 3). The unit root properties of the series are also examined to avoid problem of nonsensical regression. The CO_2 emission by all sectors, crude oil price, and the US industrial production index are integrated of order 1, with only measure pandemic, EMV, turning to be stationary at level. This property was observed in specifying our SVAR model in Eq. (14.1).

14.4.2 Main Results

We present the results of the contemporaneous effects in Table 14.2 with the distribution of shocks (that is, shocks 1–4; pandemic shock, oil price shock, production shock, and emission shock), following the ordering of the variables in Eq. (14.1). We show that the pandemic shock has significant negative contemporaneous effect on changes in oil price and US industrial productivity. This implies that COVID-19 pandemic tends to cause reduction in crude oil price and the US industrial productivity. Importantly, we observe that pandemic shock does not have significant contemporaneous impact on the sectoral CO_2 emission of the United States. This is apparent as the effect of shock 1 on changes in CO_2 emission is not significant across all sectors. This supports conclusion from multitude of recent studies that COVID-19 pandemic does not have direct impact on CO_2 emission (see for example [1, 5–9]).

Given the weak direct impact, we trace the indirect impact of pandemic shock to sectoral CO_2 emission through industrial productivity and crude oil price. The

Table 14.2 Coefficients of the contemporaneous effects.

	Shock 1	Shock 2	Shock 3	Shock 4	Shock 1	Shock 2	Shock 3	Shock 4
	Commercial sector CO_2				*Industry sector CO_2*			
EMV	1.0000	0.0000	0.0000	0.0000	1.0000	0.0000	0.0000	0.0000
Change in oil price (DOILP)	−0.3419***	1.0000	0.0000	0.0000	−0.3490***	1.0000	0.0000	0.0000
Change in industrial production index (DIPI)	−0.0209*	0.0000	1.0000	0.0000	−0.0189*	−0.0018	1.0000	0.0000
DCO_2	0.0311	−0.0188	1.3171***	1.0000	−0.0050	−0.0423	1.5938***	1.0000
	Power sector CO_2				*Residential sector CO_2*			
EMV	1.0000	0.0000	0.0000	0.0000	1.0000	0.0000	0.0000	0.0000
DOILP	−0.3355***	1.0000	0.0000	0.0000	−0.3508***	1.0000	0.0000	0.0000
DIPI	−0.0206*	0.0006	1.0000	0.0000	−0.0187*	−0.0018	1.0000	0.0000
DCO_2	−0.0139	0.1992	2.3877*	1.0000	−0.0321	−0.1660	2.8401**	1.0000
	Transport sector CO_2							
EMV	1.0000	0.0000	0.0000	0.0000				
DOILP	−0.3383***	1.0000	0.0000	0.0000				
DIPI	−0.0212**	0.0003	1.0000	0.0000				
DCO_2	−0.0401	0.0542	1.3073**	1.0000				

*, **, and *** indicate 1%, 5%, and 10% level of statistical significance. Shock 1 represents pandemic shock, shock 2 is oil price shock, shock 3 indicates US productivity shock, and shock 4 means environmental quality shock.

results show that contemporaneous change in the sectoral CO_2 emission of the United States is positively and significantly influenced by the US productivity shock. This result is confirmed across all sectors. This appears to confirm evidence from the previous studies that COVID-19 pandemic influence CO_2 emission indirectly (see [3, 10–16]). Apparently, this suggests that shutting down of industrial productivity due to pandemic causes significant reduction in sectoral CO_2 emission of the United States. The response of the residential CO_2 emission to direct pandemic shock is second to that of the transportation sector, and its response to the indirect pandemic shock (through industrial productivity shock) is the highest among all sectors. This tends to confirm our hypothesis that the stay at home policy would impact on residential CO_2 emission in a relatively unique way compared to any

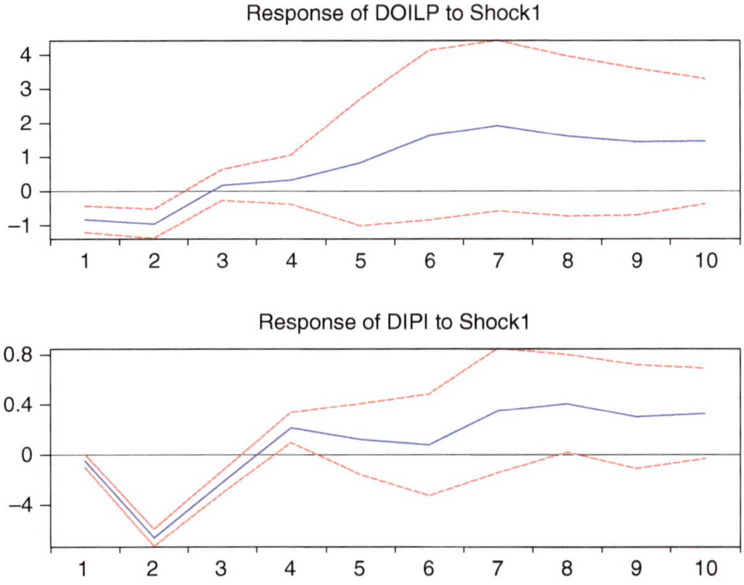

Figure 14.5 Responses of crude oil price and US industrial production to pandemic and epidemic shock.

other sector. This is related to the findings by Espejo et al. [6], Somani et al. [12], and Zambrano-Monserrate et al. [4], where negative environmental effect of COVID-19 pandemic was ascertained by low recycling and increase in waste.

In furtherance of the findings justifying the indirect impact of the pandemic on sectoral emissions via the oil price and industrial productivity changes, we highlight the impulse responses of crude oil price and the US industrial productivity to pandemic shock (see Figure 14.5). The impulse response graphs indicate that pandemic shock has temporary negative effect on crude oil price and US industrial productivity. This suggests that COVID-19 pandemic will reduce crude oil price and US industrial productivity within a short time period. Figure 14.5 shows that the negative response of change in crude oil price to pandemic shock halts after 2.5 months, while the negative response of change in industrial productivity halts after 3 months. This suggests that crude oil price would recover from pandemic before industrial productivity in the United States. This finding is consistent with the observation from the stylized facts of this study (see Section 14.2). Furthermore, the result reinforces the view that production activity increases GHG emissions via the indirect channel (see [29–31]).

Notwithstanding the insignificance of the contemporaneous direct effect of pandemics on the sectoral CO_2 emissions of the United States, we examine the dynamic effect using the impulse responses function. Hence, in Figure 14.6, we present the impulse responses of sectoral CO_2 emission to pandemic shock. The highlights of the impulse responses of sectoral CO_2 emission to pandemic shock that explain direct effect of pandemic on sectoral CO_2 emission are provided as follows. First, pandemic shock has temporary negative effect on changes in CO_2 emission across all sectors.

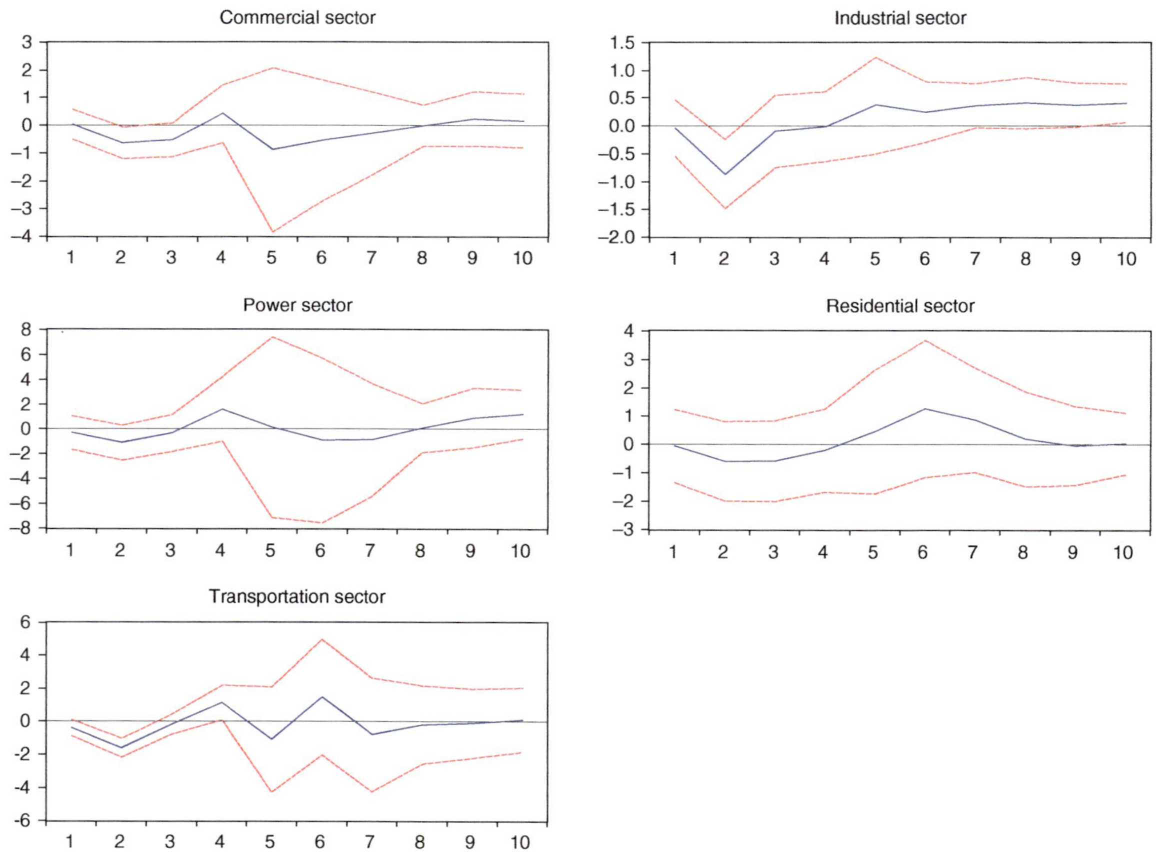

Figure 14.6 Responses of the US sectoral CO_2 emission to pandemic and epidemic shock (direct effect).

This suggests that CO_2 emission tends to reduce in all sectors because of COVID-19 pandemic. Second, the negative response of the CO_2 emission of the commercial, power, and transportation sectors halts after a three month period, while the negative response of the CO_2 emission of the industrial and residential sectors halts after a four month period. This suggests that COVID-19 pandemic tends to reduce CO_2 emission in the industrial and residential sectors for a longer period compared to the commercial, power, and transportation sectors. In other words, CO_2 emission of the industrial and residential sectors reduces at a slower rate.

We now turn to the indirect channel of transmission of pandemic shock to the sectoral CO_2 emissions. The analysis of the indirect channel between pandemic and sectoral CO_2 emission is necessary as our contemporaneous effect analysis suggests that the direct channel is weaker compared to the indirect channel. We explore the impulse responses of sectoral CO_2 emission to industrial productivity shock to explain the indirect effect of pandemic on sectoral CO_2 emission in Figure 14.7, given the argument that the emission rise due to pursuance of greater economic development (see [29–34]). The results show that the US industrial productivity shock has temporary positive effect on changes in CO_2 emission across all the five sectors. This suggests that CO_2 emission tends to reduce in all sectors as industrial productivity reduces due to COVID-19 pandemic. In terms of size, CO_2 emissions of all sectors respond maximally to productivity shock in the first period of the pandemic shock, and the effects reduce over time. Comparatively, industrial productivity shock (occasioned by the pandemic) reduces CO_2 emission of the residential sector more than any other sector (by 1.5). This supports evidence from our contemporaneous effect analysis, confirming that the CO_2 emission of the residential sector responds swiftly to COVID-19 pandemic compared to any other sector.

Meanwhile, the positive response of the CO_2 emission of the commercial, power, and transportation sectors to industrial productivity shock halts after a one month period, while the positive response of the CO_2 emission of the industrial and residential sectors halts after a three month period. This suggests that COVID-19 pandemic tends to reduce CO_2 emission in the industrial and residential sectors for a longer period compared to the commercial, power, and transportation sectors. In other words, CO_2 emission of the industrial and residential sectors reduces at a slower rate. This is consistent with the result obtained through direct channel analysis. Thus, there is ample justification to confirm our hypothesis that the stay at home policy would impact on residential CO_2 emission in a relatively unique way compared to any other sector.

14.5 Conclusion

In this paper, we examine the relationship between pandemics and CO_2 emission in the United States. The interest is motivated by the "stay at home" implication of the various stringency measures meant to contain the outbreak of COVID-19 pandemic. We utilize a new dataset by Baker et al. [28] as a proxy for uncertainty due to pandemics. This has a similar trend with the COVID-19 stringency measures employed

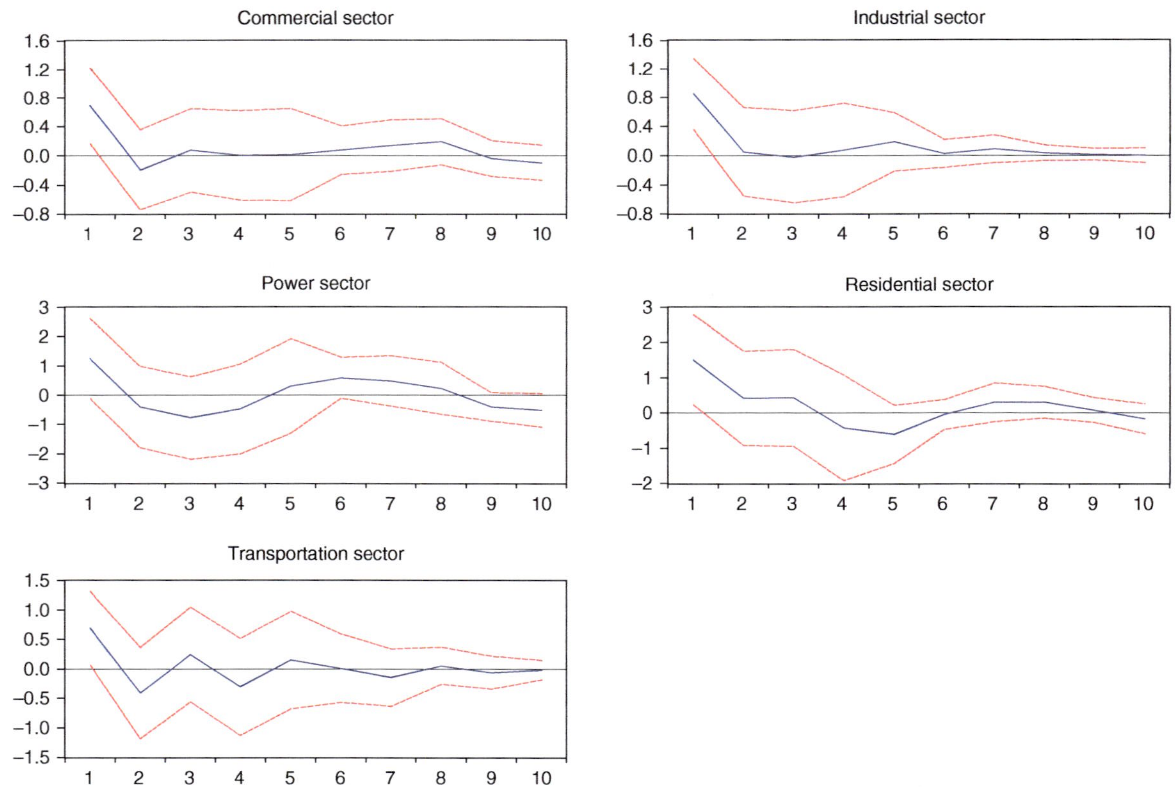

Figure 14.7 Responses of the US sectoral CO_2 emission to shocks to industrial production (indirect effect).

by government over the COVID-19 period, suggesting that EMV is a good measure of shock to economic uncertainty due to the pandemic. Other variables include WTI crude oil price, US industrial production index as a measure of economic productivity, and the CO_2 emission by sector of the United States. A SVAR approach, which allows for the analysis of the direct and indirect effects of pandemics on sectoral CO_2 emission, is constructed and estimated over the period of 1985M1–2020M06.

Our SVAR model relies on recursive assumption such that sectoral CO_2 emission can be influenced directly by pandemic shock and indirectly by the global oil price and the US productivity shocks. The results from the SVAR contemporaneous effects reveal that pandemic shock does not have significant contemporaneous impact on the sectoral CO_2 emission of the United States. This supports evidence from previous studies that COVID-19 pandemic does not have direct impact on CO_2 emission. The contemporaneous change in CO_2 emission across sectors is, however, positively and significantly influenced by the US productivity shock. This confirms evidence from the previous studies that COVID-19 pandemic influences CO_2 emission indirectly. In other words, industrial productivity shock due to pandemic causes significant reduction in CO_2 emission of the United States across all sectors. Contemporaneously, also, the response of the residential CO_2 emission to direct pandemic shock is second to that of the transportation sector, and its response to the indirect pandemic shock (through industrial productivity shock) is the highest among all sectors. This suggests that the CO_2 emission in the residential sector responds more swiftly to COVID-19 pandemic compared to any other sector. This tends to confirm our hypothesis that the "stay at home" policy of the government tends to increase residential CO_2 emission relatively higher than any other sector. Other results relating to the direct and indirect channels of transmission of the pandemic to CO_2 emission are highlighted in the paper.

References

1 Shakil, M.H., Munim, Z.H., Tasnia, M., and Sarowar, S. (2020). COVID-19 and the environment: a critical review and research agenda. *Sci. Total Environ.* https://doi.org/10.1016/j.scitotenv.2020.141022.

2 Atalan, A. (2020). Is the lockdown important to prevent the COVID-19 pandemic? Effects on psychology, environment and economy-perspective. *Ann. Med. Surg.* https://doi.org/10.1016/j.amsu.2020.06.010.

3 Paital, B. (2020). Nurture to nature via COVID-19, a self-regenerating environmental strategy of environment in global context. *Sci. Total Environ.* 730: 139088.

4 Zambrano-Monserrate, M.A., Ruano, M.A., and Sanchez-Alcalde, L. (2020). Indirect effects of COVID-19 on the environment. *Sci. Total Environ.* 728: 138813.

5 Dantas, G., Siciliano, B., França, B.B. et al. (2020). The impact of COVID-19 partial lockdown on the air quality of the city of Rio de Janeiro, Brazil. *Sci. Total Environ.* 729: 139085.

6 Espejo, W., Celis, J.E., Chiang, G., and Bahamonde, P. (2020). Environment and COVID-19: pollutants, impacts, dissemination, management and recommendations for facing future epidemic threats. *Sci. Total Environ.* https://doi.org/10.1016/j.scitotenv.2020.141314.

7 Kanniah, K.D., Zaman, N.A.F.K., Kaskaoutis, D.G., and Latif, M.T. (2020). COVID-19's impact on the atmospheric environment in the Southeast Asia region. *Sci. Total Environ.* https://doi.org/10.1016/j.scitotenv.2020.139658.

8 Lal, P., Kumar, A., Kumar, S. et al. (2020). The dark cloud with a silver lining: assessing the impact of the SARS COVID-19 pandemic on the global environment. *Sci. Total Environ.* 732 (2020): 139297.

9 Mahato, S., Pal, S., and Ghosh, K.G. (2020). Effect of lockdown amid COVID-19 pandemic on air quality of the megacity Delhi, India. *Sci. Total Environ.* 730 (2020), https://doi.org/https://doi.org/10.1016/j.scitotenv.2020.139086.

10 Nakada, L.Y.K. and Urban, R.C. (2020). COVID-19 pandemic: impacts on the air quality during the partial lockdown in São Paulo state, Brazil. *Sci. Total Environ.* 730: 139087.

11 Saadat, S., Rawtani, D., and Hussain, C.M. (2020). Environmental perspective of COVID-19. *Sci. Total Environ.* 728: 138870.

12 Somani, M., Srivastava, A.N., Gummadivalli, S.K., and Sharma, A. (2020). Indirect implications of COVID-19 towards sustainable environment: an investigation in Indian context. *Bioresour. Technol. Rep.* 11: 100491.

13 Tobías, A., Carnerero, C., Reche, C. et al. (2020). Changes in air quality during the lockdown in Barcelona (Spain) one month into the SARS-CoV-2 epidemic. *Sci. Total Environ.* 726: 138540.

14 Yao, Y., Pan, J., Wang, W. et al. (2020). Association of particulate matter pollution and case fatality rate of COVID-19 in 49 Chinese cities. *Sci. Total Environ.* 741: 140396.

15 Yongjian, Z., Jingu, X., Fengming, H., and Liqing, C. (2020). Association between short-term exposure to air pollution and COVID-19 infection: evidence from China. *Sci. Total Environ.* 727: 138704.

16 Zhang, Z., Xue, T., and Jin, X. (2020). Effects of meteorological conditions and air pollution on COVID-19 transmission: evidence from 219 Chinese cities. *Sci. Total Environ.* 741: 140244.

17 Xu, H., Yan, C., Fu, Q. et al. (2020). Possible environmental effects on the spread of COVID-19 in China. *Sci. Total Environ.* 731: 139211.

18 Fattorini, D. and Regoli, F. (2020). Role of the chronic air pollution levels in the COVID-19 outbreak risk in Italy. *Environ. Pollut.* 264: 114732.

19 Filippini, T., Rothman, K.J., Goffi, A. et al. (2020). Satellite-detected tropospheric nitrogen dioxide and spread of SARS-CoV-2 infection in Northern Italy. *Sci. Total Environ.* 739: 140278.

20 Zoran, M.A., Savastru, R.S., Savastru, D.M., and Tautan, M.N. (2020). Assessing the relationship between surface levels of PM2.5 and PM10 particulate matter impact on COVID-19 in Milan, Italy. *Sci. Total Environ.* 738: 139825.

21 JHU CCSE (2020). Novel Corona Virus (COVID-19) epidemiological data since 22 January 2020. Johns Hopkins University Center for Systems Science and Engineering (JHU CCSE).

22 Bashir, M.F., Bilal, B.M., and Komal, B. (2020). Correlation between environmental pollution indicators and COVID-19 pandemic: a brief study in Californian context. *Environ. Res.* 187: 109652.

23 Wu, X., Nethery, R.C., Sabath, B.M., et al. (2020). Exposure to air pollution and COVID-19 mortality in the United States. medRxiv. https://doi.org/10.1101/2020.04.05.20054502.

24 Hale, T., Webster, S., Petherick, A., et al. (2020). Oxford COVID-19 government response tracker. Blavatnik School of Government 25.

25 Ashraf, B.N. (2020). Economic impact of government interventions during the COVID-19 pandemic: international evidence from financial markets. *J. Behav. Exp. Finance* 27: 100371.

26 Gillingham, K.T., Knittel, C.R., Li, J. et al. (2020). The short-run and long-run effects of COVID-19 on energy and the environment. *Joule* https://doi.org/10.1016/j.joule.2020.06.010.

27 Wang, Q. and Su, M. (2020). A preliminary assessment of the impact of COVID-19 on environment – a case study of China. *Sci. Total Environ.* 728 https://doi.org/10.1016/j.scitotenv.2020.138915.

28 Baker, S.R., Bloom, N., Davis, S.J., and Terry, S.J. (2020). COVID-induced economic uncertainty. No. w26983. USA: National Bureau of Economic Research.

29 Wang, Z., Jiang, Q., Dong, K. et al. (2020). Decomposition of the US CO_2 emissions and its mitigation potential: an aggregate and sectoral analysis. *Energy Policy* 147: 111925.

30 Gozgor, G., Tiwari, A.K., Khraief, N., and Shahbaz, M. (2019). Dependence structure between business cycles and CO_2 emissions in the US: evidence from the time-varying Markov-switching copula models. *Energy* 188: 115995.

31 Xu, B., Zhong, R., Hochman, G., and Dong, K. (2019). The environmental consequences of fossil fuels in China: national and regional perspectives. *Sustain. Dev.* 27: 826–837.

32 Guo, J., Zou, L., and Wei, Y. (2010). Impact of inter-sectoral trade on national and global CO_2 emissions: an empirical analysis of China and US. *Energy Policy* 38: 1389–1397.

33 Li, Y. and Hewitt, C.N. (2008). The effect of trade between China and the UK on national and global carbon dioxide emissions. *Energy Policy* 36: 1907–1914.

34 Peters, G.P. and Hertwich, E.G. (2008). CO_2 embodied in international trade with implications for global climate policy. *Environ. Sci. Technol.* 42: 1401–1407.

15

Theranostic Approach for Coronavirus

Anushree Pandey, Asif Ali, and Yuvraj S. Negi

Indian Institute of Technology, Department of Polymer and Process Engineering, Roorkee, Uttarakhand 247667, India

15.1 Introduction

Coronavirus disease 2019 (COVID-19) was discovered in the Province of Hubei, China, in December 2019 [1]. Patient clusters were hospitalized with fever, cough, breath shortness, and other signs. There were patients scanned by means of computed tomography (CT) that showed various (solider, more abundant, and confluent) opacities in comparison to photos of healthy lungs [2]. This discovery resulted in the original pneumonia diagnosis. Additional analysis of the nucleic acid consuming real-time polymerase chain reaction (PCR) multiplex and the established pathogen sections have resulted in negative outcomes, implying that it was of uncertain origin that caused pneumonia [1]. Through 10 January 2020, extracts from the bronchoalveolar lavage (BAL) of patients fluid have been studied to discover a pathogen of a genetic link series to the lineage of β-coronavirus B. It was noticed that this novel pathogen had a similarity of 80%, 50%, and 96% to the severe acute respiratory syndrome virus genome (SARS-CoV), Middle East respiratory syndrome corona virus (MERS-CoV), RaTG13, and the bat coronavirus, respectively [1, 3].

The person-to-person transmission of severe acute respiratory syndrome coronavirus-2 (SARS-CoV-2) is planned to occur predominantly by respiratory droplets developed while coughing, sneezing, speaking, and speaking; most of it resembles the outbreak of influenza [4]. Transmission will, however, occur while an infected person is infected. The person, may or may not have symptoms, is in adjacent contact with communication with a strong one, or when one comes in contact with a healthy person one the surface is tainted and then touches his or her pupils, mouth, or nose he contracts the virus upon himself [5]. The time between infection and the onset of symptoms can stretch from 2 to 14 days [6]. It is primarily assumed that droplets of SARS-CoV-2 do not stay mobile for more than 2 min usually and do not move and stay in the air. Van Doremalen N, however. Notified that under experimental circumstances, that up to three SARS-CoV-2 aerosols can

Detection and Analysis of SARS Coronavirus: Advanced Biosensors for Pandemic Viruses and Related Pathogens, First Edition. Edited by Chaudhery Mustansar Hussain and Sudheesh K. Shukla.

stay sustainable for aerosol hours: 4 hours on silver; 24 hours on silver, plastic, and stainless steel cardboard; and 2–3 days on [7].

Studies have also shown that patients with basic disorders, such as arthritis, cardiovascular cancer, cardiovascular disorder, and disorder, may have higher levels of chance of mortality compared with other patients [8, 9]. Another such analysis has shown that additional threat factors are linked to acute respiratory distress syndrome (ARDS) progress and proceeding to death include age, neutrophilia, loss of tissues, and coagulation disorders [10]. Nevertheless, these findings are also restricted because of the absence of appropriate details on this novel disease and the small number of COVID-19 disease patients that have been involved in these trials.

Nanoparticles (NPs) are tiny but have a great surface-to-volume ratio, which makes them remarkable and exclusive. Because of these properties, nanoparticles have been employed in the fields of biotechnology, biology, distribution of medicines, and sensors, and they are viewed as a bridge between bulk and DNA labeling substances [11]. Nanotechniques have been widely used in order to enhance the delivery and efficacy of antiviral drugs, especially in combination with distribution mechanisms and nucleoside analogues, which have potential applications for drug-resistant infections of the human immunodeficiency virus (HIV) [12, 13].

This book chapter would first review the proposed conventional therapies, which are being treated extensively in clinical trials and analysis. The benefits of theranostic nanoparticles with a distinct nanoparticle with emphasis on optimum intranasal formulations, and supervision of different medicinal agents are to be addressed. Lastly, a specific priority would be dedicated to the development of treatment based on nanoparticle modalities that are expected to dramatically change therapy with COVID-19.

15.2 Conventional Medicines

COVID-19 is a highly contagious disease. It spreads when a healthy person comes in direct contact with a COVID-19 patient's respiratory droplets (Figure 15.1). There are no validated vaccinations or specific COVID-19 antiviral therapies at present. Most treatments are presently employed with supportive cardiovascular, hemodynamic, or respiratory medicine. They help people who have the infection. These therapies are, sadly, given to relieve side consequences and complications but not effectively destroy the antivirus. Therefore, there are also systematic analysis and research trials testing possible treatments, which are imminently required.

To develop an effective COVID-19 procedure, the mode of operation of the device must be well known about the infection. Analogous to severe acute respiratory syndrome virus (SARS) and Middle East respiratory syndrome virus (MERS) coronaviruses, a "Lock and Key" is included in this SARS-CoV-2 book. Mechanism of transforming angiotensin enzyme II (ACE2) serves as a "key" to specialize the entry cells that keep their "lock" [9]. It can be these target sites that originate in the lungs, kidneys, heart, vessels, and cells for the intestines. The virus will apply the host until

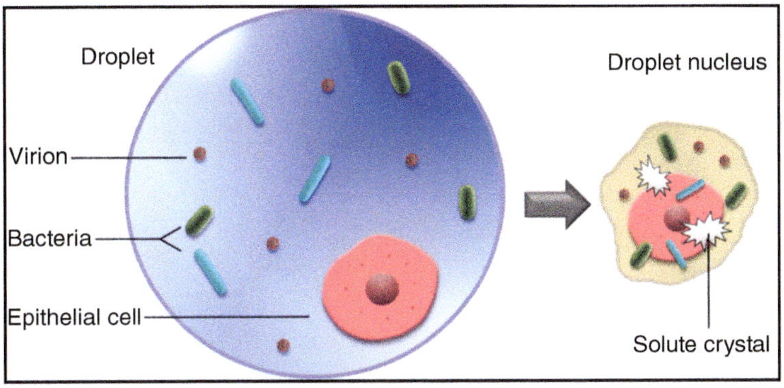

Figure 15.1 Droplet and droplet nuclei are a mechanism to transfer the infection at a very high speed. These droplets originate in respiratory tracts and when emitted in air carry bacterial cells and epithelial cells with them. Evaporation of droplets results in the formation of small nuclei. Source: Adapted from Weiss et al. [14].

it is inside in order to duplicate and contaminate other cells and cell organelles. Based on that, a therapy that avoids the virus can be helpful for entering into the cell.

What is a commonplace about all target cells for the adaptor protein complex 2 (AP2)-associated protein kinase 1 SARS-CoV-2 (AAK1), it is a main endocytosis regulator. From P. Richardson usage of machine learning tools, et al. suggested benevolent artificial intelligence (AI), the AAK1-related drugs could eliminate viral entry to target cells [15]. High doses of these antagonists, however, for example, drugs used in oncology (sunitinib and erlotinib), can be expected but could regrettably contribute the drugs being on bad side of the expected results [16]. In addition, the simulations also revealed that not all AAK1 inhibitors can show serious side effects. The baricitinib is a Janus kinase (JAK), inhibitor that can bind to another regulator of endocytosis, cyclin G-associated kinase that inhibits AAK1, thus stopping the entry of a virus into a cell. Besides that, it should be seen in cases of rheumatoid arthritis and researched as a possible cure for battling COVID-19 [17].

Other possible targets for combating COVID-19 with the corresponding use of human immunodeficiency requires protease regulators of viruses (HIV) such as lopinavir and ritonavir, as seen, suppress SARS protease-like 3-chymotrypsin MERS [18]. Multiple clinical phase III and phase IV studies have been conducted to determine the utility of these antiviral medicines. For example, open phase IV, prospective/retrospective, controlled by randomization in the study of cohorts was planned to test the effectiveness of antiviral therapies for ritonavir/lopinavir in the COVID-19 viral pneumonia therapy (from NCT04255017). It is projected that a scientific review would assess the safety and efficacy of baricitinib vs. lopinavir/ritonavir (accompanied by two other medicinal products) in people with mild to serious COVID-19 sickness (NCT04321993).

Other proposed substitute therapies under the widespread analysis include virally focused inquiries. For example, remdesivir is particularly, a nucleoside analog, it targets the polymerase and RNA-dependent suppresses a wide range of viral RNA synthesis RNA viruses, human coronaviruses included. Remdesivir is a licensed HIV reverse transcriptase inhibitor that has found applications in broad-spectrum operations in cell cultures and against RNA coronaviruses models for animals [19, 20]. M. Holshue et al. declared the good treatment of an untreated SARS-CoV-2 patient getting administration of remdesivir intravenously and no adversative reactions [21]. To better determine the security and effectiveness, multiple clinical trials were performed in phase 3 initiated in COVID-19 patients (http://ClinicalTrials.gov Identifier NCT04292899, NCT04292730, and NCT04252664, respectively).

A very encouraging option for treatment that has begun in order to be introduced in many countries, the use of chloroquine, antiviral and hydroxy-chloroquine extensive distribution of drugs beforehand employed against malaria [22] and autoimmune diseases. A protocol which includes reinforced by the use of hydroxy-chloroquine azithromycin revealed optimistic results for treating COVID-19 effectively. Potential, however, in the early years, efficacy was mainly observed. Contagiousness impairment and medication should be carried out under close supervision by doctors because of fear over the probability of arrhythmic death. A drug dependent on chloroquine was identified as inhibiting the cell-based fusion of SARS coronavirus with cells acidifying and thus inhibiting the lysosomes cathepsin, which necessitate a low pH for optimum performance of SARS-CoV-2 spike protein cleavage [23].

It is presumed that either Chloroquine can modify the SARS-CoV-2's molecular crosstalk with its target cells by suppression of kinases (i.e. Mitogen-activated protein kinases [MAPK]), or by suppression of kinases (i.e. impedes the M protein proteolytic processing and influence the assembly and budding of Virion. Additionally, by reducing the chloroquine material, chloroquine will work indirectly. Through the production of cytokines that are pro-inflammatory and/or by activating CD8+ T-cells anti-SARS-CoV-2 [24]. Recently, A. Cortegiani et al. have studied the effectiveness and security of chloroquine for the management of COVID-19 [25].

In spite of these proposed treatments, however, currently under comprehensive study and exploring other options drug research, the mortality toll in untreated patients this innovative coronavirus is still growing. Accordingly, in parallel, efforts should focus on alternative options. Approaches aimed at delivering optimal care side effects, thus reducing them. Mucosal, although vaccination, ordinarily intranasal, is the preferred route for infectious disease vaccination and treatment, and the present investigation modalities utilize the structural route. This is primarily due to intranasal barriers which need to be resolved by medication, because for some drugs, low intrinsic permeability, limited administration rate, rapid mucociliary administration before their clearance, and enzymatic degradation, hit the targeting site characteristics of the climate in the mucosal. Accordingly, the delivery of most recommended therapies to particular body sites by nanoparticles (NP) ensures that the concentration of these agents is several times greater than those acquired in the targeted infection sites, thus preventing normal cells from any side effects of the conventional techniques.

15.3 Role of Nanoparticles in COVID-19 Detection

The outbreak of COVID-19 placed global pressure on modern societies, especially health-related infrastructure, care, and diagnostic measures specific to this infection; thus, there is an immediate need for positive cases to be confirmed, for which patients need to be screened and viral surveillance should be executed. Diagnostics play a major part in function in COVID-19 containment, enabling speedy deployment control measures that limit dissemination with sensing, sensing case separation, and by touch tracing (that is detection of persons who came into direct contact with a diseased person). The opaqueness on the CT scans of lungs of patients with COVID-19 varies from that of patients who have healthy lungs: there are more solid, more abundant, and concomitant lungs [26]. Presently, diagnosis of COVID-19 is made by reverse transcription (RT)-PCR and separated through CT scans, but there are disadvantages to each technique. Molecular approaches can more reliably detect diseases. Syndromic testing or CT scans can be used because they can screen and categorize multiple pathogens [27]. Nanotechnology brings with new possibilities for cheap and flexible output methods of identification, stable personal protection devices, and new medicines that are successful. Nanosensors are a reality now and it is possible to detect very low concentrations of bacteria and viruses. Therefore, well before signs have shown up, notify clinicians with very low viral loads in patients. In line with the joint taskforce and China, from the end of December 2019 to the middle of December 2020, 104 SARS-CoV-2 viral strains were secluded and sequenced with nanopore sequencing from Illumina and Oxford in February 2020 [3]. Illumina Sequencing is a method of sequence-by-synthesis using amplification of the solid-phase bridge, while nanopore sequencing includes translocating a molecule of DNA into a molecule of a protein pore and evaluation of the following voltage changes to the order is calculated. Production of science and technology deployment is our finest tool in the battle against COVID-19 and tools for nanotechnology may be updated for the diagnosis, treatment, and prevention of the disease [28]. Various nano-based policies that have benefits over molecular methods in order to diagnose COVID-19 methods have been established. The methods currently being developed are discussed as follows.

15.4 Reverse Transcription Loop-Mediated Isothermal Amplification (RT-LAMP) Coupled with a Nanoparticle-Based Biosensor (NBS) Assay

COVID-19 is an exceptionally fatal respiratory illness spreading quickly and has triggered international fear. At present, it is identified by observing the nucleic SARS-CoV-2 acid via real-time polymerase chain reaction (RT-PCR) [29]. Also with this method, supplementary methods of gene amplification have been employed, but their methods are the key drawbacks that they are difficult and require competent skills and they are time-consuming. To prevent these restrictions, a reverse transcription loop-mediated isothermal amplification (RT-LAMP) assay to sense triple genes was developed by Yang et al. to diagnose SARS-CoV-2 easily. This was the approach in 2003 used for the coronavirus SARS test and was deemed fast and basic [27].

Since then, an RT-LAMP nanoparticle based biosensor (NBS) assay has been established to identify COVID-19 reliably and correctly [30]. Such an RT-LAMP NBS assay incorporates the amplification of loop-mediated isothermal amplification (LAMP), reverse transcription, study of nanoparticle-based biosensors and multiplex study to promote the one-step diagnosis of COVID-19, single-tube reaction. For nucleic acid amplification at a steady 60–65 °C, this technique takes 30 minutes using various primers to recognize the genes N, E, and ORF1ab. Roughly 10 collections of as seen in Figure 15.1 primers were built for each gene.

The LAMP technique is very accurate since it classifies the target sequence via six or eight unlike regions, straight employing the RNA as a template [26]. The detection consequences can be grasped macroscopically over the color alteration. These techniques comprise amplification of recombinase polymerase, helicase-based amplification, and loop-mediated isothermal amplification. Numerous academic labs have established and clinically verified RT-LAMP tests for SARS-CoV-2 [26, 27].

To establish a one-step, one-tube RT-LAMP-NBS assay, COVID-19 diagnosis, this strategy is useful and it is easy to run and requires only rudimentary, cheap operations and equipment to maintain (e.g. a water bath or a heating block) constant temperature for 30–40 minutes (63 °C) [15]. To establish a one-step, one-tube RT-LAMP-NBS assay, COVID-19 Diagnosis. This strategy is useful and it is useful and it is easy to run and requires only rudimentary, cheap operations. Equipment to maintain (e.g. a water bath or heating block) continuous temperature for 30–40 minutes (63 °C) [31]. Compared to previously established studies, which have shown that a quick and easy-to-use COVID-19 is RT-LAMP assay, NBS. The RT-LAMP can be physically and scientifically indicated by the instrument findings, thus removing the essential complex protocols (e.g. electrophoresis), unusual reagents (for example, pH metrics) and luxurious tools (e.g. PCR in real time) [31, 32]. Two target goal sequences are concurrently expanded in an isothermal series reaction and observed at the point of examination. Future research should be focused on determining the fundamental principle of COVID-19 RT-LAMP, refine the parameters for reaction (e.g. temperature of amplification), and prove the viability. The features of the COVID-19 RT-LAMP – the NBS test were investigated by detecting derived templates different diseases, such as viruses, fungi, and bacteria [30]. COVID-19 RT-LAMP-NBS mechanical overview the assay in Figure 15.2 is seen.

15.5 Point-of-care Testing

Point-of-care assessments are employed to give patients without surgery, send their samples to centralized labs, generating findings without a laboratory network being needed to classify infected patients. Lateral low detection of antigen for SARS-CoV-2

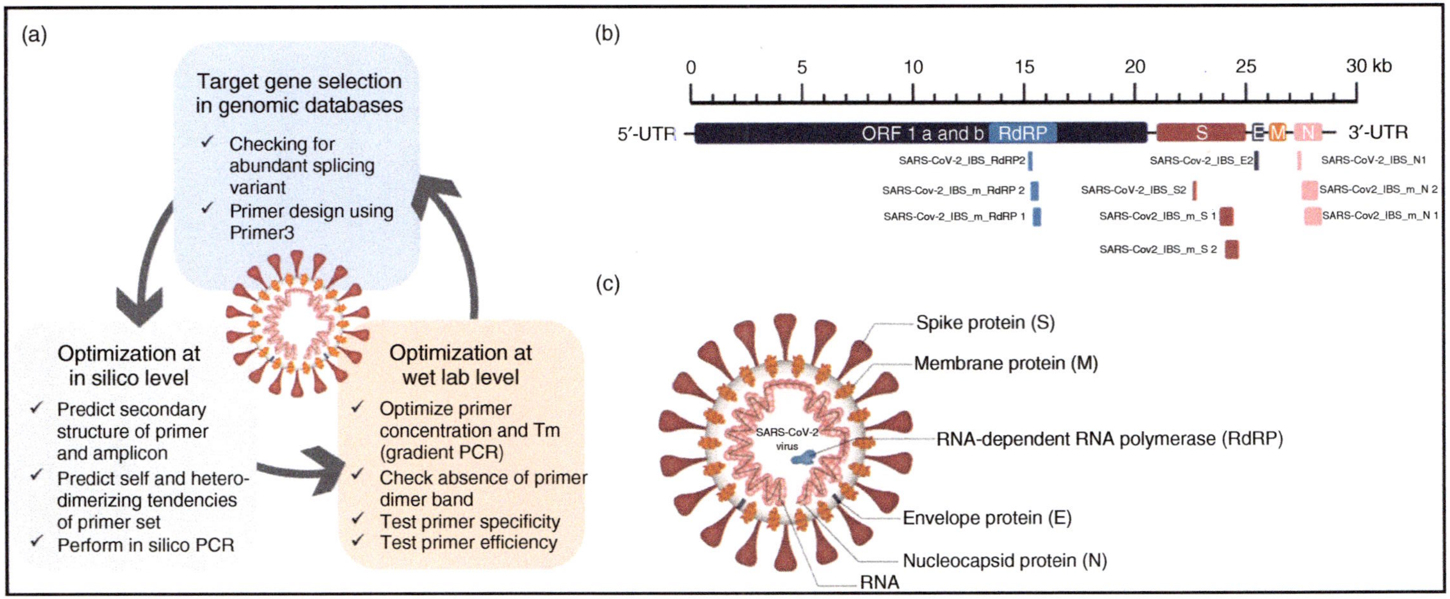

Figure 15.2 PCR primer architecture and optimization protocol, SARS-CoV-2 genome map, and primer set targets. (a) Architecture of the three-step PCR primer and instructions for optimization. (b) Place of the target genes and the selected primer sets in the SARS-CoV-2 genome. (c) The structure of SARS-CoV-2, which reveals every protein and its name. Source: Adapted from Yu et al. [31].

is a lateral detection of antigen. Point of treatment is considered for COVID-19 diagnosis [34]. Low lateral commercial assays produce a paper-like membrane two lines of strip: one includes a gold nanoparticle antibody conjugates; the remaining nonbonding antibodies are caught by the other. Patient's samples (e.g. blood and urine) are obtained from the membrane pulls the protein through the membrane and the capillary action for the rows. The antigens bind to the antibodies as the first line passes, gold conjugate of nanoparticle antibody, and the complex lows via a membrane. The antibodies that are captured immobilize the complex as the second side is reached where the red or the blue side is obvious. Specific nanoparticles of gold are red, and the combined plasmon bands allow the solution comprising the nanoparticles of clustered gold to turn blue [35]. Academics in many platforms, for example, electrochemical labs, Raman's cameras, paper-based structures, and surface-enhanced systems based on scattering, are being created.

15.6 Optical Biosensor Nanotechnology

A modern device based on the nanotechnology of optical biosensors will allow the coronavirus to be detected in about 30 minutes directly from the samples of patients without the prerequisite of centralized experiments at the clinic. It will be quick for this new technology to determine if the patient has a coronavirus infection or the virus of influenza. Potentially, this project will be used for other than the new pandemic and for the treatment of people. To evaluate different shapes, new biosensor methods can also be used. The existence of coronavirus in reservoir species, such as bats, identifies and tracks the future evolution of these viruses and prevent potential human diseases [30].

15.7 Nanopore Target Sequencing (NTS)

At the same time, the nanopore target sequencing (NTS) approach detects SARS-CoV-2 along with only 6–10 hours, like 10 other respiratory viruses. It is ideal for a clinical diagnosis of COVID-19; the system will, however, be extended to detect other cases. NTS is grounded on the enhancement of 11 SARS-CoV-2 fragments of virulence-related and special genes using an existing primary panel accompanied (e.g. orf1ab) sequencing on a nanopore of the bigger section platform [36]. This project employs a network of nanopores for sequencing, which can sequence long fragments of nucleic acid in real time and concurrently evaluate the data output. This confirmation of SARS-CoV-2 infections in minutes sequencing by mapping the sequence to the SARS-CoV-2 gene and identification analysis, validity, and reading of the output sequence number. Depending on the virulence region (genome 21 563–29 674 bp; NC 045512.2) ORF3a (275 AA), encoding S (1273 amino acids; AA), M (222 AA), E (75 AA), ORF7a (121 AA), ORF6 (66 AA), ORF8 (121 AA), ORF10 (38 AA), and N (419 AA), proteins to detect the pivotal cause of SARS-CoV-2 virulence genes. On one MinION, the NTS is carried out for all test samples, and the sequence data, the sequencer chip an in-house bioinformatics pipeline, are analyzed at the routine cycles. Both high-identity, to improve plasmid concentrations reads, is calculated by the mapping of read output on genome of SARS-CoV-2. As a quantitative polymerase chain reaction (qPCR) norm, NTS cannot regulate by checking whether a sample is positive for infection just one or two locations; signals from all target areas should be available [37, 38].

15.8 Role of Nanotechnology in the Treatment

Effective therapeutics are needed for the rapid expansion of COVID-19 battle tactics for this lethal pneumonia. Nonetheless, for the treatment of COVID-19 [29, 39], special medications have been authorized. Centered on both clinician and pharmacological expertise, numerous immunomodulatory and antiviral medicines have been proposed as possible tentative pathway therapies and those are in research tests today. The immunomodulatory and antiviral medicines encompass oseltamivir, remdesivir, mycophenolate, favipiravir, darunavir/cobicistat, teicoplanin, interferon, tocilizumab, convalescent plasma, tocilizumab ribavirin, etc. [40].

On 28 March 2020, however, the Health and Food Systems – Drug Administration issued an emergency use permit for the oral formulation of sulfate of hydroxychloroquine and for the treatment of COVID-19 [41, 42] chloroquine phosphate. Pursuant of the most recent literature, the recommendations issued in Belgium, as well as the recommendations were issued by the Netherlands, Switzerland, recommended in France and Italy: remdesivir, lopinavir/ritonavir, chloroquine, and tocilizumab in chloroquine or hydroxychloroquine in addition to compassionate treatment, artificial ventilation, including oxygen and [43]. Favipiravir, an antiviral agent that has been permitted in China and Japan, is under scrutiny for influenza care usage vs. COVID-19 [44].

In China's medications, in addition to these proposed medications, the treatment of COVID-19 plays a significant role. In the Chinese Versions of the Academy of Sciences, along with the Institute of Shanghai, it was proposed by Materia Medica that SARS-CoV-2 may be the use of Shuanghuanglian oral liquid was prohibited [45, 46]. These compounds, for example, baicalin, chlorogenic acid, and forsythin liquid, show antiviral and antibacterial properties [47]. No vaccines are currently approved for COVID-19; any recently designed vaccines are in clinical trials, though.

Significantly, nanotechnology may lead to the treatment of COVID-19. Various nanoparticles have antiviral and antibacterial properties [28]. The use of these nanoparticles may be to decrease the incidence and risk of diseases. No nano-based drugs for the treatment of COVID-19 are available to date. Research is, however, in the development stage, and hopefully, nano-based medicines will be out in near future among which few are summarized in Table 15.1. Zinc is a nanoparticle, for instance, that can be substantially used against COVID-19 [48].

Table 15.1 Summary of nanoparticle-based therapy approaches toward COVID-19.

Nanoparticle-based therapy	Drug	Details
RNA vaccines	mRNA-1273	Lipid nanoparticles encoding the SARS-CoV-2 S protein
	BNT162	Lipid nanoparticles encoding the SARS-CoV-2 S protein
Protein subunit	NVX-CoV2373	Full-length recombinant SARS-CoV-2 glycoprotein nanoparticle vaccine adjuvanted with Matrix M™ adjuvant
DNA vaccines	INO-4800	DNA plasmid encoding the SARS-CoV-2 S protein delivered through electroporation
Viral vector-based vaccines	Ad5-nCoV	Nonreplicating adenovirus type 5 vector carrying the gene for the SARS-CoV-2 S protein
	ChAdOx1 nCoV-19	ChAdOx1 construct (an adenovirus vaccine vector) carrying the gene for the SARS-CoV-2 S protein

Source: Based on Dong et al. [48] and Adapted from Liu et al. [49].

15.9 Conclusion

Although the cause of COVID-19 is still being studied, COVID-19 has characteristics typical of the family of Coronaviridae and has been identified as 2b linage β-coronavirus. It is possible to relay COVID-19 between individuals. Interventions that require extensive touch tracking accompanied by quarantine as well as isolation will, with the impact of travel restrictions, significantly minimize the spread of COVID-19. Wearing gloves, washing hands, and disinfecting surfaces help prevent contamination. Companies like Moderna and Pfizer are actively trying to get vaccines, and with the advancement of technology, we can hope for a cure soon.

References

1 Zhou, P., Yang, X.L., Wang, X.G. et al. (2020). A pneumonia outbreak associated with a new coronavirus of probable bat origin. *Nature* 579: 270–273. https://doi.org/10.1038/s41586-020-2012-7.

2 Ai, T., Yang, Z., Hou, H. et al. (2020). Correlation of chest CT and RT-PCR testing for coronavirus disease 2019 (COVID-19) in China: a report of 1014 cases. *Radiology* 296: E32–E40. https://doi.org/10.1148/radiol.2020200642.

3 Lu, R., Zhao, X., Li, J. et al. (2020). Genomic characterisation and epidemiology of 2019 novel coronavirus: implications for virus origins and receptor binding. *Lancet* 395: 565–574. https://doi.org/10.1016/S0140-6736(20)30251-8.

4 Zhao, W., Zhong, Z., Xie, X. et al. (2020). CT scans of patients with 2019 novel coronavirus (COVID-19) pneumonia. *Theranostics* 10: 4606–4613. https://doi.org/10.7150/thno.45016.

5 Liu, F., Zhang, Q., Huang, C. et al. (2020). CT quantification of pneumonia lesions in early days predicts progression to severe illness in a cohort of COVID-19 patients. *Theranostics* 10: 5613–5622. https://doi.org/10.7150/thno.45985.

6 Rothan, H.A. and Byrareddy, S.N. (2020). The epidemiology and pathogenesis of coronavirus disease (COVID-19) outbreak. *J. Autoimmun.* 109. https://doi.org/10.1016/j.jaut.2020.102433.

7 Van Doremalen, N., Bushmaker, T., Morris, D.H. et al. (2020). Aerosol and surface stability of SARS-CoV-2 as compared with SARS-CoV-1. *N. Engl. J. Med.* 382: 1564–1567. https://doi.org/10.1056/NEJMc2004973.

8 Yang, J., Zheng, Y., Gou, X. et al. (2020). Prevalence of comorbidities and its effects in coronavirus disease 2019 patients: a systematic review and meta-analysis. *Int. J. Infect. Dis.* 94: 91–95. https://doi.org/10.1016/j.ijid.2020.03.017.

9 Yu, Q., Wang, Y., Huang, S. et al. (2020). Multicenter cohort study demonstrates more consolidation in upper lungs on initial CT increases the risk of adverse clinical outcome in COVID-19 patients. *Theranostics* 10: 5641–5648. https://doi.org/10.7150/thno.46465.

10 Wu, C., Chen, X., Cai, Y. et al. (2020). Risk factors associated with acute respiratory distress syndrome and death in patients with coronavirus disease 2019 pneumonia in Wuhan, China. *JAMA Intern. Med.* 180: 934. https://doi.org/10.1001/jamainternmed.2020.0994.

11 Gnanasangeetha, D. and SaralaThambavani, D. (2013). One pot synthesis of zinc oxide nanoparticles via chemical and green method. *Res. J. Mater. Sci.* 1: 1–8.

12 Agarwal, H.K., Chhikara, B.S., Doncel, G.F., and Parang, K. (2017). Synthesis and anti-HIV activities of unsymmetrical long chain dicarboxylate esters of dinucleoside reverse transcriptase inhibitors. *Bioorg. Med. Chem. Lett.* 27: 1934–1937. https://doi.org/10.1016/j.bmcl.2017.03.031.

13 Chhikara, B.S., Rathi, B., Singh, J., and Fnu, P. (2020). Corona virus SARS-CoV-2 disease COVID-19: infection, prevention and clinical advances of the prospective chemical drug therapeutics: a review on corona virus disease COVID-19, epidemiology, prevention, and anticipated therapeutic advances. *Chem Biol Lett* 7: 63–72.

14 Weiss, C., Carriere, M., Fusco, L. et al. (2020). Toward nanotechnology-enabled approaches against the COVID-19 pandemic. *ACS Nano* 14: 6383–6406. https://doi.org/10.1021/acsnano.0c03697.

15 Richardson, P., Griffin, I., Tucker, C. et al. (2020). Baricitinib as potential treatment for 2019-nCoV acute respiratory disease. *Lancet* 395: e30–e31. https://doi.org/10.1016/S0140-6736(20)30304-4.

16 Pu, S.Y., Xiao, F., Schor, S. et al. (2018). Feasibility and biological rationale of repurposing sunitinib and erlotinib for dengue treatment. *Antiviral Res.* 155: 67–75. https://doi.org/10.1016/j.antiviral.2018.05.001.

17 Chen, N., Zhou, M., Dong, X. et al. (2020). Epidemiological and clinical characteristics of 99 cases of 2019 novel coronavirus pneumonia in Wuhan, China: a descriptive study. *Lancet* 395: 507–513. https://doi.org/10.1016/S0140-6736(20)30211-7.

18 Li, G. and De Clercq, E. (2020). Therapeutic options for the 2019 novel coronavirus (2019-nCoV). *Nat. Rev. Drug Discovery* 19: 149–150. https://doi.org/10.1038/d41573-020-00016-0.

19 Wang, M., Cao, R., Zhang, L. et al. (2020). Remdesivir and chloroquine effectively inhibit the recently emerged novel coronavirus (2019-nCoV) in vitro. *Cell Res.* 30: 269–271. https://doi.org/10.1038/s41422-020-0282-0.

20 de Wit, E., Feldmann, F., Cronin, J. et al. (2020). Prophylactic and therapeutic remdesivir (GS-5734) treatment in the rhesus macaque model of MERS-CoV infection. *Proc. Natl. Acad. Sci. U.S.A.* 117: 6771–6776. https://doi.org/10.1073/pnas.1922083117.

21 Holshue, M.L., DeBolt, C., Lindquist, S. et al. (2020). First case of 2019 novel coronavirus in the United States. *N. Engl. J. Med.* 382: 929–936. https://doi.org/10.1056/NEJMoa2001191.

22 Savarino, A., Boelaert, J.R., Cassone, A. et al. (2003). Effects of chloroquine on viral infections: an old drug against today's diseases? *Lancet Infect. Dis.* 3: 722–727. https://doi.org/10.1016/S1473-3099(03)00806-5.

23 Simmons, G., Bertram, S., Glowacka, I. et al. (2011). Different host cell proteases activate the SARS-coronavirus spike-protein for cell-cell and virus-cell fusion. *Virology* 413: 265–274. https://doi.org/10.1016/j.virol.2011.02.020.

24 Devaux, C.A., Rolain, J.M., Colson, P., and Raoult, D. (2020). New insights on the antiviral effects of chloroquine against coronavirus: what to expect for COVID-19? *Int. J. Antimicrob. Agents* 55. https://doi.org/10.1016/j.ijantimicag.2020.105938.

25 Cortegiani, A., Ingoglia, G., Ippolito, M. et al. (2020). A systematic review on the efficacy and safety of chloroquine for the treatment of COVID-19. *J. Crit. Care* 57: 279–283. https://doi.org/10.1016/j.jcrc.2020.03.005.

26 Chotiwan, N., Brewster, C.D., Magalhaes, T. et al. (2017). Rapid and specific detection of Asian- and African-lineage Zika viruses. *Sci. Transl. Med.* 9. https://doi.org/10.1126/scitranslmed.aag0538.

27 Yang, W., Dang, X., Wang, Q. et al. (2020). Rapid detection of SARS-CoV-2 using RT-LAMP method. *medRxiv*. https://doi.org/10.1101/2020.03.02.20030130.

28 Chan, W.C.W. (2020). Nano research for COVID-19. *ACS Nano* 14. https://doi.org/10.1021/acsnano.0c02540.

29 Law, S., Leung, A.W., and Xu, C. (2020). Severe acute respiratory syndrome (SARS) and coronavirus disease-2019 (COVID-19): from causes to preventions in Hong Kong. *Int. J. Infect. Dis.* 94: 156–163. https://doi.org/10.1016/j.ijid.2020.03.059.

30 Zhu, X., Wang, X., Han, L. et al. (2020). Reverse transcription loop-mediated isothermal amplification combined with nanoparticles-based biosensor for diagnosis of COVID-19. *medRxiv*. https://doi.org/10.1101/2020.03.17.20037796.

31 Yu, L., Wu, S., Hao, X. et al. (2020). Rapid colorimetric detection of COVID-19 coronavirus using a reverse transcriptional loop-mediated isothermal amplification (RT-LAMP) diagnostic plat-form: iLACO. *medRxiv*. https://doi.org/10.1101/2020.02.20.20025874.

32 Lamb, L., Bartolone, S., Ward, E., and Chancellor, M. (2020). Rapid detection of novel coronavirus (COVID-19) by reverse transcription-loop-mediated isothermal amplification. *SSRN Electron. J.* https://doi.org/10.1101/2020.02.19.20025155.

33 Park, M., Won, J., Choi, B.Y., and Lee, C.J. (2020). Optimization of primer sets and detection protocols for SARS-CoV-2 of coronavirus disease 2019 (COVID-19) using PCR and real-time PCR. *Exp. Mol. Med.* 52: 963–977. https://doi.org/10.1038/s12276-020-0452-7.

34 Xiang, J., Yan, M., Li, H. et al. (2020). Evaluation of enzyme-linked immunoassay and colloidal gold-immunochromatographic assay kit for detection of novel coronavirus (SARS-Cov-2) causing an outbreak of pneumonia (COVID-19). *medRxiv*. https://doi.org/10.1101/2020.02.27.20028787.

35 Udugama, B., Kadhiresan, P., Kozlowski, H.N. et al. (2020). Diagnosing COVID-19: the disease and tools for detection. *ACS Nano* 14: 3822–3835. https://doi.org/10.1021/acsnano.0c02624.

36 Wang, M., Fu, A., Hu, B. et al. (2020). Nanopore target sequencing for accurate and comprehensive detection of SARS-CoV-2 and other respiratory viruses. *medRxiv*. https://doi.org/10.1101/2020.03.04.20029538.

37 Liu, R., Fu, A., Deng, Z. et al. (2020). Promising methods for detection of novel coronavirus SARS-CoV-2. *View* 1. https://doi.org/10.1002/viw2.4.

38 Yu, L., Tong, Y., Shen, G. et al. (2020). Immunodepletion with hypoxemia: a potential high risk subtype of coronavirus disease 2019. *medRxiv*. https://doi.org/10.1101/2020.03.03.20030650.

39 Peeri, N.C., Shrestha, N., Rahman, M.S. et al. (2020). The SARS, MERS and novel coronavirus (COVID-19) epidemics, the newest and biggest global health threats: what lessons have we learned? *Int. J. Epidemiol.* 49: 717–726. https://doi.org/10.1093/ije/dyaa033.

40 Ziaie, S., Koucheck, M., Miri, M.M. et al. (2020). Review of therapeutic agents for treatment of COVID-19. *J. Cell. Mol. Anesth.* 5: 32–36. https://doi.org/10.22037/jcma.v5i1.29760.

41 Song, Z., Hu, Y., Zheng, S. et al. (2020). Hospital pharmacists' pharmaceutical care for hospitalized patients with COVID-19: recommendations and guidance from clinical experience. *Res. Soc. Adm. Pharm.* https://doi.org/10.1016/j.sapharm.2020.03.027.

42 Tobaiqy, M., Qashqary, M., Al-Dahery, S. et al. (2020). Therapeutic management of patients with COVID-19: a systematic review. *Infect. Prev. Pract.* 2: 100061. https://doi.org/10.1016/j.infpip.2020.100061.

43 Donders, F., Lonnée-Hoffmann, R., Tsiakalos, A. et al. (2020). ISIDOG recommendations concerning COVID-19 and pregnancy. *Diagnostics* 10. https://doi.org/10.3390/diagnostics10040243.

44 Sanders, J.M., Monogue, M.L., Jodlowski, T.Z., and Cutrell, J.B. (2020). Pharmacologic treatments for coronavirus disease 2019 (COVID-19): a review. *JAMA* 323: 1824–1836. https://doi.org/10.1001/jama.2020.6019.

45 Wang, L., Wang, Y., Ye, D., and Liu, Q. (2020). Review of the 2019 novel coronavirus (SARS-CoV-2) based on current evidence. *Int. J. Antimicrob. Agents* 55: 105948. https://doi.org/10.1016/j.ijantimicag.2020.105948.

46 Allahbadia, G., Human, F., and Singh, M. (2020). A comprehensive review on epidemiology, aetiopathogenesis, diagnosis and treatment of the novel coronavirus syndrome – COVID-19. *Iberoam. J. Med.* 2: 1–15.

47 Cui, W., Cheng, L., Hu, C. et al. (2013). Electrospun poly(L-lactide) fiber with ginsenoside Rg3 for inhibiting scar hyperplasia of skin. *PLoS One* 8: e68771. https://doi.org/10.1371/journal.pone.0068771.

48 Dong, L., Hu, S., and Gao, J. (2020). Discovering drugs to treat coronavirus disease 2019 (COVID-19). *Drug Discov. Ther.* 14: 58–60. https://doi.org/10.5582/ddt.2020.01012.

49 Liu, X., Liu, C., Liu, G. et al. (2020). COVID-19: progress in diagnostics, therapy and vaccination. *Theranostics* 10: 7821–7835. doi:https://doi.org/10.7150/thno.47987.

Index

Detection and Analysis of SARS Coronavirus: Advanced Biosensors for Pandemic Viruses and Related Pathogens, First Edition. Edited by Chaudhery Mustansar Hussain and Sudheesh K. Shukla.

d

e

o

p

q

r

s